TRANSGENIC ANIMAL TECHNOLOGY

TRANSGENIC
ANIMAL
TECHNOLOGY

A Laboratory Handbook

Edited by

Carl A. Pinkert

Department of Comparative Medicine
Schools of Medicine and Dentistry
The University of Alabama at Birmingham
Birmingham, Alabama

ACADEMIC PRESS, INC.

A Division of Harcourt Brace & Company

San Diego   New York   Boston

London   Sydney   Tokyo   Toronto

Front cover photograph : Fertilized mouse ova, from pronuclear zygotes to blastocysts, obtained between 12 and 84 hr after fertilization. Original magnification 100X. Photomicrograph by Carl A. Pinkert.

This book is printed on acid-free paper. ∞

Academic Press, Inc.
525 B Street, Suite 1900, San Diego, California 92101-4495

United Kingdom Edition published by
Academic Press Limited
24–28 Oval Road, London NW1 7DX

Library of Congress Cataloging-in-Publication Data

Transgenic animal technology : a laboratory handbook / edited by Carl
  A. Pinkert.
      p.   cm.
    Includes bibliographical references and index.
    ISBN 0-12-557165-8
    1. Transgenic animals--Laboratory manuals. 2. Animal genetic
  engineering--Laboratory manuals. I. Pinkert, Carl A.
  QH442.6.T69   1993
  599'. 015' 0724--dc20
                                                    93-8793
                                                    CIP

PRINTED IN THE UNITED STATES OF AMERICA
93 94 95 96 97 98   EB   9 8 7 6 5 4 3 2 1

Contents

Contributors

Numbers in parentheses indicate the pages on which the authors' contributions begin.

Charles J. Bieberich (221, 235), Department of Virology, Jerome H. Holland Laboratory, American Red Cross, Rockville, Maryland 20855

Howard Y. Chen (279), Merck, Sharp and Dohme Research Laboratories, Rahway, New Jersey 07065

Linda C. Cioffi (279), Progenitor, Inc., Athens, Ohio 45701

Thomas Doetschman (115), Department of Molecular Genetics, Biochemistry, and Microbiology, University of Cincinnati College of Medicine, Cincinnati, Ohio 45267

Harold W. Hawk (339), United States Department of Agriculture, Agricultural Research Service, Beltsville Agricultural Research Center, Livestock and Poultry Sciences Institute, Gene Evaluation and Mapping Laboratory, Beltsville, Maryland 20705

Jan K. Heideman (265), Transgenic Animal Facility, University of Wisconsin, Madison, Wisconsin 53706

Gilbert Jay (221, 235), Department of Virology, Jerome H. Holland Laboratory, American Red Cross, Rockville, Maryland 20855

John J. Kopchick (279), Edison Animal Biotechnology Center, Ohio University, Athens, Ohio 45701

Michael J. Martin (315), DNX Corporation, Princeton, New Jersey 08540

Glenn M. Monastersky (177), Charles River Laboratories, Wilmington, Massachusetts 01887, and Transgenic Alliance, St. Germain sur l'Arbresle, France

Lien Ngo (235), Department of Virology, Jerome H. Holland Laboratory, American Red Cross, Rockville, Maryland 20855

Paul A. Overbeek (69), Department of Cell Biology and Institute for Molecular Genetics, Baylor College of Medicine, Houston, Texas 77030

Carl A. Pinkert (3, 15, 315), Department of Comparative Medicine, Schools of Medicine and Dentistry, The University of Alabama at Birmingham, Birmingham, Alabama 35294

H. Greg Polites (15), Department of Molecular Neurobiology, Hoechst–Roussel Pharmaceutical, Inc., Somerville, New Jersey 08876

Caird E. Rexroad, Jr. (339), United States Department of Agriculture, Agricultural Research Service, Beltsville Agricultural Research Center, Livestock and Poultry Sciences Institute, Gene Evaluation and Mapping Laboratory, Beltsville, Maryland 20705

James M. Robl (265), Department of Veterinary and Animal Sciences, University of Massachusetts, Amherst, Massachusetts 01003

Brad T. Tinkle (221), Department of Virology, Jerome H. Holland Laboratory, American Red Cross, Rockville, Maryland 20855

Philip A. Wood (147), Department of Comparative Medicine, Schools of Medicine and Dentistry, The University of Alabama at Birmingham, Birmingham, Alabama 35294

Preface

Transgenic animal technology and the ability to introduce functional genes into animals are powerful and dynamic tools for dissecting complex biological processes. The questions to be addressed span the scientific spectrum from biomedical and biological applications to production agriculture. Transgenic methodologies have influenced a cross section of disciplines. They are recognized as instrumental in our expanding understanding of gene expression, regulation, and function. There are many general reviews on transgenic animals that are indeed very useful and timely. However, aside from the manuals devoted to mouse embryology, a single text illustrating the methodologies employed by leading laboratories in their respective disciplines has not previously been compiled. This handbook covers the technical aspects of gene transfer in animals—from molecular methods to whole animal considerations across a host of species. With this in mind, this handbook is envisioned as a bridge for researchers and as a tool to facilitate training of students and technicians in the development of various transgenic animal model systems.

I thank all of the contributing authors and those individuals in my laboratory, both past and present, who have assisted in the course of development of this handbook.

This handbook is dedicated to the memory of Norman L. Pinkert, my first mentor and role model.

Carl A. Pinkert

PART I

Overview

1

Introduction to Transgenic Animals

Carl A. Pinkert

Department of Comparative Medicine
Schools of Medicine and Dentistry
The University of Alabama at Birmingham
Birmingham, Alabama 35294

I. INTRODUCTION

The 1970s and 1980s have witnessed a rapid advance of the application of genetic engineering techniques for increasingly complex organisms, from single-cell microbial and eukaryotic culture systems to multicellular whole animal systems. The whole animal is generally recognized as an essential tool for biomedical and biological research, as well as for pharmaceutical development and toxicological/safety screening technologies. Moreover, an understanding of the developmental and tissue-specific regulation of gene expression is achieved only through *in vivo* whole animal studies.

Today, transgenic animals embody one of the most potent and exciting research tools in the biological sciences. Transgenic animals represent unique models that are custom-tailored to address specific biological questions. Hence, the ability to introduce functional genes into animals provides a very powerful tool for dissecting complex biological processes and systems. Gene transfer is of particular value in those animal species where long life cycles reduce the value of classical breeding practices for rapid genetic modification. For identification of interesting new models, genetic screening and characterization of chance mutations remains a long and arduous task. Furthermore, classical genetic monitoring cannot engineer a specific genetic trait in a directed fashion.

II. HISTORICAL BACKGROUND

In the early 1980s, only a handful of laboratories possessed the technology necessary to produce transgenic animals. With this in mind, this text is envisioned as a bridge to the development of various transgenic animal models. The gene transfer technology that is currently utilized in laboratory and domestic animals was pioneered using the mouse model. Today, the mouse continues to serve as a starting point for implementing gene transfer procedures and is the standard for optimizing experimental efficiencies for other species. Inherent species differences are frequently discounted by researchers who are planning studies with a more applicable species model. However, when one attempts to compare experimental results generated in mice to those obtained in other species, not surprisingly, many differences become readily apparent. Therefore, an objective of this text will be to address the adaptation of relevant protocols.

When initiating work related to gene transfer, it is important to look at the rapid advancement of a technology that is still primitive by many standards. From an historical perspective, one readily contemplates potential technologies and methods that lie just ahead. Whereas modern recombinant DNA techniques are of primary importance, the techniques of early mammalian embryologists were crucial to the development of gene transfer technology. While we can look at just over a decade of transgenic animal production, the preliminary experiments leading to this text go back millennia to the first efforts to artificially regulate or synchronize embryo development. More recently, we observed the centennial of the first successful embryo transfer experiments, dating back to the efforts first published in the 1880s and to Heape's success in 1891. By the time the studies by Hammond were reported in the late 1940s, culture systems were developed that sustained ova through several cleavage divisions. Such methods provided a means to systematically investigate and develop procedures for a variety of egg manipulations. These early studies led to experiments that ranged from mixing of mouse embryos and production of chimeric animals, to the transfer of inner cell mass cells and teratocarcinoma cells, to nuclear transfer and the first injections of nucleic acids into developing ova. Without the ability to culture or maintain ova *in vitro,* such manipulations or the requisite insights would not be possible (see Brinster and Palmiter, 1986).

In 1977, Gurdon transferred mRNA and DNA into *Xenopus* eggs and observed that the transferred nucleic acids could function in an appropriate manner. This was followed by a report by Brinster *et al.* (1980) of similar studies in a mammalian system, using fertilized mouse ova in initial experiments. Here, using rabbit globin mRNA, an appropriate translational product was obtained.

Major turning points in science continue to accelerate at an incredible pace. The technology at hand today in many areas will appear antiquated in a few short years. It is amazing to look back at the major events related to genetic engineering of animals and how our ability to manipulate the genome has come so far.

The production of transgenic mice has been hailed as a seminal event in the development of animal biotechnology. In reviewing the early events leading to the first genetically engineered mice, it was fascinating to note that the entire procedure for DNA microinjection was described over 25 years ago. While some progress seems extremely rapid, it is still difficult to believe that, following the first published report of a microinjection method in 1966 (Lin, 1966; see Figs. 1 and 2), it took 15 years before transgenic animals were created. The five pioneering laboratories that reported success at gene transfer (Gordon *et al.*, 1980; Wagner *et al.*, 1981a,b; Harbers *et al.*, 1981; Brinster *et al.*, 1981; Costantini and Lacy, 1981; Gordon and Ruddle, 1981) would not have been able to do so, were it not for the recombinant DNA technologies necessary to develop protocols or document results. In

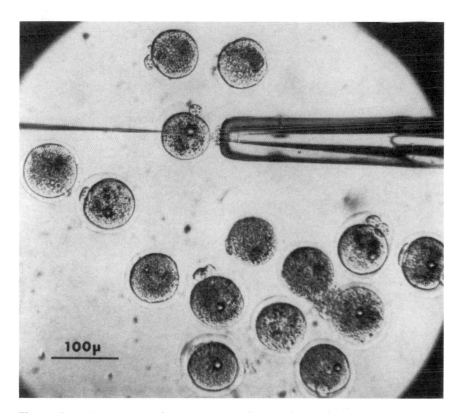

Figure 1. Microinjection of murine zygotes. The initial procedures for DNA microinjection were outlined in 1966. Here, zygotes are being injected with oil droplets. The zygotes survived this mechanical trauma, from use of holding pipettes to insertion of an injection pipette. [Reprinted with permission from Lin (1966).]

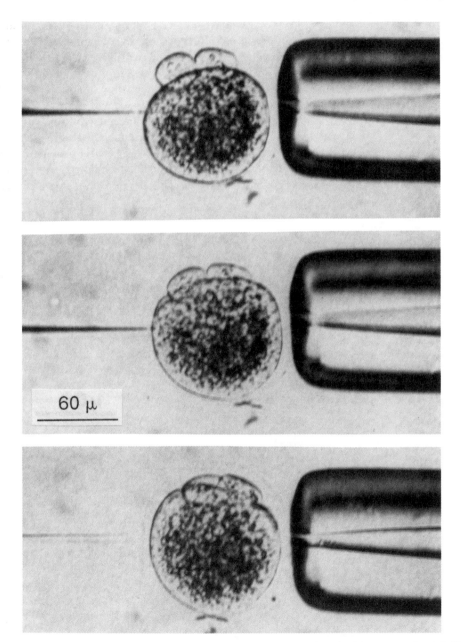

Figure 2. Microinjection of murine zygotes. As described in the 1966 paper by T. P. Lin, zygotes survived not only the mechanical trauma associated with the rudimentary injection procedures, but the injection of a bovine γ-globulin solution as well. [Reprinted with permission from Lin (1966).]

gene transfer, animals carrying new genes (integrating foreign DNA segments into their genome) are referred to as "transgenic," a term first coined by Gordon and Ruddle (1981). As such, transgenic animals are recognized as specific variants of species following the introduction and/or integration of a new gene or genes into the genome.

There are now well over 100 excellent reviews that detail the production of transgenic animals [in addition to a journal, *Transgenic Research,* which is dedicated to this field, readers are referred to reviews by Brinster and Palmiter (1986), Bürki (1986), Camper (1987), Cordaro (1989), First and Haseltine (1991), Grosveld and Kollias (1992), Hogan *et al.* (1986), Palmiter and Brinster (1986), Pattengale *et al.* (1989), Pinkert (1987), Pinkert *et al.* (1990), Pursel *et al.* (1989), Rusconi (1991), Scangos and Bieberich (1987), and Van Brunt (1988)]. To me, the singular effort with the greatest influence on this technology would not be among the initial reports just described. Rather, the work of Richard Palmiter and Ralph Brinster related to growth performance, and the dramatic phenotype of growth hormone transgenic mice (Fig. 3), subsequently influenced the emerging field in a most compelling manner for both basic and applied sciences (Palmiter *et al.*, 1982, 1983).

III. APPLICATIONS AND OVERVIEW OF TEXT

Scientists have envisioned many potential studies and applications, if an animal genome could be readily modified. Therefore, the realization of the many technologies at hand today has opened new avenues of research promise. Production of transgenic mice marked the convergence of previous advances in the areas of recombinant DNA technology and manipulation and culture of animal cells and embryos. Transgenic mice provide a powerful model to explore the regulation of gene expression as well as the regulation of cellular and physiological processes. The use of transgenic animals in biomedical, agricultural, biological, and biotechnological arenas requires the ability to target gene expression and to control the timing and level of expression of specific genes. Experimental designs have taken advantage of the ability to direct specific expression (including cell types, tissue, organ type, and a multiplicity of internal targets) and ubiquitous, whole-body expression *in vivo*.

From embryology to virology, the applications of transgenic mice provide models in many disciplines and research areas. Examples include the following:

• *Genetic bases of human and animal disease and the design and testing of strategies for therapy.* Many human diseases either do not exist in animals or are developed only by "higher" mammals, making models scarce and expensive. Many times, an animal model does not exist and the rationale for development is limited.

• *Disease resistance in humans and animals.* From a basic research and ethical

Figure 3. Production of transgenic mice harboring a growth hormone (GH) fusion construct. Animals harboring the GH transgene and expressing the GH gene product grew at a rate of 2- to 4-fold greater than control littermates, reaching a mature weight that was twice that of controls. This dramatic phenotype led the way for the exponential development of gene transfer technology. [Reprinted with permission from Palmiter *et al.* (1982).]

standpoint, it is imperative that we develop models for enhancing characteristic well-being.

• *Gene therapy.* Models for growth, immunological, neurological, reproductive, and hematological disorders have been developed. Circumvention and correction of genetic disorders are now possible to address using a variety of experimental methods.

• *Drug and product testing/screening.* Toxicological screening protocols are already in place in trials which utilize transgenic animal systems. For preclinical drug development, from a fundamental research perspective, a whole animal model for screening is essential in relation to understanding disease etiology, investigating drug pharmacokinetics, and evaluating therapeutic efficacy. A comparable need is crucial to product safety testing as well.

• *Novel product development through "molecular farming."* In domestic animals, biomedical proteins have been targeted to specific organs and body fluids with reasonable production efficiencies. Tissue plasminogen activator (TPA), factor IX, and human hemoglobin are a few products produced in transgenic animals that are currently in different stages of validation and commercialization.

• *Production agriculture.* Long term, it may become possible to produce animals with enhanced characteristics that will have profound influences on the food we eat, influences ranging from production efficiency to the inherent safety of our food supply.

The numerous strategies for producing genetically engineered animals extend from mechanistic (e.g., DNA microinjection, embryonic stem cell- or retrovirus-mediated transfer) as well as molecular (cloning) techniques. As the chapters of this text unfold, it will be apparent that the technology has extended to a variety of animal species in addition to the mouse, including the production of transgenic rats, rabbits, swine, ruminants (sheep, goats, and cattle), poultry, and fish. Although genetically engineered amphibians, insects, nematodes, lower eukaryotes and prokaryotes, and members of the plant kingdom have been acknowledged in the literature, such models are beyond the scope of this text.

However, advances in the understanding of promoter–enhancer sequences and external transcription-regulatory proteins involved in the control of gene expression continue to evolve using different model systems. In the systems explored in this text, gene transfer technology is a proven asset in science as a means of dissecting gene regulation and expression *in vivo*. However, the primary question that is addressed concerns the particular role of a single gene in development or in a given developmental pathway. With this caveat, considerations include the ramifications of gene activity—from intracellular to inter- and extracellular events within a given tissue or cell type milieu. Normally, gene function is influenced by *cis*-acting elements and *trans*-acting factors. For transferred genes, the *cis*- and *trans*-activators in conjunction with the gene integration/insertion event within the host genome influence regulation of both endogenous and transferred genes. Using genes that code for (or are composed of) reporter proteins (e.g., growth hormone or *lacZ* constructs), analysis of transgenic animals has revealed the importance of those three factors in determining the developmental timing, efficiency, and tissue distribution of gene expression. Additionally, transgenic animals have proved quite useful in unraveling *in vivo* artifacts of other model systems or techniques.

Although gene transfer technology continues to open new and unexplored biological frontiers, it also raises questions concerning regulatory and commercialization issues. It is not within the scope of this text, however, to fully address these issues. Suffice it to say that a number of issues exist and will continue to plague the development of many of the systems which are described herein. Major aspects of the regulation of this technology will focus on the following issues as we enter the twenty-first century:

- Environmental impact following "release" of transgenic animals
- Public perceptions
- Ethical considerations
- Legislation
- Safety of transgenic foodstuffs
- Patent aspects and product uniformity/economics

Contrary to the early prospects related to mainstreaming of this technology, there are numerous societal challenges regarding potential risks that are still ahead. The risks at hand can be defined by scientific evidence but also in relation to public concern (whether perceived or real). Therefore, the central questions will revolve around the proper safeguards to employ and the development of a coherent and unified regulation of the technology. Can new animal reservoirs of fatal human diseases be created? Can more virulent pathogens be artificially created? What is the environmental impact of the "release" of genetically engineered animals? Do the advantages of a bioengineered product outweigh potential consequences of its use? These are but a few of the questions that researchers cannot ignore and must approach. They are not alone, however, as the many regulatory hurdles that exist today will challenge not only scientists and policymakers, but sociologists, ethicists, and legal scholars as well.

The chapters in this text outline the basic techniques that various laboratories currently use to develop transgenic animals. The methods used to initiate experiments, develop vector systems, maintain animals (and the associated husbandry and experimental needs), and analyze and evaluate animals, with the requisite strategies to enhance experimental efficiency, are described at each step of experimentation. Discussion of all interlaboratory variations for each procedure is not feasible. As the chapter authors have learned, the strategies associated with the production of transgenic animals are quite variable, even between laboratories that utilize the same systems. Therefore, in some instances, alternatives to published and commonly used techniques are presented. However, most of the techniques for extensions to other systems are unique and timely for new investigators. The overall efficiency of many procedures will vary, as will the cost–benefit ratios. However, do not let the mechanics of experimentation outweigh the most important reason that one enters into these studies (unless you are a postdoctoral fellow looking for a niche and eventual job placement), which is the development and characterization of a biological model with specific utility.

Our goal is to illustrate a number of variations or novel methods which differ from the standard protocols outlined in detail for the mouse [some of the earlier references on embryology and micromanipulation of ova are listed in the references and include Rafferty (1970), Daniel (1971), Bürki (1986), and Hogan *et al.* (1986)]. The organization of this text is designed in a manner to assist those interested in developing an understanding of the basic species differences in transgenic animal research.

For the novice or new trainee, as well as for the experienced researcher, this text should influence proficiency and ultimately help provide an increase in overall productivity. For those wishing to develop a transgenic animal research program, this text will provide an overview of the requirements needed for development of a comprehensive gene transfer program.

There is only one take-home message to readers beyond the development of a desired biological model. An appreciation for the effort involved in each step of experimentation is most important in order to see a project through, from its design and implementation to the validation of a defined animal model. From a personal standpoint, one cannot discount the equal importance of the many unrelated disciplines, from molecular to whole animal biology, and the necessary training to ensure the overall success of transgenic animal technology.

REFERENCES

Brinster, R. L., and Palmiter, R. D. (1986). Introduction of genes into the germ line of animals. *Harvey Lect.* **80**, 1–38.

Brinster, R. L., Chen, H. Y., Trumbauer, M. E., and Avarbock, M. R. (1980). Translation of globin messenger RNA by the mouse ovum. *Nature (London)* **282**, 499–501.

Brinster, R. L., Chen, H. Y., Trumbauer, M. E., Senear, A. W., Warren, R., and Palmiter, R. D. (1981). Somatic expression of herpes thymidine kinase in mice following injection of a fusion gene into eggs. *Cell (Cambridge, Mass.)* **27**, 223–231.

Brinster, R. L., Chen, H. Y., Trumbauer, M. E., Yagle, M. K., and Palmiter, R. D. (1985). Factors affecting the efficiency of introducing foreign DNA into mice by microinjecting eggs. *Proc. Natl. Acad. Sci. U.S.A.* **82**, 4438–4442.

Bürki, K., ed. (1986). Experimental embryology of the mouse. *In* "Monographs in Developmental Biology," Vol. 19. Karger, New York.

Camper, S. A. (1987). Research applications of transgenic mice. *BioTechniques* **5**, 638–650.

Chen, H. Y., Trumbauer, M. E., Ebert, K. M., Palmiter, R. D., and Brinster, R. L. (1986). Developmental changes in the response of mouse eggs to injected genes. *In* "Molecular Developmental Biology" (L. Bogorad, ed.), pp. 149–159. Alan R. Liss, New York.

Cordaro, J. C. (1989). Transgenic mice as future tools in risk assessment. *Risk Anal.* **9**, 157–168.

Costantini, F., and Lacy, E. (1981). Introduction of a rabbit β-globin gene into the mouse germ line. *Nature (London)* **294**, 92–94.

Daniel, J. C. (1971). "Methods in Mammalian Embryology." Freeman, San Francisco, California.

First, N., and Haseltine, F. P. (1991). "Transgenic Animals." Butterworth-Heinemann, Stoneham, Massachusetts.

Gordon, J. W., and Ruddle, F. H. (1981). Integration and stable germ line transmission of genes injected into mouse pronuclei. *Science* **214**, 1244–1246.

Gordon, J. W., Scangos, G. A., Plotkin, D. J., Barbosa, J. A., and Ruddle, F. H. (1980). Genetic transformation of mouse embryos by microinjection of purified DNA. *Proc. Natl. Acad. Sci. U.S.A.* **77**, 7380–7384.

Grosveld, F., and Kollias, G. (1992). "Transgenic Animals." Academic Press, San Diego.

Gurdon, J. B. (1977). Egg cytoplasm and gene control in development. *Proc. R. Soc. London B* **198**, 211–247.

Hanahan, D. (1988). Dissecting multistep turmorigenesis in transgenic mice. *Annu. Rev. Genet.* **22**, 479–519.

Harbers, K., Jahner, D., and Jaenisch, R. (1981). Microinjection of cloned retroviral genomes into mouse zygotes: Integration and expression in the animal. *Nature (London)* **293**, 540–542.

Heape, W. (1891). Preliminary note on the transplantation and growth of mammalian ova within a uterine foster mother. *Proc. R. Soc. London* **48**, 457–458.

Hogan, B., Costantini, F., and Lacy, E. (1986). "Manipulating the Mouse Embryo: A Laboratory Manual." Cold Spring Harbor Laboratory, Cold Spring Harbor, New York.

Lin, T. P. (1966). Microinjection of mouse eggs. *Science* **151**, 333–337.

Palmiter, R. D., and Brinster, R. L. (1986). Germ-line transformation of mice. *Annu. Rev. Genet.* **20**, 465–499.

Palmiter, R. D., Brinster, R. L., Hammer, R. E., Trumbauer, M. E., Rosenfeld, M. G., Birnberg, N. C., and Evans, R. M. (1982). Dramatic growth of mice that develop from eggs microinjected with metallothionein–growth hormone fusion genes. *Nature (London)* **300**, 611–615.

Palmiter, R. D., Norstedt, G., Gelinas, R. E., Hammer, R. E., and Brinster, R. L. (1983). Metallothionein–human GH fusion genes stimulate growth of mice. *Science* **222**, 809–814.

Pattengale, P. K., Stewart, T. A., Leder, A., Sinn, E., Muller, W., Tepler, I., Schmidt, E., and Leder, P. (1989). Animal models of human disease. *Am. J. Pathol.* **135**, 39–61.

Pinkert, C. A. (1987). Gene transfer and the production of transgenic livestock. *Proc. U.S. Anim. Health Assoc.* **91**, 129–141.

Pinkert, C. A., Dyer, T. J., Kooyman, D. L., and Kiehm, D. J. (1990). Characterization of transgenic livestock production. *Domest. Anim. Endocrinol.* **7**, 1–18.

Pursel, V. G., Pinkert, C. A., Miller, K. F., Bolt, D. J., Campbell, R. G., Palmiter, R. D., Brinster, R. L., and Hammer R. E. (1989). Genetic engineering of livestock. *Science* **244**, 1281–1288.

Rafferty, K. A. (1970). "Methods in Experimental Embryology of the Mouse." Johns Hopkins Press, Baltimore, Maryland.

Rusconi, S. (1991). Transgenic regulation in laboratory animals. *Experientia* **47**, 866–877.

Scangos, G., and Bieberich, C. (1987). Gene transfer into mice. *Adv. Genet.* **25**, 285–322.

Van Brunt, J. (1988). Molecular farming: Transgenic animals as bioreactors. *Bio/Technology* **6**, 1149–1154.

Wagner, E. F., Stewart, T. A., and Mintz, B. (1981a). The human β-globin gene and a functional viral thymidine kinase gene in developing mice. *Proc. Natl. Acad. Sci. U.S.A.* **78**, 5016–5020.

Wagner, T. E., Hoppe, P. C., Jollick, J. D., Scholl, D. R., Hodinka, R. L., and Gault, J. B. (1981b). Microinjection of a rabbit β-globin gene into zygotes and its subsequent expression in adult mice and their offspring. *Proc. Natl. Acad. Sci. U.S.A.* **78**, 6376–6380.

PART II

Transgenic Animal Production Focusing on the Mouse Model

DNA Microinjection and Transgenic Animal Production

H. Greg Polites
Department of Molecular Neurobiology
Hoechst–Roussel Pharmaceuticals, Inc.
Somerville, New Jersey 08876

Carl A. Pinkert
Department of Comparative Medicine
Schools of Medicine and Dentistry
The University of Alabama at Birmingham
Birmingham, Alabama 35294

I. INTRODUCTION

Over the last decade, DNA microinjection has become the most widely applied method for gene transfer in mammals. While relative efficiencies are debated and new methodologies explored, DNA microinjection has been the workhorse providing the bulk of useful animal models for study. In the previous chapter, the two most practiced methods, DNA microinjection and embryonic stem (ES) cell transfer were mentioned. However, to identify the characteristic function of dominant genes in a host of applications, following *in vitro* studies, DNA microinjection can be the first step in delineating possible research directions.

In the course of this chapter, it is our intention to identify important criteria for the selection of this methodology, principles and requirements of the technology, and what expectations one might develop in the course of experimentation. It is not the intent to illustrate the *only* way to proceed, but rather to establish alternatives and ideas to complement the existing texts and handbooks related to microinjection technology that describe specific methods/techniques in great detail. All animals utilized in these procedures or in the experiments that follow are cared for according to National Institutes of Health (NIH) Office for Protection from Research Risks (OPRR) guidelines for appropriate husbandry under strict barrier adherence.

II. GENERAL METHODS

A. *DNA Preparation and Purification*

1. DNA Construct/Fragment Structure

In comparison with other gene transfer methods (particularly in relation to retroviral packaging), it seemed that DNA microinjection was one of the only methods where DNA fragment size was not at issue. However, fragment size constraints were originally imposed by plasmid (12 kb) and cosmid (45 kb) cloning vectors. As related technology developed, the ability to clone larger and *stable* constructs has been extended to yeast artificial chromosomes (YAC) and P1 vectors (Sternberg, 1992; Schedl *et al.*, 1993), as well as to the transfer of large chromosomal fragments (megabase lengths; Richa and Lo, 1989).

Although genomic fragments can be rather large and difficult to isolate and clone, recent studies demonstrate that homologous recombination can be effected using DNA injection into mouse zygotes. In the first example, using injection of major histocompatibility complex (MHC) class II gene, homologous recombination was effected in 1 of 500 mice incorporating the transgene (Brinster *et al.*, 1989). In a second example (Pieper *et al.*, 1992), three DNA fragments in the 30- to 40-kb range with 2.5–3.0 kb of overlapping ends were coinjected and homologously recombined before inserting into the mouse chromosome. Treating fragment ends or heterologous ends with phosphatase was not necessary to prevent concatamer formation. Apparently, recombination was a faster event compared to DNA end ligation. As such, use of overlapping fragments may prove to be a convenient method to circumvent cloning and construction of large gene constructs.

2. Factors That Influence Transgenic Mouse Production

Many of the criteria for successful production of transgenic mice have been defined (Brinster *et al.*, 1985; Brinster and Palmiter, 1986; Chen *et al.*, 1986; and many, many others since the mid-1980s). Briefly, to optimize experimental efficiencies, the following considerations are important.

a. Linear DNA fragments integrate with greater efficiency than supercoiled DNA. The DNA fragment size/length does not affect integration frequency.

b. A low ionic strength microinjection buffer should be prepared (10 mM Tris, pH 7.4, with 0.1–0.3 mM EDTA).

c. The efficiency of producing transgenic mice (DNA integration and development of microinjected eggs to term) appears most efficient at a DNA concentration between 1.0 and 2.0 ng/μl.

d. Linear DNA fragments with blunt ends have the lowest chromosomal integration frequency, whereas dissimilar ends are more efficient.

e. Injection of DNA into the male pronucleus is slightly more efficient than injection into the female pronucleus for producing transgenic mice.

f. Nuclear injection of foreign DNA is dramatically more efficient than cytoplasmic injection.

We have used an elongation factor 1-α promoter (Kim *et al.*, 1990) driving the *lacZ* marker gene (EF-GAL) to confirm these findings. This construct allows the assessment of conditions affecting integration efficiency and minimizes the confounding losses during *in vivo* development (H. G. Polites, A. McNab and G. Vogeli, unpublished data, 1991). After microinjection of EF-GAL into pro nuclear eggs, eggs are cultured to the hatched blastocyst stage and stained for β-galactosidase (β-gal) activity.

The EF-GAL construct was used to analyze the effect of DNA concentration and the type of DNA ends on foreign gene integration. Our results confirmed that integration increases rapidly with higher DNA concentrations; however, DNA is toxic to the egg, and viability diminishes as the DNA concentration increases. The integration frequency at 2 ng/μl equaled 22%, and that at 6 ng/μl was 54%. In another series of experiments we allowed the blastocysts to attach and the inner cell mass (ICM) to grow out. When they were stained we observed a surprisingly high degree of mosaicism in the positive-staining ICMs.

In addition, we have used an elastase–EJ *ras* fusion construct (Quaife *et al.*, 1987; Pinkert, 1990) for trainee preparation as well as foreign gene integration studies. The latter construct produces a visible phenotype (abdominal enlargement, pancreatic tumor formation, and ascites accumulation) at days 18–20 of gestation. While obviating the need for biochemical or molecular analyses, this construct affords the trainee an opportunity to evaluate the entire spectrum of procedures necessary to produce transgenic mice (described in greater detail in Chapter 3).

3. Preparation of DNA for Microinjection

a. Construction and Isolation of DNA   A given DNA fragment, after verification of correct cloning and confirmation of the nucleotide sequence, is purified through several steps to ensure that the DNA fragment does not contain nicks or strand breaks and is as pure as possible. For these final steps one should use the best reagents available. All solutions are made up from tissue culture grade, 18 MΩ water. Commercially available water systems, in combination with media preparations described below, have been used since the early 1980s. Ultrafiltration or deionization followed by reverse osmosis and distillation provides water that is also free of viruses and mycoplasmas. Good results can be obtained using water systems produced by most manufacturers (e.g., Millipore, Bedford, MA, Milli-Q; Barnstead, Dubuque, IA, NANOpure; Corning, Corning, NY, Mega-Pure). After preparation, solutions are also processed using a 0.45 or 0.2-μ filter.

Plasmid DNA can be replicated in any standard *Escherichia coli* host, but the DH5 methylase-defective strain CPLK-17 (Cat. No. 200292, Stratagene, La Jolla, CA) is both MCR$^-$ and MRR$^-$ and helps maintain the methylation state of the gene construct during cloning. It also allows transgene rescue to analyze flanking regions of the chromosomal insertion for toxicology applications or to localize insertional mutants (Short *et al.*, 1989). Methods for basic cloning or assembly of gene constructs and methods for large-scale preparation of DNA are well documented (Sambrook *et al.*, 1989).

Before a given gene fragment has been linearized and is readied for microinjection, the plasmid DNA is first purified by isolating supercoiled DNA from a cesium chloride (CsCl) gradient. The fragment is then isolated from vector sequences by restriction enzyme digestion, removing as much of the vector sequences as possible. There are numerous studies pointing to the relative severity and/or influence of vector DNA sequences on the function of foreign genes after chromosomal integration (reviewed in Brinster and Palmiter, 1986; Rusconi, 1991). For the most part, we routinely minimize the amount of flanking vector sequences for injection fragments. Additionally, the construct should be designed so that comigrating bands of DNA are easily separated on agarose gels.

One should avoid any preparative steps or conditions that might introduce nicks or contaminants into the purified DNA. Overdigestion or partial digestion of DNA with restriction enzymes, use of excessive heat to dissolve dried DNA, vigorous vortexing or pipetting, use of unsterilized pipette tips or microcentrifuge tubes, exposure to short-wavelength UV light, exposure to varying temperatures and DNase activity, and exposure to equipment that had not been cleaned/sterilized are examples of standard problems that should be carefully controlled in preparing DNA for microinjection.

b. DNA Purification   Once cut from a given vector, the DNA sample can be isolated by electrophoresis through agarose (SeaKem GTG grade, FMC, Rockland, ME). After adequate separation of the DNA bands in agarose, we generally purify fragments by either electroelution, followed by desalting and concentration by running the DNA on a column of DEAE-Sephacel (Sambrook *et al.*, 1989) (e.g., Elutip D, Schleicher and Schuell, Keene, NH), or glass bead adsorption (e.g., Geneclean, Bio 101 Inc., La Jolla, CA). After using either method, purification is completed as follows:

1. Resuspend the dried DNA pellet in microinjection buffer (TE, 10 mM Tris, 0.1 mM EDTA, pH 7.4) at a concentration greater than or equal to 10 ng/μl.

2. Perform two phenol/chloroform extractions (with chloroform containing 5% isoamyl alcohol).

3. Perform two chloroform extractions.

4. Add 1/10 volume of 2 M NaCl and 3 volumes of 100% ethanol and precipitate at $-20°$C overnight.

5. Centrifuge the DNA at 12,000 g for 10 min.

6. Wash twice in 70%(v/v) ethanol at room temperature and dry.

7. Resuspend in microinjection buffer.

8. Warm in a 37°C water bath for 5 min.

9. Gently vortex to dissolve the pellet.

10. Spin at 12,000 g for 10 min.

11. Transfer the aqueous volume to a new tube, leaving behind a small volume of liquid in the bottom of the tube.

12. Measure the DNA in a 1-μl volume and then dilute to 2 ng/μl in microinjection buffer; determine the DNA concentration of the microinjection aliquot.

13. The microinjection aliquot, sealed with Parafilm, may be stored at 4°C or frozen at $-20°$C.

There are numerous possible modifications of the purification protocol, and all are practical if they ensure that the final DNA preparation provides DNA that is (a) free of salts, organic solvents, or traces of agarose, (b) in the correct and sterile buffer, and (c) not sheared or nicked.

c. Purification and Quantification of DNA   Routinely, we cut enough plasmid DNA, containing the fragment of interest, to yield 5 to 20 μg of insert and use a large preparative agarose gel (Horizon 20–25, BRL, Gaithersburg, MD) to isolate the DNA fragment. This allows for large losses common to long purification protocols and produces enough DNA for accurate quantification and dilution into the microinjection buffer. We usually try to obtain a final dilution ratio of 1:20 or greater and aliquot a minimal volume of 300 μl in a 500-μl microcentrifuge tube to prevent evaporation from changing the concentration.

Quantification of DNA can be performed using either a fluorometer with calf thymus standards or comparative size standards of known concentration on an agarose gel containing ethidium bromide. There are DNA-specific fluorometers (e.g., TKO 100, Hoeffer, San Francisco, CA), and the fluorometric analysis is the more accurate method. However, DNA comparisons on agarose gels allow a final quality control check on a given sample to verify both fragment size and purity.

With either method precision is very important. Integration frequencies for foreign genes are concentration dependent (Brinster *et al.*, 1985), but egg viability is inversely related to DNA concentration. At DNA concentrations of 1–2 ng/μl injection buffer, viability averages 20% and still allows on average 30% live-born transgenic mice. At higher concentrations (e.g., 6 ng/μl) integration frequencies can increase to 60%, but viability drops to 10% or less and reaches the point where there are insufficient uterine implantations for pregnancy maintenance.

B. Superovulation, Egg Culture, and Harvest

1. Mouse Strains for Microinjection and Transgenic Models

a. Strains   Several different hybrids are currently popular for DNA microinjection, and the C57BL/6 × SJL F1 hybrid has been shown to be efficient in the generation of transgenic mice (Brinster *et al.*, 1985). The majority of the hybrids used for microinjection utilize the C57BL/6 inbred strain as one of the parental stocks because of favorable genetic and embryological characteristics. Hybrid mice are popular because hybrid vigor not only imparts desirable reproductive characteristics but also enhances the egg quality, hence leading to desirable microinjection characteristics. However, one should be concerned with the uniformity of the genetic background in which the transgene will be functioning, particularly in experimental designs where large populations or many generations of transgenic mice will be required.

Concepts related to the genetic background of mouse strains, including isogenicity (the degree to which individuals of a strain are genetically identical) and homogenicity (the degree to which individuals are homozygous at all genetic loci), are now influencing the characterization of transgenic models as lines are expanded. The initial selection of hybrid strains for gene transfer work focused on the efficiency of maintenance and reproduction as well as known embryology and response to experimental manipulations. However, current applications necessitate transgenic mouse models produced in specific background strains. Recent reports (Harris *et al.*, 1988; Chisari *et al.*, 1989) illustrate that transgene expression can be modulated or suppressed by "background" genetics of particular strains. In these examples, transgenic mice backcrossed to particular strains show characteristic repression or enhancement of transgene expression. In experiments where other strains are utilized, losses in experimental efficiencies will be encountered. In some instances (e.g., using FVB or C57BL/6 strains) efficiency losses may be minimal, but conditions for manipulation and timing of specific biological end points (e.g., pronuclear egg formation, response to gonadatropins) using various strains must be identified. As with strains used in different fields (e.g., other inbreds or congenics), efficiency losses can range from a few percent to more than a 100-fold difference (C. A. Pinkert, unpublished data, 1984; using DBA/2 congenic mice) when compared to C57BL/6 × SJL hybrids.

b. Genetic Variability and Transgene Expression Significant variation in transgene expression between individuals, litters, or generations can doom a transgenic model at several stages. It can complicate the initial characterization of the model if the genetic background severely influences the transgene expression. Alternatively, as the model is bred through several generations, inbreeding can alter transgene expression slowly or bring out new recessive phenotypes unrelated to the transgene.

With hybrid strains, isogenicity and phenotypic uniformity are high in the F1 generation but rapidly drop as further generations are produced. Homozygosity in hybrids is low at the F1 generation and drops to very low in the F2 (Festing, 1979). However, inbred strains such as the C57BL/6 and FVB have very high characteristics for all these traits at any generation. Although efficiencies in inbred mice (related to superovulation, microinjection, and reproduction) are reduced, the utility associated with genetic characterization adds a dimension that may be required for particular projects. However, the "ease" related to production of transgenic mice using hybrid donors may warrant production of founder transgenics from hybrid stock, *then* backcrossing to strains of choice.

2. Superovulation of Mice

a. Pregnant Mare's Serum Gonadotropin and Human Chorionic Gonadotropin For superovulation we use pregnant mare's serum gonadotropin (PMSG) (Ayerst, Montreal, Quebec; Diosynth, Chicago, IL; Equitech, Atlanta, GA; Sigma, St. Louis, MO) and human chorionic gonadotropin (HCG) (ICN, Costa Mesa, CA; Organon, West Orange, NJ; Sigma) both at 5.0 to 7.5 units i.p. per female mouse (3–8 weeks of age). Stocks are resuspended or diluted to 25 units/ml in phosphate-buffered saline (PBS) or water, then stored at −20°C (or lower temperatures) until thawed for use. Hormones have been stored in excess of 5 years at −70°C with no apparent loss of biological activity.

With a 12- to 14-hr light cycle (6 am to 6 pm or 7 am to 9 pm light), we administer PMSG at noon followed at 24 hr with HCG for C57BL/6 × SJL, ICR, or FVB mice or at 26 hr for C57BL/6 mice. Natural variation in responsiveness to administration of exogenous hormones requires that the time and dosage be "titered" for any new strain in the colony, supplier of hormones, or change in colony conditions. A biological assay related to the quality and quantity of eggs as well as mating performance (plug formation) are evaluated following hormone batch preparation.

Excessive PMS will lead to hormone refractoriness or increased proportions of nonfertilized, crenated, or abnormal eggs. At a proper dosage, one can reasonably expect to obtain 20–35 eggs per female mouse [dependent on age and strain in our experience, including C57BL/6 × SJL, C57BL/6 × DBA/2, and C57BL/6 × C3H hybrids; C57BL/6 and BL/10 (and congenics), C3H, and FVB inbreds; and outbred Swiss mice (Swiss-Webster, ICR, CD-1, and ND-4)]. Hybrid strains show the lowest percentage of abnormal egg development, whereas inbreds have the highest. Normal eggs and some abnormal mouse eggs produced following super-ovulation with PMS are illustrated below in Fig. 6C and include the following: (a) one-cell eggs with a degenerative cytoplasmic appearance (perivitelline space with multiple, fragmented polar bodies or devoid of polar bodies); (b) highly frag-mented, one-cell eggs with multiple unequal fragments in the perivitelline space; (c) precociously matured, fragmented two-cell eggs containing unequally divided blastomeres, with cytoplasmic fragmentation; and (d) well-developed one-cell pro-nuclear stage zygotes having a clear single or double polar body, two well-expanded pronuclei, and no cytoplasmic fragmentation.

b. Follicle-Stimulating Hormone and Luteinizing Hormone with Osmotic Pumps Use of follicle-stimulating hormone (FSH) (Vetrapharm, London, Ontario) delivered via an implantable osmotic minipump (Model 1007D, Alza Corp, Palo Alto, CA) and induction of ovulation with either HCG or luteinizing hormone (LH) can significantly improve egg quality and quantity in poorly responding strains. A method for mice was adapted from a protocol published by Leveille and Armstrong (1989) for superovulation in rats. For a 6-week-old, 20-g mouse (e.g., C57BL/6), FSH is resuspended at 1 mg/ml and loaded into the pump, which is implanted subcutaneously following the manufacturer's recommendations. The female donors are anesthesized with Avertin before the pumps are implanted. The osmotic pumps supply a constant infusion of FSH, and the degree of superovulation can be regu-lated by changing the pump rate or the concentration of FSH. The pumps are im-planted the morning of day −2 (7 am), and LH (Vetrapharm; 0.1 mg/ml i.p.) or HCG (Organon; 5.0 units i.p.) is administered on day 0 at 1 pm. The donors are then mated, and the pumps remain implanted in the females until the eggs are harvested the following day.

The FSH/pump protocol for superovulation is significantly more expensive and time-consuming than the PMSG/HCG regimen. The FSH osmotic pump method has been used successfully to induce superovulation in nonovulating (natural or PMSG/HCG-induced) transgenic females from a variety of lineages. In several at-tempts we have found that old, obese females which do not respond to PMSG can be successfully superovulated with the osmotic pump protocol. However, the FSH concentration is adjusted to a milligram per kilogram body weight level; for a 60-g female mouse, the FSH concentration is increased to 2 mg/ml.

3. Production of Eggs from Superovulated Females

a. Colony   The continuous production of superovulated females is dependent on a healthy colony with good management and environmental regulation. Additional factors are discussed in great detail in *The Mouse in Biomedical Research* (Foster *et al.*, 1983) and include the number and training of animal care technicians handling the mice, feed, bedding, water, noise, and housing density. The reproductive performance of both donor females and stud males is the first characteristic to degenerate when husbandry conditions are suboptimal. It is imperative that all personnel are familiar with the daily operations and acceptable conditions in the mouse colony so that any aberrations or problems can be readily identified and rectified.

b. Influence of Litter Number   Interestingly, in setting up a breeding colony to produce C57BL/6 × SJL F1 donor stock, we found a strong correlation between the frequency of abnormal eggs and parity of the C57BL/6 female dams. Initial superovulation was attempted exclusively with donors derived from first parity females. A high frequency (35–45%) of abnormal eggs was obtained. As females derived from second and subsequent parity dams became available, the frequency of abnormal eggs declined to a fairly consistent range around 15%. This may relate to the superior reproductive performance from second through fourth parity females (particularly with inbred strains). However, variations in hormone treatments, breeding pairs, and especially colony environment are additional components to be carefully evaluated.

4. *In Vitro* Culture of Eggs

a. Equipment   A dissecting stereo zoom microscope (6.5–40×) with wide-field (20×) eyepieces (Fig. 1) offers excellent resolution and working distances (e.g., Wild M3Z, Leica, Deerfield, IL; or Zeiss SV8, Thornwood, NY) for the harvest and manipulation of eggs. The binocular microscope is mounted on a mirrored transmitted light base for transillumination and egg contrast (e.g., Diagnostic Instruments Inc., Sterling Heights, MI).

There are several warming plates that precisely control the egg culture temperature at the laboratory bench to within 0.1°C (e.g., HT-400, Mini Tube of America, Cambridge, IA; 700 series, PMC Corp., San Diego, CA). Whether custom-configured or commercially available, such plates will significantly enhance egg viability during manipulations (and, with their adjustable range, readily function as do slide warmers to keep mice warm following surgery). In addition, the

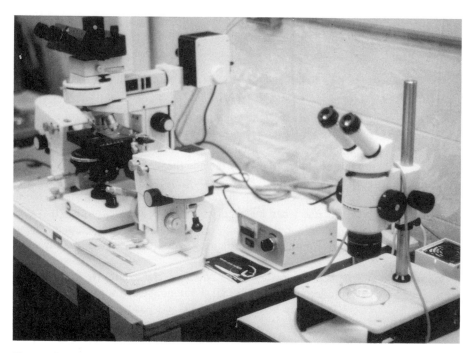

Figure 1. Microinjection station with binocular dissecting microscope at right and micro-injection microscope with manipulators on an antivibration table at left. The station is located in a HEPA-filtered clean room (note frame and heavy gauge plastic enclosure behind station).

HT-400 can be equipped with a stage warming plate that is independently controlled to avoid temperature drops during egg manipulation under the microscope.

   b. Atmosphere for Culturing Eggs   Eggs and media are maintained in a 5% O_2, 5% CO_2, and 90% N_2 atmosphere (by volume) that is controlled with a flow-meter (E500, Manostat). The mixture is passed through a gas humidifier, then into a small plastic incubator (e.g., a pipette tip box lid with a hole cut to accommodate tubing for gas flow) covering the microdrop dishes containing eggs. Finally, equipment for basic surgical requirements can be obtained from several suppliers and are well documented elsewhere.

   c. Glass Pipettes for Manipulating Eggs   Pipettes for manipulating eggs can be prepared from 9-in. Pasteur pipettes or from 4-mm-diameter glass tubing (Baxter, Muskegon, MI, G6100-4) cut at 5-in. lengths. The taper of the Pasteur pipette or the center of the 5-in. tube is heated over a flame, and the tube is rapidly pulled apart when the glass becomes pliable. The smaller the area of tubing that is heated, the smaller the final tube diameter. The pulled end is pinched off to a length of 3 in., placed in a clean container, and sterilized. With a little practice one can learn how to pull pipettes/tubes of consistent diameter. Small diameters [i.e., slightly greater than one egg ($\sim$80–90 μm) in internal diameter] are appropriate for egg transfers. However, a larger inner diameter ($\sim$200 μm) of the final taper facilitates more rapid collection and transfer of eggs during the microinjection procedure.

Once the glass is prepared, the wide end of the glass pipette is attached with latex tubing to a capillary adapter (with cotton plug or filter; Fig. 2). Using larger bore latex tubing, the adapter is then attached to a plastic mouthpiece. This mouth-controlled pipetting device, with a little practice, greatly enhances control of egg manipulation and is superior to hand-held micrometer-type devices (Rafferty, 1970).

d. Media Two different egg culture media are used for manipulation and microinjection. First, a bicarbonate-buffered medium such as modified BMOC-3, requiring a 5% O_2, 5% CO_2, 90% N_2 atmosphere, may be used. Additionally, a HEPES-buffered medium is used to maintain pH when eggs are removed from

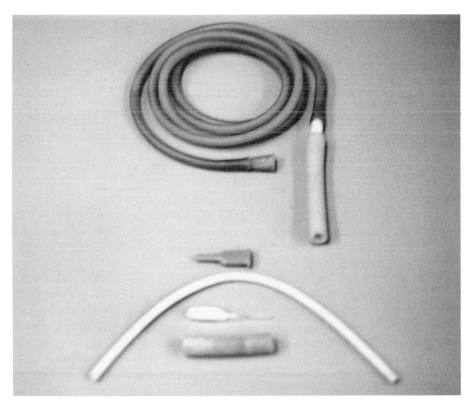

Figure 2. Mouth pipette assembly. Parts include large gauge tubing to attach to the Pasteur pipette, a cotton-filled glass capillary adapter (produced from a 4-cm tapered portion of a Pasteur pipette), small gauge tubing, and a mouthpiece. [See Rafferty (1970).]

TABLE 1
Media Recipes Suitable for the Culture of Mouse Ova[a]

	Amount (g/liter)				
Component	BMOC-3	M16	SECM	BMOC-3 + HEPES	M2
NaCl	5.200	5.533	5.540	5.200	5.533
KCl	0.356	0.356	0.356	0.356	0.356
$CaCl_2 \cdot 2H_2O$	0.252	0.252	—	0.252	0.252
Calcium lactate pentahydrate	—	—	0.527	—	—
KH_2PO_4	0.162	0.162	0.162	0.162	0.162
$MgSO_4 \cdot 7H_2O$	0.294	0.293	0.294	0.294	0.293
$NaHCO_3$	2.112	2.101	2.112	—	0.349
Sodium pyruvate	0.028	0.036	0.028	0.028	0.036
Sodium lactate	2.520	2.610	2.416	2.520	2.610
BSA	5.000	4.000	1.000	5.000	4.000
EDTA	0.037	—	—	0.037	—
Glucose	1.000	1.000	1.000	1.000	1.000
HEPES	—	—	—	5.950	4.969
Penicillin G potassium salt[b]	0.070	0.060	0.060	0.070	0.060
Streptomycin sulfate[b]	0.050	0.050	0.050	0.050	0.050
Phenol red[b]	0.010	0.010	0.005	—	0.010
Distilled water added to a final volume of 1 liter					

[a]BMOC-3 was defined in Brinster (1972) and Brinster *et al.* (1985). M2 and M16 were defined in Hogan *et al.* (1986) [original references: Quinn *et al.* (1982) and Whittingham (1971), respectively]. SECM was defined in Burki (1986) [original reference: Biggers *et. al.* (1971)].
[b]Optional.

the controlled atmosphere (see recipes in Table 1). Egg viability is higher in the bicarbonate-buffered medium, which is the preferred medium for culture, whereas the HEPES-buffered medium is used only for extended periods when the triple-gas or controlled atmosphere is unavailable.

The bicarbonate-buffered medium can be equilibrated with CO_2 and stored frozen for up to 6 months, with bovine serum albumin (BSA) added on thawing. The phenol red concentration can be reduced or omitted if feasible. Antibiotics are routinely omitted from media preparations to minimize toxicity and to allow quick identification of possible contamination. All reagents should be cell-culture tested and preferably "hybridoma-quality" reagents. Preparing media in large volumes (500 ml) allows for greater ease in accurate measurement of components while decreasing time-consuming preparation steps in the laboratory. Media have been stored at 4°C for over 3 years and provide egg efficiencies similar to those of freshly prepared aliquots (our only precondition was to exclude aliquots where contamination was grossly evident).

e. Water Sources   Water quality is a critical component, and the importance of testing several sources to ensure the quality is stressed. We have found that acceptable sources are in-house double processing systems (distilled then reverse osmosis filter systems) and commercial cell-culture suppliers (e.g., Sigma; Gibco, Grand Island, NY) that also screen for viral and mycoplasmal contamination.

f. Culture Dishes   Tissue culture dishes are routinely used for egg culture procedures. Small 35-mm tissue culture dishes (e.g., Corning 35 × 10-mm poly-styrene, No. 25000) are filled with about 3 ml of media and used for egg collection and "washing." Larger dishes (e.g., Corning 60 × 15-mm polystyrene, sterile tissue culture dishes, No. 25010) are used for holding large numbers of eggs before/after microinjection and for long-term culture. Small volumes of media (20–50 μl in microdrops) are overlaid with silicone oil and placed in the dishes to maintain eggs during manipulation. The drops are best identified by marking quadrants/sections on the bottom of dishes *before* media or oil is added to the dishes, similar to labeling quadrants on petri dishes before pouring plates (Fig. 3). Silicone oil (e.g.,

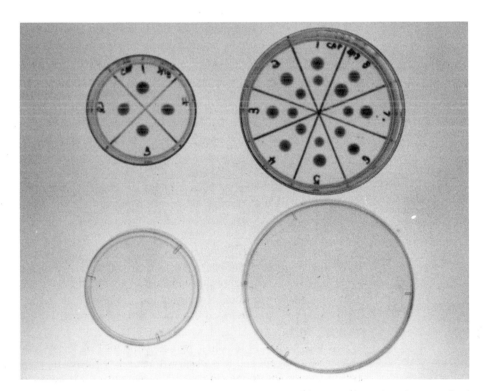

Figure 3.   Tissue culture dishes (35 and 60 mm) with 10- to 15-μl drops of medium overlaid with silicone oil. Before medium and oil are added, dishes are appropriately labeled on the bottom surface.

Dow Corning, Midland, MI, 200 fluid) is used routinely and readily minimizes diffusion and evaporation of the microdrops. In our experience, washing or equilibrating oil has not proved to be necessary. However, to wash silicone oil, omit BSA from the medium, mix a 1:1 volume of oil and medium by shaking vigorously for 15 min., and allow overnight separation. Washed oil can then be separated from the top layer as needed. After microdrop dishes are prepared, they are placed in the temperature- and gas-controlled environment. A minimum of 10 min is required to equilibrate conditions before use.

5. Harvesting Superovulated Eggs

a. Mating   After female egg donors receive HCG, they are placed in cages with male mice. The next morning, the females are checked for copulatory plugs. Bred females are sacrificed for egg recovery.

b. Oviduct Dissection   Harvesting of the oviducts and eggs should be done within 8–9 hr of the midpoint of the last dark cycle. This allows one to easily isolate the eggs as a mass from the oviduct, obviating the need to flush the individual eggs. After killing donor females, the oviducts are obtained though a midventral or dorsolateral approach. Either method is satisfactory, and personal preference will likely dictate the method of choice (note: the authors have mixed preferences).

After cutting through the skin and body wall is complete, the fat pad by the ovary is located, then the uterotubal junction (where the uterus and oviduct join) is grasped with a pair of forceps. The uterus is then severed just below the forceps. A second cut between the oviduct and the ovary frees the oviduct, which is still grasped with forceps. The oviduct is then placed into a sterile petri or tissue culture dish filled with bicarbonate buffer. The dissections from donor females continue as rapidly as possible to remove all the oviducts, but care is taken not to excessively manipulate oviducts, avoiding possible rupture and release or loss of egg masses.

c. Release of Eggs from the Oviduct   The oviducts are collected in a barrier hood, then brought to the microinjection laboratory and transferred to a clean dish. Some laboratories collect or place oviducts into a hyaluronidase-containing medium at this point. We postpone hyaluronidase treatment until the eggs, contained in cumulus masses, are liberated from oviducts in order to minimize enzyme exposure to eggs.

At this point, the swollen section (ampulla) of the oviduct should contain a large mass of eggs, and it is observed using low-power magnification under a dissecting stereomicroscope (Fig. 4). The oviduct or adjacent membrane/tissue, *but not the ampulla,* is first grasped with a pair of fine forceps or held in place with a

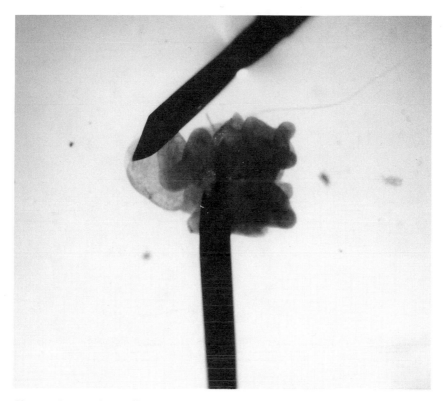

Figure 4. Oviduct collection. Low-power magnification (20×) of dissected oviduct; the needle at top is pointed to the ampulla where the cumulus mass is collected.

small gauge needle. The ampulla is then ruptured by tearing or pricking, with another pair of forceps or a needle, in proximity to the egg mass (Fig. 5). the intra-oviductal pressure is such that the egg mass is expelled without further manipulation. Using a large bore pipette (e.g., Pasteur pipette with bulb suction filler or mouth pipette), the egg mass is then placed into a dish containing 3 ml of bicarbonate buffer supplemented with hyaluronidase (type IV-S, 100 units/ml; Sigma). The eggs remain until the cumulus cells surrounding the eggs are digested and the zonae pellucidae are free of any attached cells or debris (Fig. 6). Practice is necessary to keep the exposure of eggs to hyaluronidase to a minimum (~3–6 min in total).

To remove the eggs from the hyaluronidase-containing medium and for routine handling of eggs, a glass pipette (200 μm final inner diameter) is first filled by capillary action with medium. Then two small air bubbles are drawn into the narrow bore to enhance fine control of egg manipulation (Fig. 7).

The isolated eggs are carefully picked up, avoiding rough vacuuming action that will damage the egg. The eggs are placed in a new dish of hyaluronidase-free medium,

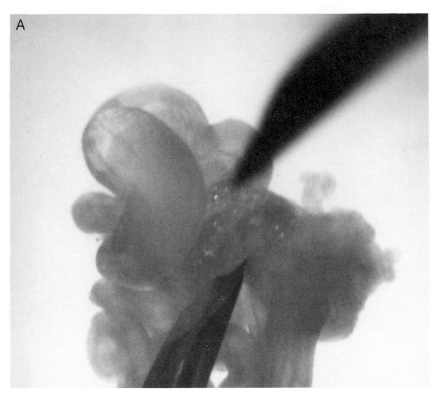

Figure 5. Cumulus mass collection. The sequence of events necessary to harvest cumulus masses is depicted under high-power magnification (40×). (A) The oviduct is positioned with two 25-gauge needles. (B) One needle is used to prick or cut the ampulla. (C) The cumulus exudes from the ampulla.

minimizing the transfer of any remaining extraneous cells/debris. The eggs are then pipetted into a second wash and counted. Good quality eggs are selected and placed in a dish containing microdrops of medium until readied for microinjection.

   d. Egg Development and Microinjection Timing   During any manipulation with bicarbonate buffer, attention is paid to the color of the medium and egg morphology to ensure pH balance. As the CO_2 diffuses, egg viability decreases rapidly. If eggs are harvested early, before pronuclei are well formed, the microdrop dish can be stored in a modular incubator chamber (Billups-Rothenberg, Del Mar, CA) and placed in a standard temperature-controlled incubator in order to maintain temperature, humidity, and atmospheric conditions relatively inexpensively. The optimal windows for microinjection range from the time when the male pronuclei at the periphery of the egg membrane are identifiable until the male and female pronuclei

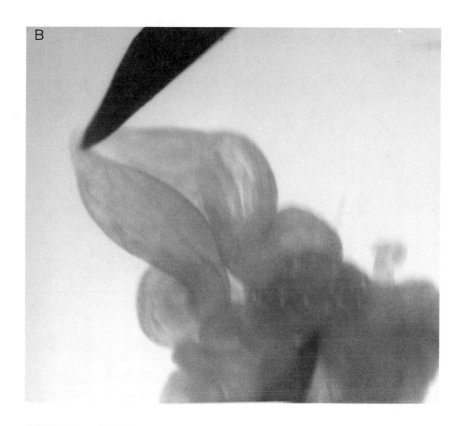

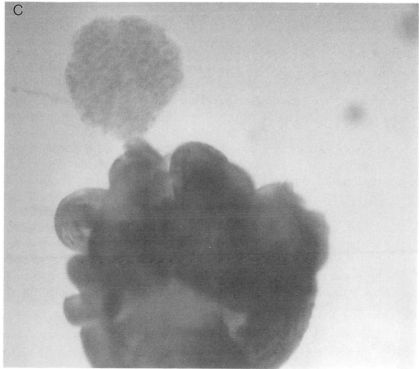

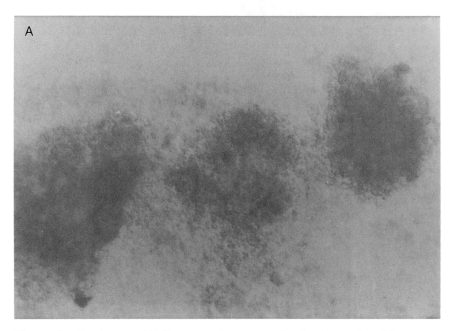

Figure 6. Egg harvest. (A) Three cumulus masses shortly after hyaluronidase treatment. Here, cumulus cells are just beginning to dissociate from the eggs. (B) After 5 min in medium containing hyaluronidase, eggs are well dissociated from cumulus cells and debris. (C) After washing eggs in fresh medium (without enzyme), the cumulus cells and other debris are removed. Note here that, in obtaining eggs from an outbred donor, a proportion of degenerating or abnormal ova are obtained in addition to the pronuclear eggs.

merge just prior to the first cleavage division. The time of this window varies between strains with outbred and hybrid strains having a more uniform and a smaller time window compared to inbred mice. Typically the injection window is 3–4 hr, but it is affected by *in vitro* culturing methods and care during routine handling.

e. Efficiency of Egg Culture Media conditions and strains have been chosen to obviate the 2-cell block in egg development. This allows one to test the quality of all media components and manipulations by culturing the eggs *in vitro* for several days. Under optimal *in vitro* culturing conditions, we have obtained 95% or greater development of C57BL/6 × SJL hybrid derived zygotes to the blastocyst stage in 4 days, with 70% then capable of growing out inner cell masses within 3 days. Any suboptimal components will lower these efficiencies or delay development. It is important to run quality checks of media periodically. Inbred and outbred strains of mice will exhibit lower developmental rates *in vitro,* and one should choose a representative strain with reasonable developmental uniformity for quality control checks of *in vitro* culturing conditions.

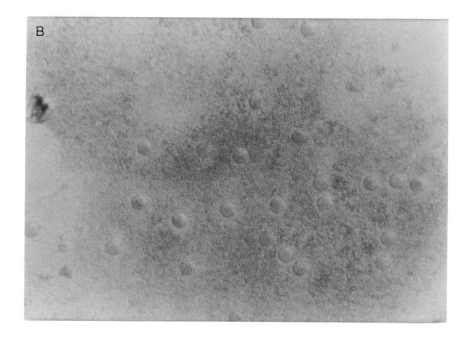

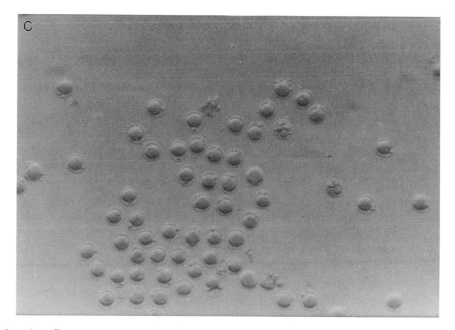

Figure 6 (*continued*).

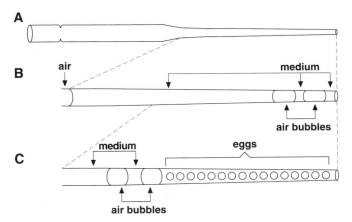

Figure 7. Egg handling pipette. After heating the tapered portion of a 9-in. Pasteur pipette to a molten state (orange color), the pipette can be pulled to an appropriate diameter (90–200 μm, depending on particular needs). After cooling (~3 sec), the two ends can again be pulled (without heat), breaking the glass evenly. The diagrams illustrate the procedure from pulling a pipette (A) to loading pipette (B and C represent enlargements of the distal end of the pipette). (B) Medium is drawn up the pipette by capillary action. When the medium reaches equilibrium, the pipette is attached to the mouth pipette assembly, and two air bubbles are drawn into the tip of the pipette. The addition of air bubbles provides a greater precision in egg movement. (C) Following the air bubbles, eggs are drawn into the pipette. For egg transfers to recipient females, a tip diameter slightly larger than the eggs is used. As shown here, the smallest amount of medium is used as the eggs are drawn, side by side, into the pipette. For washing eggs or moving them from one treatment dish to another, a wider bore pipette (i.e., 200 μm diameter at the tip) is more convenient.

C. Microinjection Needles and Slides

1. Types of glass

In actual practice, either borosilicate or aluminosilicate glass is used to make microinjection needles. Additionally, thick-walled (ID:OD < 75%) or thin-walled (ID:OD > 75%) glass may be used for needle production. The ID:OD ratio we prefer is just under 75% (inner diameter 0.027 inches, outer diameter 0.037 inches). This measurement is important because it determines whether the tip of the microinjection needle will be fused shut or left open after it is pulled. Thick-walled glass will tend to shut at the tip. An excellent reference for micropipette pullers and microneedles was written by Flaming and Brown (1982). There are two types of glass tubing configurations as well that are used for microinjection, namely, hollow capillary tubes (e.g., KG-33; Garner Glass, La Jolla, CA) and tubes with an internal filament (e.g., TW100F-4; World Precision Instruments, Sarasota, FL). The hollow microcapillaries require backfilling (from the tip back), which is performed after the needles are attached to oil-filled lines and an opening has been created at the tip. The filament-containing tubes can be filled prior to attachment to hydraulic lines and before the tip is opened/cracked.

2. Preparation of Glass

Many protocols have been developed for pretreatment of washing of glass tubing to be used for forming microinjection needles, but we have bypassed such procedures without a loss in overall efficiency. We prefer to simplify the process of making transgenic animals, minimizing some procedures for the sake of expediency when possible.

3. Pipette Pullers

Pipette pullers are available from several manufacturers and are of vertical (e.g., Kopf, Tujunga, CA; Narishige, Greenvale, NY) or horizontal design (Sutter, San Raphael, CA; Fig. 8). These units have distinct variations in filament and con-

Figure 8. Pipette puller. Uniform tapers and shapes of microneedles can be programmed using horizontal (depicted) or vertical pullers.

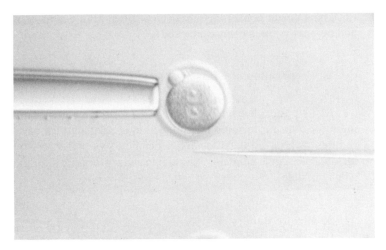

Figure 9. Injection and holding pipettes. With an egg in place (~70 μm diameter), the relative size of the injection and holding pipettes becomes evident. The holding pipette may range from 15 to 50 μm in diameter, depending on style (not polished to very polished), whereas the injection pipette is approximately 0.75 μm in diameter. Generally, the injection pipette is aligned with the pronucleus before insertion into the egg. (Magnification, 20×)

trol options. In our experience, the Sutter pipette puller (P-87) has proved to be very reliable, delivering a high degree of uniformity and consistency. Its controls include five programmable options, with the ability to cycle several combinations of parameters during the pulling of a single pipette. With a preset ramp program, one can evaluate the heating of any filament arriving at a "ramp" setting, allowing cross-referencing and use of different micropipettes and filaments.

4. Injection Pipette Shapes

We have used three basic shapes of pipettes/needles for microinjection that differ in two regions, namely, the tip, which is the final taper that directly enters the egg during microinjection, and the shank, which is the taper from the tip back to the full diameter of the glass tube (Fig. 9).

Sutter Settings for Injection Pipettes[a]

Type	Heat	Pull	Velocity	Time	Pressure
1	750	170	60	250	800
2	765	175	60	250	800
3	780	180	70	250	800

[a]Ramp = 729, box filament 3 × 3 mm, single cycle.

The three basic shapes and applications are as follows. Type 1 injection pipettes have fast tapering shanks and tips that produce a wide diameter bore at all sections of the micropipette. They can be used for large pronuclei at a late stage in development. Type 2 pipettes have shanks and tips with longer tapers and more gradual changes in tube diameter. The majority of eggs are injected with type 2 pipettes, which are fine enough to inject small pronuclei. Type 3 injection pipettes have extremely fine tips with very long shanks. They can be used for injecting 2-cell eggs, or eggs with very small pronuclei just after fertilization.

It takes considerable time to optimize the parameters for the pipette shape one desires, but the uniformity of each pipette from the Sutter unit is worthwhile. One should understand the effects of altering individual parameters before trying to arrive at an optimal shape. In general, finer tips are achieved by increasing the heat or by increasing the pull. The taper of the shank is increased by decreasing the pressure, decreasing the time, or increasing the velocity. The most difficult "skill" to develop in arriving at a pipette shape involves working with parameters that are interdependent but without equal influence on pipette shape.

5. Beveling Hollow Capillary Microinjection Needles

After the microinjection needle is pulled, one method for creating a sharp tip is to break the tip of the injection needle on the holding pipette. This method, while expedient, provides tips that are not as reproducible as those preformed on a beveler. With practice of either method, however, one can easily produce a sharp tip that will penetrate the zona pellucida and pronuclear membranes without lysing or damaging the egg.

If beveling of the needle is desired, a reproducible way to create sharp pipettes is to grind the tips with a beveler (e.g., Sutter BV-10) using a 0.5 to 5-μm grit size diamond wheel. A fiber optic illuminator and a stereomicroscope with wide-field eyepieces allow adequate monitoring of the tip beveling. A positive-pressure gas supply (nitrogen; 60 psi) is attached through narrow-diameter tubing (Tygon, ID 1/32, OD 3/32, wall 1/32 in.) to the microinjection needle. This pressure helps prevent grit from accumulating in the tip.

To bevel the microinjection needle, first attach it to the nitrogen source, then secure it in the beveler elevator. The pipette is positioned so that the grinding direction is parallel and rotating away from the pipette and not crosscutting the tip. Next, the needle tip is lowered onto the wheel with the coarse adjustment. The shadow of the pipette on the wheel is then brought into the viewing field of the binocular microscope, and the injection needle is carefully lowered with the coarse adjustment (then the fine control knob) until it just touches its shadow on the wheel.

The diameter of the tip is controlled by the time in contact with the wheel and by how far it's lowered onto the wheel. If the tip is ground too far, even with nitrogen pressure, it will accumulate grit within and make control of fluid flow difficult. The optimal bevel is observable with the microscope at $120\times$ magnification and takes 3–5 sec to form with the minimal amount of pressure on the grinding wheel. Practice is needed to create uniform tips. Once the tips are beveled, the microinjection pipettes may be bent to a 30° degree angle as the holding pipette if desired.

6. Microforge Shaping of Pipettes

a. Manufacturers   There are several microforge manufacturers (e.g., De-Fonbrune, TPI, St. Louis, MO; Narishige; Nikon, Melville, NY; Mini Tube). The instruments have common features, including a heating element, cooling capability, glass holders, and stereomicroscope assembly (either attached or separate). As recommended, with the DeFonbrune-type unit (Fig. 10), we use a 31-gauge heating filament made of platinum and iridium. For magnification we use either a stereo zoom or compound microscope head, with either type equipped with wide-field eyepieces. If desired, an eyepiece can be equipped with a micrometer/reticle to measure injection tip diameter and the overall dimensions of pipettes.

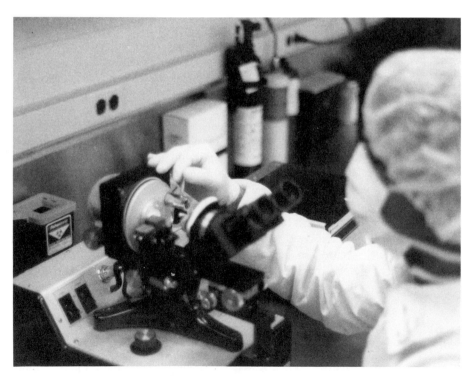

Figure 10.   Microforge. Needles may be fire polished or bent to specification using the microforge mounted with either a compound or dissecting microscope head.

b. Heating Filament The heating filament is bent to a U or V shape at the tip and held between two electrodes. A small glass bead is melted on the inside curve or tip of the filament. This allows uniform heating and will provide for uniform pipette breakage for production of holding pipettes as well as performing other shaping steps.

c. Holding Pipette Either the Kopf or the Sutter pipette pullers can be used to form holding pipettes. The Kopf puller has a coiled platinum heating element that forms much wider diameter and longer tapering pipettes. For holding pipettes the Kopf settings are approximately 13.2 for the filament heater and 2.1 for solenoid, although these settings do require fine-tuning that is element dependent.

The holding pipette is formed by angling the pipette 30° off the horizontal plane above the glass bead on the microforge filament. The rheostat is turned on and adjusted until the bead just begins to glow (orange in color). The pipette is lowered to gently contact the bead and is held for about 5 sec; a minimal deformation of the pipette should be observed. Then the heat is turned off, and, as the filament contracts on cooling, the fused bead and pipette break to form the tip of the pipette. The break will produce a flat face if the pipette angle was correct and the heat or fusing time was not excessive.

The holding pipette is then brought vertically over the filament glass bead, the electrode is turned on until the bead is again orange, and the pipette is lowered until the open tip just begins to melt. This polishes the glass tip, removing any sharp edges. (This step, however, may be omitted.) If an additional bend is desired so that the tip of the needle is in a plane horizontal to the injection slide, the pipette is again treated. The glass bead and filament are turned on until the bead is heated. Positioned near the side of the pipette, the filament is used to bend the pipette to a 30° angle (Fig. 11). The pipette is then stored in a dust-free container.

d. Injection Pipette After several microinjection needles are pulled and beveled, they, too, may be bent to a 30° angle should an elevated angle or straight approach to a microinjection slide or dish be desired. These needles may be stored in the same container as the holding pipettes, but sections of the container should be labeled for the types of needles stored to expedite selection.

7. Microinjection Slides for Eggs

a. Slide Designs DNA and eggs can be held in depression slides or in cut-out slides (Fig. 12; for cut-out slides, rather than a depression in the center, the equivalent area is cut out of a flat glass slide). The cut-out slides make use of disposable glass coverslips for holding microdrops overlaid with oil. The coverslips are attached to the bottom of the cut-out slide with a ring of paraffin (outside the periphery of the hole). (Use of 1- to 2-mm blocks/risers on the bottom of both ends of the slide keep the coverslip from contacting surfaces.) This configuration offers

Figure 11. Microneedle with 30° bend. Bends in the needles allow better access to cut-out slides and facilitate movements of needles parallel to the surface of slides or dishes. (Magnification, 20×)

an advantage to depression slides, because the coverslips are disposable and because there is less light refraction. However, because a greater needle angle is needed to manipulate eggs, as required for working with dishes on a microscope, the needle tips must be bent (30° angle, see Fig. 11) prior to microinjection. Again, personal preference will dictate the method of choice.

b. Microinjection Media For microinjection, a HEPES-buffered medium such as BMOC-3 plus HEPES (recipes in Table 1) supplemented with cytochalasin B (5 μg/ml; Brinster *et al.*, 1985) is used. If the egg membrane is dragged into the cytoplasm during injection, then the egg will likely lyse rapidly. Cytochalasin B stiffens the membranes during microinjection and helps prevent lysis of the egg. Alternatively, 7% (v/v) ethanol has a similar effect (P. Hoppe, personal communication, 1991). The membranes of one-cell eggs can also be stiffened by lowering the slide temperature to 10°C with a KT Model 5000 stage cooler (Technology Inc., Whitehouse Station, NJ). This will not alter egg survival or the percentage of transgenic mice produced, as long as the eggs are returned to 37.5°C within approximately 45 min (H. G. Polites and C. A. Pinkert, unpublished data, 1992). The

Figure 12. Cut-out slide. Custom slides use disposable coverslips for maintaining micro-drops (with eggs and possibly DNA samples) overlaid with oil; however, the fragility of cut-out slides may favor the use of alternative depression slides, which suffer from greater light refraction.

cooling effect takes up to 10 min before the egg membranes are noticeably stiffer; therefore, the simpler route using cytochalasin B is preferred.

    c. Arrangement of DNA and Media Drops   On the slide two drops are formed, one large drop (40 μl) of HEPES-buffered medium and a small drop (1- to 2-μl) of DNA. The large drop of medium is spread out from 1 to 1.5 cm so that the top surface is flattened to prevent refraction of light. The drops are then covered with silicone oil, making sure all the surfaces are submerged to prevent evaporation or optical distortion.

D. Microinjection Equipment

1. Microscopes

    A number of microscopes and micromanipulators are marketed with different advantages and disadvantages. Rather than sell one model or another, we highlight equipment representative of our experience.

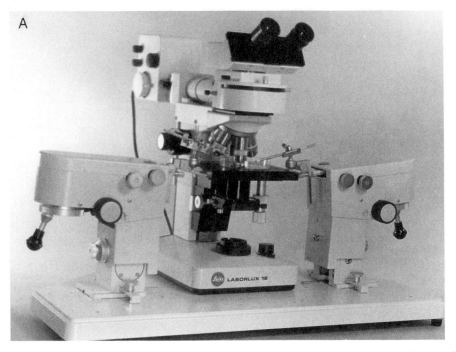

Figure 13. Inverted and upright DIC microscopes and micromanipulator assemblies used for DNA microinjection. (Leitz courtesy of Leica Inc., Deerfield IL; Narishige courtesy of Narishige USA, Greenvale, NY; Nikon courtesy of Nikon Inc., Melville, NY.)

Various microscope configurations (Fig. 13) from upright to inverted styles (Leitz Laborlux and Labovert, Leica Axioplan and Axiovert, Zeiss; Optiphot, Nikon; etc.) afford excellent differential interference contrast (DIC; either Nomarski or Smith). Magnification between 180 and $400\times$ is commonly used, and a decision should be evaluated before purchase. Final DIC magnification may be limited by the microscope brand or configuration, and desired field diameter is generally dictated by personal experiences or past training. Usually, DIC is only required at the microinjection magnification ($200\times$). Low-power magnification ($50\times$), while providing a reasonably large working field for priming pipettes and sorting eggs, does not require DIC optics. Thus, the final configuration and necessary adjustments (e.g., light filament, lens, diaphragms, and focusing lenses) should be tested preferably when the manufacturer's representative and a few eggs are available at the same time. The representative can set an "optimal" DIC configuration, but this does not necessarily provide the best pronuclear images. He or she should be consulted to ensure that all optical adjustments are demonstrated effectively and can be handled properly in the representative's absence.

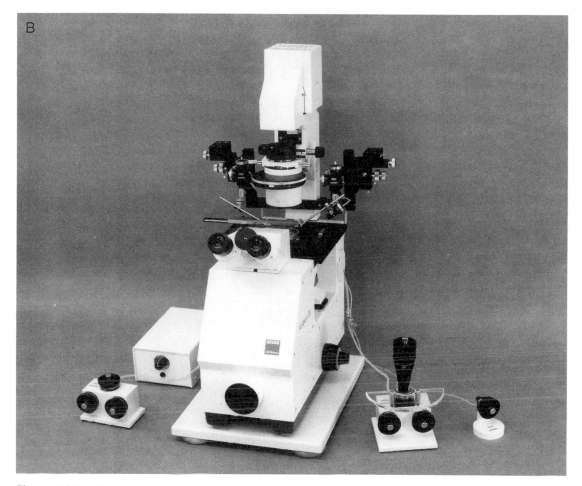

Figure 13 (continued).

2. Micromanipulators

2. Micropipette Holders   The pipette holders for the Leitz micromanipulators have two brass collars that hold together a plastic washer. The assembly is covered with a metal, threaded cap which adjusts the assembly around the micropipette to create a leak-proof seal. The tubing used for the washer should be of an appropriate diameter to match the outer diameter of the micropipette and needs frequent replacement to function effectively.

b. Micromanipulator Adjustments   The micromanipulators have several adjustable dials for regulating the sensitivity and tension on the joysticks. An autho-

C

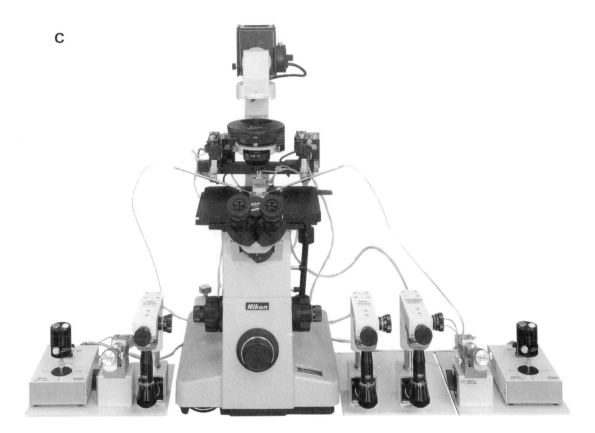

Figure 13 (*continued*).

rized technician should be consulted to demonstrate the various adjustments that can be made on the micromanipulators. Not all the adjustments are obvious from the visible knots and collars.

2. Antivibration Tables   When the microinjection needle is inserted into the egg, the slightest vibration will disrupt the egg, influencing lysis (immediately or delayed), and will make pronuclear placement rather difficult. To avoid frustration, the use of vibration-free tables is necessary. We have used a number of antivibration bases (Kinetic Systems, Roslindale, MA; Micro-G, Woburn, MA; Barry Wright, Burbank, CA), which support solid stainless steel surfaces on four nitrogen-driven pistons. In addition, and at lower cost, marble balance tables work well in most circumstances. Laboratory location and environment will play a role in defining specific needs.

3. Microinjection Systems

a. General   Systems for controlling microinjection of DNA and holding eggs are grouped into either air-driven systems (Narishige; Nikon; Zeiss; and Eppendorf, Brinkmann, Westbury, NY) or oil-driven hydraulic systems (Fig. 14). The air-driven systems can also be attached to electronically controlled delivery systems (e.g., Eppendorf 5221; PLI-100 Medical Systems Corp., Great Neck, NY; Narishige IM-200), that allow preprogrammed regulation of fluid delivery volume, time, and pressure. Excellent results may be obtained with or without such systems, and the choice is dependent more on injector experience or bias. The technique, skill, and training of the injector are factors that must ultimately be weighed in the choice of injection systems.

b. Oil-Driven System for Injection   The system that we currently use is a custom hydraulic unit (oil-driven) that is very inexpensive compared to other automated injector systems. The P-3 syringe assembly (C. A. Pinkert; Fig. 14B) is a low-cost, low-maintenance microsyringe assembly that can be used with different syringes (from 100 μl to 2 ml) for a number of applications. The P-3, or other microsyringe assembly (e.g., Eppendorf; 4095 Alcatel, Micro Instruments Ltd., Oxford, UK; Stoelting, Chicago, IL), is attached to the injection pipette and holder via flexible tubing (PE-100; Clay Adams/BD, Parsippany, NJ). Additionally, a double Luer fitting tube (Bio-Rad, Richmond, CA, No. 732-8202) can be attached to a three-way Luer stopcock value (Bio-Rad, No. 732-8103) for greater ease of handling (i.e., purging air in lines). Ultimately, the micrometer assembly will allow precise positive and negative pressure as required.

The syringes are loaded with Dow Corning 200 fluid. The tubing, the microinjection needle holder, and the needle itself are filled with fluid to dampen control and allow continuous flow of DNA without repeated syringe adjustments. The degree of control in the injection system can be reduced by using lower viscosity oil or greatly increased by higher viscosity oil (oils of 200–300 centistokes give very fine control). One advantage in this system is that purging air bubbles and filling the system are easy using Luer tip syringes filled with the appropriate oil. Air or any obstructions (broken pipettes, dirt, shaved pieces of tubing) should always be checked for and cleared from the system before setting up to perform microinjections.

4. Oil-Driven Holding Pipette System

The holding pipette system we use is controlled by the same microsyringe assembly. The syringe is connected with PE-100 or Tygon tubing (ID 1/8, OD 1/4, and wall 1/16 in.) directly to the back of the pipette holder (Fig. 14B). Dow Corning 200 fluid fills the entire system. Any air bubbles near the pipette holder or inside

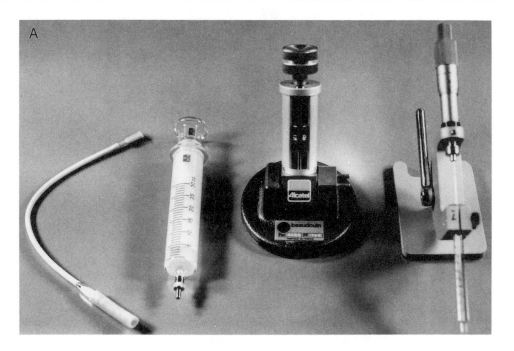

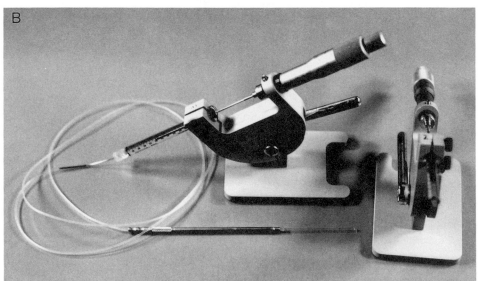

Figure 14. (A) Microsyringe assemblies (from left to right): mouth pipette control (length of tubing shortened here); glass syringe; and two, more elaborate, microsyringe assemblies. Whether air or hydraulic systems are used, these assemblies provide a low-cost alternative to electronic systems, while providing different degrees of precision in delivery. (B) P-3 microsyringe assembly with tubing from a micrometer-controlled Hamilton syringe to the manipulator collar and microneedle. The manipulator collar is mounted on the micromanipulator, and the needle is aligned before use.

the barrel can be removed by holding the pipette holder vertically, removing the two brass collars, and gently tapping on the side of the pipette holder. The bubbles will slowly move up the tube and be visible once they emerge from the pipette holder. A few extra turns on the syringe will flush the bubble from the tube.

To recharge the holding pipette system with more oil, remove the brass collars, push out some fresh oil, and quickly plunge the tip into a beaker of oil. Fresh oil can then be pulled up by reversing the syringe pressure while making certain that air bubbles are not introduced.

5. Electronic Delivery Systems

Automatic microinjectors have been developed for controlled injection of cultured cells and have been applied to DNA microinjection techniques. Of several manufacturers, Narishige, Nikon, and Eppendorf produce the most widely available systems. There are similar controls on all models that allow regulation of fluid delivery volume, time, and pressure. The pressure controls include separate regulators that allow a static holding pressure, an increased pressure for injection, and a higher pressure for clearing the injection tip. Advantages of automatic injectors over manually controlled micrometer syringe systems include the use of a foot pedal to activate the injection pressure and allowance for two fluid pressures during injection for easier egg penetration. For novices, an important skill to develop is an ability to control the flow of DNA through the microinjection needle. If the flow is too great, the membranes of the egg will be pushed away from the tip as the microinjection needle enters the egg; hence, the needle tip will not successfully penetrate the pronucleus or deliver DNA properly. If the flow is too slow, the microinjection needle will easily penetrate the egg, but the slow delivery of DNA will enhance the likelihood of egg lysis. With manual injection systems, a continuous flow rate between these two extremes is learned, but after much tribulation. An automated system can obviate one time-consuming stage of training.

It should be noted that not all electronically controlled systems have an easily accessible vacuum function to allow filling of the microinjection needle from the tip. Therefore, DNA is loaded from the back of the needles, as described for filament-containing tubes, or with the aide of microsyringes or capillary tubes. A nonmetallic syringe needle for filling micropipettes is available for such needs (MicroFil, World Precision Instruments).

E. Microinjection Procedure

1. Setting Up the Injection and Holding Pipettes

a. Holding Pipette   The holding pipette can be inserted directly into the pipette holder, after the system has been checked for air bubbles. A few turns on the

Hamilton syringe will push oil all the way to the tip of the pipette. The holding pipette is then oriented above the microscope lens so that the bend in the pipette is parallel to the slide. Finer adjustments are made by viewing under low power and adjusting the focus to ensure that the tip of the pipette is the lowest part of the holding needle.

b. Injection Pipette   The injection pipette is first filled with oil (the same as contained in the injection system) using a 4-in. 24-gauge needle and 5 ml Luer-lock syringe. While filling, keep the injection needle vertical and fill from the tip back, making sure to only trap one air pocket in the tip of the pipette. Once the injection needle is filled and the injection system is cleared of air bubbles, turn the microsyringe micrometer to push some oil out of the tip of the injection pipette holder. Insert the back of the injection needle into this oil drop and down into the pipette holder. The metal cap of the injection holder should be loose enough to avoid having to jam the injection needle through the brass collars or plastic washer. This procedure is done with the injection needle held vertically. The microsyringe micrometer is then turned to increase the pressure on the air trapped in the tip of the injection needle, which will keep it in place until it is all pushed through the opening at the tip.

c. Aligning the Injection Needle   The injection needle and holder are then placed on the micromanipulator and adjusted to orient it parallel to the slide. As with the holding pipette, the injection pipette is first oriented so that it is perpendicular to the microscope. Next, under low power adjust the angle of the injection pipette so that the full length of the bent tip section is in focus. This adjustment is sometimes time-consuming but is important because it ensures that the pipette enters the egg at a straight angle and avoids shearing the egg as one pushes the needle into the pronucleus. This adjustment will greatly increase the ease, speed, and precision of microinjection. It is also important to put the same degree of bend into each injection pipette to reduce the amount of time that is needed to realign each injection pipette in the micromanipulator.

The final adjustment is to check the tracking of the injection needle pipette under high power. With the tip of the needle in focus on one side of the field, move the needle directly across to the opposite side. If the tip goes out of focus, adjust the angle knob for the micromanipulator until the needle tip stays in focus across the entire field. This adjustment also increases the speed, accuracy, and precision of microinjection by having the injection pipette tip working plane overlap the focal plane. This saves time during injection of multiple eggs by eliminating the need to realign the injection tip and pronucleus for each egg.

d. Loading DNA into the Injection Needle   If the injection needle tip has been sufficiently beveled, then oil should be flowing out of the tip by the time orientation adjustments are complete. The size of the oil beads running down the pipette will indicate the diameter of the tip. The micrometer assembly is dialed back to reduce

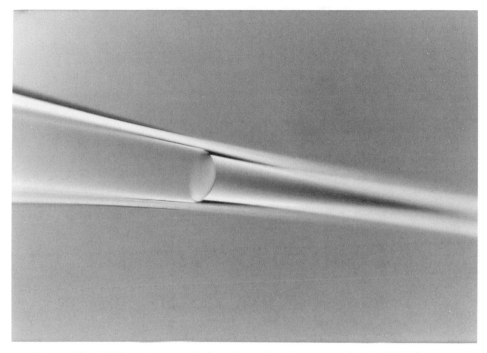

Figure 15. DNA meniscus in a hollow fiber microinjection needle. The DNA flow rate can be readily monitored when using a hollow fiber injection needle. The meniscus can be used to determine or follow DNA flow during microinjection procedures. (Magnification, 320×.)

the flow, and the injection pipette is lowered into the DNA drop. The quality of the tip can be observed at this point and the needle replaced if any problems are found.

DNA is drawn up into the injection pipette as far as possible, and filling is allowed to continue while the eggs are being selected and prepared for microinjection (Fig. 15). Once the injection pipette is loaded and the eggs are ready for microinjection, switch the stopcock valve to open the microsyringe micrometer and reverse oil flow to start the DNA/oil meniscus moving as slowly as possible down the pipette (observe under high power).

2. Preparing Eggs for Microinjection

Once the DNA is readied in the injection pipette, the eggs to be injected are selected from the holding dish. The eggs are first transferred to a drop of injection medium (containing cytochalasin B) in a dish under oil. The dish is held on a warming plate for 5–7 min or until the egg membranes recover from "blebbing" (i.e., no longer have a scalloped outline).

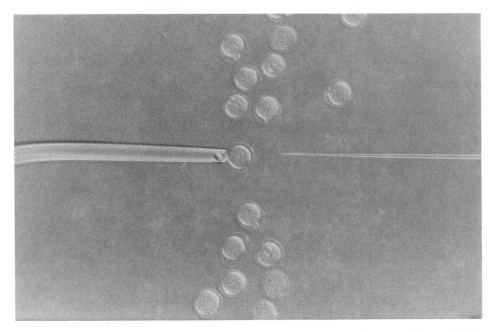

Figure 16. Egg organization during microinjection. Noninjected eggs are located in the drop of medium, covered with silicone oil, in a group above the needles (injection area), whereas eggs are placed below the needles after injection. (Magnification, 100×.)

Only eggs with visible pronuclei and normal appearance are selected and transferred to the drop of medium on the microinjection slide. There are many arrangements for the eggs on the slide, but we prefer to keep them organized in a line above and below the microinjection area (where needles are located; see Fig. 16). The slide is placed on the microscope, the holding and injecting pipettes are lowered into the drop of medium under low power, and the eggs are visualized. The high-power (DIC) objective is then moved into place, and microscope adjustments are made with the eggs in view.

3. Steps for Efficient Manipulation of Eggs during Microinjection

a. General   As with any precision skill, constant review should be focused on the efficiency of hand movements to reduce unnecessary work or strain on the operator. A major facet of good microinjection skill is keeping the movements and adjustments of the microscope or manipulators simple and quick. The faster the egg can be returned to the correct temperature and medium (even *in vivo* reimplantation), the more likely that the eggs will survive the *in vitro* manipulations.

b. Micromanipulators The joystick on the micromanipulators should be adjusted to give adequate movement of the pipette with minimal pressure from the fingers. Pulling or pushing too hard on the joystick will slightly deflect the injection or holding pipette tip and reduce the accuracy of injecting the pronucleus. Keep in mind that such joysticks provide a three-dimensional movement of the micropipettes and do not afford the same accuracy as movement in two dimensions. In some instances, particularly in training, it is worthwhile to avoid use of three-dimensional movement.

c. Microinjection Needle With the eggs on the slide next adjust the focal plane under high power to be about 1-1/2 to 2 egg diameters above the coverslip. Bring the tip of the injection needle into focus. During the injection of eggs on this slide the focus does not have to be readjusted, nor does the height of the injection needle need to be altered. The flow of the DNA can be checked by waving the injection tip in the media and looking for currents formed near the tip—dial back the flow rate if this is observed. Fine micrometer adjustments are usually sufficient during the injection process to control the DNA flow rate (and are typically made once or twice during the injection of about 50 eggs).

d. Controls and Egg Manipulation During microinjection, eggs are pulled to the holding pipette by suction (negative pressure). The focus of the pronuclei is adjusted with the height control of the holding pipette. The positions of the egg and injection needle are kept as close as possible to the center of the viewing field by use of the joystick or manual two-dimensional control knobs.

The number of manipulations for each hand are kept to a minimum. One hand controls the injection needle position with the joystick (or control knobs) and the microsyringe assembly for suction control of the holding pipette. The other hand controls the position of the holding pipette and the focus of the egg with the x-axis control on the holding pipette. The first hand also occasionally moves the slide with the stage knob and as well adjusts DNA flow. This arrangement allows a skilled operator to microinject routinely an average of 60–120 eggs per hour.

4. Visualizing and Injecting the Egg Pronuclei

a. Orientation The optimal orientation for microinjection of pronuclei is to have the male pronucleus closest to the injection needle (Fig. 17). The height control of the holding pipette is raised and lowered until one is sure the center of the pronucleus is in focus. The injection needle tip should also be in focus prior to penetrating the egg.

b. Egg Membrane Penetration with Injection Needle Because the DNA is constantly running out the tip of the injection needle, one has to work quickly to

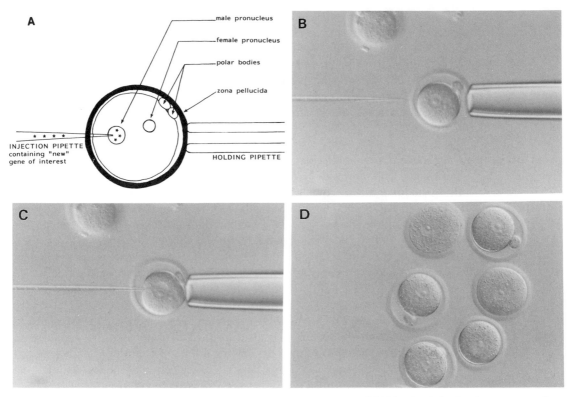

Figure 17. DNA microinjection. (A) Diagram of DNA microinjection into a pronuclear murine zygote [reproduced with permission from Pinkert (1987)]. A pronuclear egg is held by a large bore pipette using gentle suction. A small bore injection pipette containing the DNA solution is inserted into the male pronucleus; the DNA solution is then slowly expelled into the pronucleus, which expands approximately 2-fold. The diameter of the mouse egg is about 75 μm. (B) The male pronucleus is in focus and the injection needle aligned before DNA injection. Note the diameter of the pronucleus. (C) During injection, note the increase in diameter of the pronucleus and the uniform spherical periphery indicative of full expansion. (D) Within minutes of injection, some of the eggs will obviously lyse and "fill" the zona pellucida (e.g., eggs at 11 and 3 o'clock have lysed). (Magnification (B–D), 200×.)

penetrate the zona pellucida membrane and pierce the pronuclear membrane to prevent the buffer and DNA from "pushing" the membranes (note the cytoplasmic delivery in Fig. 18). A sharp needle without debris stuck to the tip will have the best penetrating characteristics. The DNA flow may have to be finely adjusted if bubbling is still occurring with a good pipette.

   c. *Filling the Pronuclei with DNA Solution*   The injection pipette should fill the pronucleus with DNA until the expansion of the pronucleus stops. One must

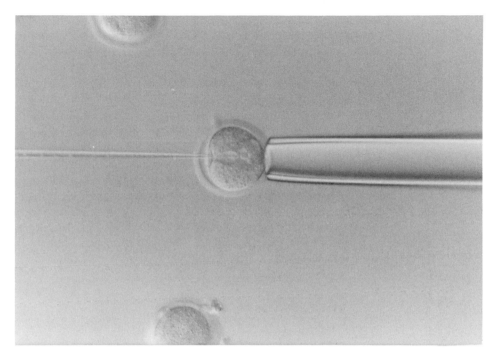

Figure 18. DNA microinjection into cytoplasm. Misdirection of injection can occur if the injection needle is not aligned with the pronucleus or is too blunt for penetration, or if the flow rate of DNA solution is too great. Cytoplasmic injection is to be avoided but can result in the appearance of a cytoplasmic vacuole that may be maintained even after the needle is withdrawn. (Magnification, 200×.)

simultaneously observe the swelling of the egg to ensure not too much DNA is delivered. When first learning microinjection, one should try several different degrees of filling and record the results. Practice is the only way to understand the correct amount of DNA to deliver. The injection pipette does not always penetrate the pronucleus perfectly, and leaking of DNA out of the pronucleus will frequently occur. In this case the pronucleus will not expand as rapidly, and the whole egg will enlarge. The egg will lyse if too much DNA is delivered into the cytoplasm.

5. Criteria for Microinjection to Yield Transgenic Mice

a. Treatment after Microinjection Regardless of the skill and speed of the injector, the eggs should be left on the slide no longer than 30 min. Once the injection slide is completed, the pipettes are raised, the slide carefully returned to the warming stage on the dissecting microscope, and the injected eggs transferred through two "wash" dishes of culture medium (without cytochalasin B) and then placed in an appropriately labeled microdrop dish.

b. Lysed Eggs   Routinely 10–25% of eggs will lyse as a result of microinjection, but this figure is highly dependent on the skill of the injector, the egg background strain, the injection pipette shape, and the quality and concentration of the DNA preparation. If less than 10% of the microinjected eggs lyse, such results would suggest that the volume of DNA delivered into the egg pronucleus was insufficient to generate transgenic mice efficiently.

c. Record Keeping   Accurate records during all of the microinjection-related procedures aid in the development of microinjection skills, help pinpoint sources of problems, and help maintain mouse identification (Fig. 19). Representative data

Figure 19.   Record keeping and animal care are paramount to transgenic experimentation. Animals are maintained in rooms under controlled conditions of temperature, humidity, and photoperiod in barrier caging that prevents introduction of bacterial, viral, or other pathogens.

sheets that we use are illustrated at the end of this chapter (see Appendix), whereas Chapter 3 details and explains specific needs and requirements.

d. In Vitro Culture of Microinjected Eggs Zygotes can be routinely cultured to the 2-cell or blastocyst stage, then transferred to the oviduct of day 1 pseudo-pregnant mice (Bronson and McLaren, 1970; Kooyman *et al.,* 1990). Using outbred (e.g., ICR) or hybrid (e.g., C57BL/6 × SJL) mice, typically 40–60% and 60–80% of injected eggs (60–80 and 75–100% of noninjected eggs), respectively, develop overnight to the 2-cell stage in our experience. A high proportion of eggs going to the 2-cell stage will usually proceed to the blastocyst stage. The significant decrease in egg viability following microinjection is due to the trauma associated with disturbing the ionic and physical environment of the zygote. Egg culture provides a reasonable assessment of microinjection proficiency. If survival frequencies approach noninjection efficiencies (culturing control, noninjected eggs simultaneously), then review of injection procedures is warranted. If survival frequencies drop way below controls, then experimental procedures should also be reevaluated. Inbred strains exhibit lower overall egg survivability and may not develop beyond the 2-cell stage with the suggested media preparations outlined.

F. Egg Transfer

1. Ova Reimplantations

Procedures and factors influencing egg transfer have been described (Adams, 1982; Hogan *et al.,* 1986). Briefly, a pool of female mice are maintained and observed for visual evidence indicative of proestrus (vulval and vaginal appearance; Champlin *et al.,* 1973). The day after mating to vasectomized males, those females exhibiting copulatory plugs are used as egg transfer recipients. (As a result of mating to vasectomized males, these females will have the ability to maintain the eggs through gestation, without uterine competition from naturally ovulated eggs. Because the mating to vasectomized males renders the females "pseudopregnant," if fertilized eggs are not transferred, the females will not return to estrus for about 11 days.)

For transfer, recipients are anesthetized and placed in ventral recumbency under a dissecting microscope. The surgical incision site is liberally swabbed with disinfectant (e.g., Betadine). Using sterile surgical instruments, a 5-mm transverse skin incision is made between 10 and 15 mm from the spine, posterior to the rib cage. By sliding the skin incision to a point located above one-third of the way between the dorsal and ventral midline, the ovarian fat pad will be located through the intact body wall. When the fat pad has been located, the body wall is gently grasped with blunt forceps, and a small incision is made in the body wall. The ovarian fat pad will then be exteriorized (the fat pad is the landmark used to locate the ovary and oviduct). The oviduct is maintained in place using paper sponges around the uterus or placing a sponge 4 cm long and 3 mm wide through the mesometrium, or by grasping the fat pad with a small surgical clamp (Hogan *et al.,* 1986). The bursa surrounding the ovary is then either cut or torn to expose the ostium of the oviduct.

Once these steps are accomplished, eggs are obtained using mouth pipetting under a second dissecting microscope. Between 25 and 30 injected ova are then expelled into the ostium of the oviduct of the recipient and are blown through to the ampulla by gentle pressure through the pipette. (Note: The number of eggs transferred may vary based on injector skill or available microinjected eggs. Up to 4 control eggs may be transferred to help ensure pregnancy maintenance if there is concern regarding proficiency or quantity of microinjected eggs.) The reproductive tract is then carefully replaced into the abdomen, and the body wall is closed with 1 or 2 sutures followed by skin closure with wound clips. Surgery is performed with the animal resting on dry, soft padding. Body temperature is maintained during the procedure. Surgical depth of anesthesia is ensured by observation of respiratory rate, tail reflex, or foot pinch reflex at frequent intervals.

2. Vasectomy

Male mice are vasectomized and used to mate proestrous-stage females in order to induce pseudopregnancy. For this procedure, each male mouse is anesthetized. The abdomen is then thoroughly swabbed with Betadine. Using sterile instruments, a 10 to 15-mm midventral skin incision is followed by an incision through the body wall musculature. The fat pad by each testis should be visible and is grasped using blunt forceps to exteriorize the testis and vas deferens. The vas deferens can then be isolated and ligated or cauterized. Generally, the removal of 15–25 mm of the vas deferens is sufficient to ensure irreversible surgery (and limit worry of reanastomosis). The procedure is repeated for the second testis. Only one abdominal incision is made for this procedure; testes are exteriorized through a single site. On completion of the vasectomy, the body wall is closed with sutures and the skin closed using wound clips. (Note: We generally place the sections of vas deferens aside following their removal to keep count of the number of sections removed. Once in the last 6 years, a trainee left one intact vas deferens but was able to identify the error while the mouse was still sedated before the wound clips were applied.)

Support care will include placement of the animal on clean, soft padding during the procedure and warming anesthetized animals. Observation of the respiratory rate and foot pinch reflex of the animal is routine throughout the procedure and postoperative period. Wound clips are removed after healing (i.e., in 7 to 10 days). At this time, the males are tested for mating ability. Copulatory plug formation and evidence of two infertile matings are required before the males are used in the vasectomized male battery. As a rule, we generally maintain twice the number of males that might be used for a given day of experiments. To arrive at this number, we estimate the number of recipients needed and assume that we will obtain copulatory plugs in approximately 50% of the paired "proestrous females," selected on vulval and vaginal appearance. The vasectomized males are used until infirmities appear or until they do not mate with six successive proestrous females.

3. Postoperative Care

After completion of any surgical procedure, animals will be kept at 30°C and carefully monitored until they regain reflex responsiveness. At that time, they are placed into clean cages and observed daily thereafter. As these surgeries are short in duration, are not invasive in nature, and aseptic techniques are employed, postoperative analgesia and antibiotic use are not favored.

G. Egg Cryopreservation

Cryopreservation of preimplantation-stage ova was first described in 1972, using DMSO as a cryoprotectant (Whittingham et al., 1972; Wilmut, 1972). Procedures for cryopreservation of murine eggs have been greatly simplified over the last few years. We have utilized two different approaches to freeze early-stage eggs. A vitrification kit is commercially available (Cryozyte CT100; TSI, Milford, MA) that is suitable for storage of 375 eggs (8-cell to early blastocyst stage). Using this kit, survival rates greater than 80% were reported for a number of mouse strains after vitrification and storage in liquid nitrogen (TSI data reports).

In addition, we have used a recently described one-step freezing method for cryopreservation of murine eggs (Leibo and Oda, 1993). As described, for freezing medium, M2 (or BMOC-3 + HEPES) is supplemented with 2M ethylene glycol (Sigma E-9129) and 7.5% polyvinylpyrrolidone (PVP; Sigma P-2307). Eggs are incubated in the freezing medium for 5 min, drawn into a sterile 0.25-ml cryopreservation straw (PETS, Canton TX; 05-079-119-1), sealed, and plunged into liquid nitrogen. To recover eggs, the straws are removed from the liquid nitrogen tank and immediately thawed by submersion in a 37°C water bath. The contents of the straw are expelled into a dish of sterile culture medium and the eggs are washed twice before readied for transfer. We have obtained a 50% or greater survival rate of 2- to 8-cell eggs; however, our results with morula stage eggs were significantly lower (C. A. Pinkert and L. W. Johnson, unpublished data, 1993).

Routine freezing of transgenic lines has several important advantages.

a. Should loss of transgene expression occur in latter generations of a given line, the original genetic background (and characterized transgene expression patterns) can be recovered.

b. For transgenic lines that will be in great demand, or where breeding spans a number of generations, genetic drift may also be problematic. This is a crucial concern in the use of transgenic models to derive clinical correlations. Genetic drift can be reduced by freezing many ova at an early or specific generation.

c. Disease or line contamination can unexpectedly destroy valuable transgenic animals. Preservation of a relatively small number of eggs can safeguard any given project.

d. In a catastrophic situation, the use of cryostorage can provide a means not only to safeguard research but to influence rapid return to productivity.

e. Cryopreserving transgenic lines can significantly reduce or eliminate the costs associated with housing and maintaining lines of mice.

f. Egg cryopreservation can be used in conjunction with rederivation of pathogen-free stocks of animals. In mouse experiments, embryo transfer of both frozen and non-frozen eggs have been used to rederive important animals. Various rodent pathogens have been "cleaned up" using embryo transfer. The classic method of obtaining rederived pathogen-free stocks involves the aseptic collection of fetuses by cesarean-section with transfer to pathogen-free foster mothers. However, while many infections can be vertically transmitted, the zona pellucida can provide a barrier to infection of preimplantation ova (see Reetz *et al.*, 1988). Several groups have successfully obtained pathogen-free offspring from heavily contaminated stock using variations of embryo washing/transfer techniques (Reetz *et al.*, 1988; Rouleau *et al.*, 1992). We have successfully used these procedures to rederive mouse lines that were contaminated by indigenous murine pathogens.

III. SUMMARY

As summarized in Fig. 20, the steps involved in producing and evaluating transgenic mice derived by DNA microinjection are straightforward, although mastering the steps can be very time consuming. In general, it takes between five and twelve months of dedicated effort to become proficient at the necessary techniques. One must begin with embryo handling and practice embryo transfer skills to develop good coordination. These skills will ultimately maximize all gene transfer efforts. Then, microinjection practice will lead to the production of the first transgenic mice, but continued commitment will be necessary in order to develop proficiency.

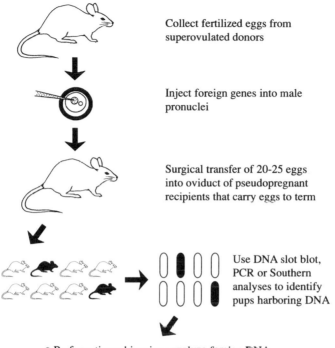

Collect fertilized eggs from
superovulated donors

Inject foreign genes into male
pronuclei

Surgical transfer of 20-25 eggs
into oviduct of pseudopregnant
recipients that carry eggs to term

Use DNA slot blot,
PCR or Southern
analyses to identify
pups harboring DNA

● Perform tissue biopsies - analyze foreign DNA
integration, mRNA transcription, and protein
production

● Establish transgenic lines to study gene regulation
in progeny

Figure 20. Gene transfer in mice. The methodology employed in the production and subsequent evaluation of transgenic mice is presented schematically.

It is hoped that this chapter and the chapters that follow will illustrate the techniques necessary to create and maintain appropriate animal models both successfully and efficiently.

APPENDIX

For record-keeping purposes there are a number of database systems and programs available for experimentation and colony-related needs. The following figures highlight some of the records or windows that we have found to be of value in order to maintain a running status of projects.

TA/ESC COLONY SUMMARY

Date: _____ By: _____

COLONY TOTALS

	Room #	# of animals	# small cages	# large cages
Expt. Section	_____	_____	_____	_____
	_____	_____	_____	_____
	_____	_____	_____	_____
	_____	_____	_____	_____
	_____	_____	_____	_____
Expt. Subtotal:		~~~~~~~~	~~~~~~~~	~~~~~~~~
Stock Section	_____	_____	_____	_____
	_____	_____	_____	_____
	_____	_____	_____	_____
	_____	_____	_____	_____
Stock Subtotal:		~~~~~~~~	~~~~~~~~	~~~~~~~~
Colony Total:		_____	_____	_____

* *

STOCK AVAILABLE

	# of animals	# of animals designated for use
S-W recipient ♀ [≤7wks/>7 wks]	_____	
C57BL/6 ♀		
3-4 weeks	_____	_____
5-8 weeks	_____	_____
>8 weeks	_____	_____
B6xSJL donor ♀		
3-4 weeks	_____	_____
5-8 weeks	_____	_____
>8 weeks	_____	_____
11 or 12	_____	_____
_____	_____	_____
_____	_____	_____

* *

WEANING

	Total	# to stock	# to mating
S-W ♀	_____	_____	_____
S-W ♂	_____	_____	_____
B6xSJL F1 ♀	_____	_____	_____
B6xSJL F1 ♂	_____	_____	_____
C57BL/6J ♀	_____	_____	_____
C57BL/6J ♂	_____	_____	_____
SJL/J ♀	_____	_____	_____
SJL/J ♂	_____	_____	_____
_____	_____	_____	_____
_____	_____	_____	_____
TOTAL WEANED:	_____		

Figure A.1. Colony summary sheet. Additional pages are also updated weekly with identification of animals on a room by room basis.

INJECTION SCHEDULE from _____ to _____

Expt day/ Initials	# Mice/ Strain	Age (B/W/R)	Fri	Sat	Sun	Mon	Tues	Wed	Thurs

Figure A.2. Injection schedule. A list is posted by the freezer with the aliquoted hormones in addition to individual cage cards identifying appropriate animals and treatments. Age abbreviations: B, birth; W, weaning; R, received date.

GENE LOG

Last Revised:

#	Name/Expt Description	Lab	Project	Rec'd	Date Injected	Eggs Trans	Trans Total	#Tg/ Total	Comments (# = transgenic animal log #)
1	El ras	CAP	DNA	5/1/91	practice/training	826	49	13/68	
2	AB1	PAW	ES cell	10/21/91	11/2/91 - 11/6/91	75	7	5/8-20	12 control eggs were used
3	MMTV-slGF.3	CAP	DNA	10/22/91	12/19/91	259	12	16/58	
4									
5									
6									
7									
8									
9									
10									
11									

Figure A.3. Gene log. All projects are kept on one central inventory. Abbreviations: #, sequential experiment log number; Eggs Trans, total of eggs transferred to all recipient females; Trans Total, total of all embryo transfer recipients; #Tg/Total, transgenic animals per all founders obtained.

MICROINJECTION

\# & NAME/EXP. DESCRIP.:_____

Notebook
Reference:_____

DATE:_____   STRAIN:_____   CLIENT:_____

INJECTOR:_____

\# OF DONORS OPENED:_____

\# OF EGGS OBTAINED:_____

\# OF EGGS GOOD:_____

\# OF EGGS GIVEN TO OTHER INJECTOR (NAME & #):_____

INJECTED	NOT INJ.	TOTAL	# ALIVE	START/FINISH	COMMENTS

EGGS FOR TRANSFER:_____

RECIPIENTS AVAILABLE:_____

START TRANSFER:_____

END TRANSFER:_____

Figure A.4.   Microinjection log. A detailed log is maintained for each day of injection.

TODAY'S DATE: ___/___/___

MATERNITY

EXPERIMENT (# & NAME):_____   REVIEWED: _____
DNA CONCENTRATION: _____

RECIPIENT D 1: _____/_____/_____   INJECTOR: _____
EGG STAGE (DAY): _____   TRANS DATE: _____/_____/_____   TRANSFER: _____
OPEN ON DAY*: _____   OPEN DATE: _____/_____/_____   OPENED: _____

ID #	[ep#]	Eggs transferred/side	L/C#	Pups/ fetuses	RS	Comments
TOTALS						

_____/_____   FEMALES PREGNANT   _____ %

_____/_____   # IMPLANTATION   _____ %
                   SITES (PUP + RS)

_____/_____   PUPS (OR FETUSES)   _____ %

ADDITIONAL COMMENTS:
_____
_____
_____
_____
_____

#Livebirth (L), Cesarian (C). If C, then indicate the side that the pup or resorption site located.
*Based on recipient

Figure A.5. Maternity log. After the embryo transfers are completed, a copy of the day's transfers are placed into a central book. Although all cages are checked twice daily, the maternity log book is checked daily, to be certain that no litters have gone unattended.

SAMPLE ANALYSIS

EXPERIMENT (#, name): _____ PAGE: ____ OF ____ INVESTIGATOR: _____ ANALYSIS BY: _____ REVIEW: ____

DATE SAMPLES READY: _____ DATE SAMPLES SENT: _____ DATE RESULTS RECEIVED: _____ DATE ANIMALS SHIPPED/SAC'D: _____

LOG #	ANIMAL ID	DAM	SIRE	DOB	DOD/DOS	TISSUE	DNA	RNA	METHOD	ANALYSIS/COMMENTS

Figure A.6. Sample analysis. All tail samples or other biopsies are given a log number and maintained in a common notebook for the laboratory.

TRANSGENIC ANIMAL SUMMARY

EXPERIMENT (#, name): _____   Investigator: _____   Founder Transgenic Rate: ____ / ____   Date Revised: _____

ANIMAL ID	LINE #	STRAIN	DOB	DOD	DNA		mRNA		PROTEIN		PATH-OLOGY	PROGENY	COMMENTS
					LOG	RES	LOG	RES	LOG	RES	(Y/N)		

Figure A.7. Transgenic animal summary. After founder transgenic animals are identified, a log is maintained by experiment (individual construct or a gene, and accession number) for founders (and subsequent offspring).

ACKNOWLEDGMENTS

The authors gratefully acknowledge their mentors and colleagues, and the opportunities and support from the Laboratory of Reproductive Physiology at the University of Pennsylvania, The Upjohn Company, Hoechst-Roussel Pharmaceuticals, and the University of Alabama at Birmingham. Appreciation is extended to K.D. Clay, M.H. Irwin, L.W. Johnson, C. Traina, G.J. Vergara, K.R. Marotti, and L.E. Post. This work was also supported in part by funds from the U.S. Department of Agriculture and the National Institute of Environmental Health Sciences.

REFERENCES

Adams, C. E. (1982). Factors affecting the success of egg transfer. *In* "Mammalian Egg Transfer" (C. E. Adams, ed.), pp. 176–183. CRC Press, Boca Raton, Florida.

Brinster, R. L. (1972). Culture of the mammalian embryo. *In* "Nutrition and Metabolism of Cells in Culture" (G. Rothblat and V. Cristafalo, eds.), vol. 2, pp. 251–286. Academic Press, New York.

Brinster, R. L., and Palmiter, R. D. (1986). Introduction of genes into the germ line of animals. *Harvey Lect.* **80**, 1–38.

Brinster, R. L., Chen, H. Y., Trumbauer, M. E., Yagle, M. K., and Palmiter, R. D. (1985). Factors affecting the efficiency of introducing foreign DNA into mice by microinjecting eggs. *Proc. Natl. Acad. Sci. U.S.A.* **82**, 4438–4442.

Brinster, R. L., Braun, R. E., Lo, D., Avarbock, M. R., Oram, F., and Palmiter, R. D. (1989). Targeted correction of a major histocompatibility class II Eα gene by DNA microinjected into mouse eggs. *Proc. Natl. Acad. Sci. U.S.A.* **86**, 7087–7091.

Bronson, R. A., and McLaren, A. (1970). Transfer to the mouse oviduct of eggs with and without the zona pellucida. *J. Reprod. Fertil.* **22**, 129–137.

Bürki, K. (1986). Experimental embryology of the mouse. *In* "Monographs in Developmental Biology," Vol. 19. Karger, New York.

Champlin, A. K., Dorr, D. L., Gates, A. H. (1973). Determining the stage of the estrous cycle in the mouse by the appearance of the vagina. *Biol. Reprod.* **8**, 491–494.

Chen, H. Y., Trumbauer, M. E., Ebert, K. M., Palmiter, R. D., and Brinster, R. L. (1986). Developmental changes in the response of mouse eggs to injected genes. *In* "Molecular Developmental Biology" (L. Bogorad, ed., pp. 149–159. Alan R. Liss, New York.

Chisari, F. V., Klopchin, K., Moriyama, T., Pasquinelli, C., Dunsford, H. A., Sell, S., Pinkert, C. A., Brinster, R. L., and Palmiter, R. D. (1989). Molecular pathogenesis of hepatocellular carcinoma in hepatitis B virus transgenic mice. *Cell (Cambridge Mass.)* **59**, 1145–1156.

Festing, M. F. W. (1979). "Inbred Strains in Biomedical Research." Oxford Univ. Press, New York.

Flaming, D. G., and Brown, K. T. (1982). Micropipette puller design: Form of the heating filament and effects of filament width on tip length and diameter. *J. Neurosci. Methods* **6**, 91–102.

Foster, H. L., Small, J. D., and Fox, J. G. (1983). "The Mouse in Biomedical Research." Academic Press, New York.

Harris, A. W., Pinkert, C. A., Crawford, M., Langdon, W. Y., Brinster, R. L., and Adams, J. M. (1988). The Eμ-*myc* transgenic mouse: A model for high incidence spontaneous lymphoma and leukemia of early B cells. *J. Exp. Med.* **167**, 353–371.

Hogan, B., Costantini, F., and Lacy, E (1986). "Manipulating the Mouse Embryo: A Laboratory Manual." Cold Spring Harbor Laboratory, Cold Spring Harbor, New York.

Kim, D. W., Uetsuki, T., Kaziro, Y., Yamaguchi, N., and Sugano, S. (1990). Use of the human elongation factor 1α promoter as a versatile and efficient expression system. *Gene* **91**, 217–223.

Kooyman, D. L., Baumgartner, A. P., and Pinkert, C. A. (1990). The effect of asynchronous egg transfer in mice on fetal weight. *J. Anim. Sci.* **68** (Suppl. 1), 445.

Leibo, S. P., and Oda., K. (1993). High survival of mouse zygotes and embryos cooled rapidly or slowly in ethylene glycol plus polyvinylpyrrolidone. *Cryoletters.* In **14**, 133–144.

Leveille, M.-C. and Armstrong, D. T. (1989). Preimplantation embryo development and serum steroid levels in immature rats induced to ovulate or superovulate with pregnant mare's serum gonadotropin injection or follicle-stimulating hormone infusions. *Gamete Res.* **23**, 127–138.

Pieper, F. R., deWit, I. C. M., Pronk, A. C. J., Kooiman, P. M., Strijker, R., Krimpenfort, P. K. A., Nuyens, J. H., and deBoer, H. A. (1992). Efficient generation of functional transgenes by homologous recombination in murine zygotes. *Nucleic Acids Res.* **20**, 1259–1264.

Pinkert, C. A. (1987). Gene transfer and the production of transgenic livestock. *Proc. U.S. Anim. Health Assn.* **91**, 129–141.

Pinkert, C. A. (1990). A rapid procedure to evaluate foreign DNA transfer into mammals. *BioTechniques* **9**, 38–39.

Quaife, C. J., Pinkert, C. A., Ornitz, D. M., Palmiter, R. D., Brinster, R. L. (1987). Pancreatic neoplasia induced by *ras* expression in acinar cells of transgenic mice. *Cell (Cambridge, Mass.)* **48**, 1023–1034.

Quinn, P., Barros, C., and Whittingham, D. G. (1982). Preservation of hamster oocytes to assay the fertilizing capacity of human spermatozoa. *J. Reprod. Fertil.* **66**, 161–168.

Rafferty, K. A. (1970). "Methods in Experimental Embryology of the Mouse." Johns Hopkins Press, Baltimore, Maryland.

Reetz, I. C., Wullenweber-Schmidt, M., Kraft, V., and Hedrich, H. J. (1988). Rederivation of inbred strains of mice by means of embryo transfer. *Lab. Anim. Sci.* **38**, 696–701.

Richa, J., and Lo, C. W. (1989). Introduction of human DNA into mouse eggs by injection of dissected chromosome fragments. *Nature (London)* **245**, 175–177.

Rouleau, A., Kovacs, P., Kunz, H. W., and Armstrong, D. T. (1992). Decontamination of rat embryos and transfer to SPF recipients for the production of a breeding colony. *Theriogenology* **37**, 289 (abstr.).

Rusconi, S. (1991). Transgenic regulation in laboratory animals. *Experientia* **47**, 866–877.

Sambrook, J., Fritsch, E. F., and Maniatis, T. (1989). "Molecular Cloning: A Laboratory Manual." Cold Spring Harbor Press, Cold Spring Harbor, New York.

Sandgren, E. P., Quaife, C. J., Pinkert, C. A., Palmiter, R. D., and Brinster, R. L. (1989). Oncogene-induced liver neoplasia in transgenic mice. *Oncogene* **4**, 715–724.

Schedl, A., Montoliu, L., Kelsey, G., and Schütz, G. (1993). A yeast artificial chromosome covering the tyrosinase gene confers copy number-dependent expression in transgenic mice. *Nature (London)* **362**, 258–261.

Short, J. M., Blakeley, M., Sorge, J. A., Huse, W. D., and Kohler, S. W. (1989). The effects of eukaryotic methylation on recovery of a lambda phage shuttle vector from trangenic mice. *J. Cell. Biochem.* **13b**, (Suppl.), 184.

Sternberg, N. L. (1992). Cloning high molecular weight DNA fragments by the bacteriophage P1 system. *Trends Genet.* **8**, 11–16.

Whittingham, D. G. (1971). Culture of mouse ova. *J. Reprod. Fertil. (Suppl.)* **14**, 7–21.

Whittingham, D. G., Leibo, S. P., and Mazur, P. (1972). Survival of mouse embryos frozen to −195°C and −269°C. *Science* **178**, 411–414.

Wilmut, I. (1972). The effect of cooling rate, warming rate, cryoprotective agent and stage of development on survival of mouse embryos during freezing and thawing. *Life Sci.* **11**, 1071–1079.

3

Factors Affecting Transgenic Animal Production

Paul A. Overbeek

Department of Cell Biology and
Institute for Molecular Genetics
Baylor College of Medicine
Houston, Texas 77030

I. Introduction
II. Getting Started
  A. Microinjection versus Retroviral Infection versus Embryonic Stem Cells
  B. Strains of Mice: Inbred, Outbred, and Hybrid
  C. Mouse Husbandry: Caging, Mating, Pregnancy, Record Keeping, etc.
III. Troubleshooting
  A. Assaying for Successful Pronuclear Microinjection
  B. Visual Identification of Transgenic Mice
  C. Superovulation Problems
  D. Colony Problems
IV. Transgenic Phenomenology
  A. Transgene Expression: Effects of Integration Site and Copy Number
  B. Genetic Mosaicism in Founder Mice
  C. Intrafamily Variation in Expression
  D. Integration into a Sex Chromosome
  E. Multiple Sites of Integration in Founder Mice
  F. Homozygous versus Hemizygous
  G. Coinjections
  H. Genetic Instability
V. Summary
  Appendix: Standardized Nomenclature for Transgenic Animals
  References

I. INTRODUCTION

This chapter is intended to provide some general advice about generating transgenic mice. The major goal of the chapter is to help simplify some of the decisions that need to be made when a laboratory begins to do research with transgenic mice. What are good strains of mice to use? Should the transgenic mice be generated by microinjection or retroviral infection, or by genetic manipulation of embryonic stem cells? If microinjection is chosen, what are good vectors to use for the initial training stages? What about record keeping for the mice?

The first half of the chapter provides information about factors to consider when getting started, including husbandry and record keeping suggestions. The second half of the chapter discusses troubleshooting strategies and transgenic phenomenology. Most of the recommendations in this chapter are based on personal experience acquired while running a laboratory that generates transgenic mice by microinjection. The strategies suggested in this chapter have worked well for my laboratory, but they may need to be modified for other research settings. The information in this chapter represents a short introduction to inbred strains of mice, mouse husbandry, mouse breeding, and record keeping. More detailed information about these topics can be found in Hogan *et al.* (1986), Festing (1990, 1992), Hetherington (1987), Otis and Foster (1983), Lang (1983), Rafferty (1970), Green (1975), Dickie (1975), and Les (1975).

II. GETTING STARTED

A. *Microinjection versus Retroviral Infection versus Embryonic Stem Cells*

There are three general techniques for generating transgenic mice: microinjection of one-cell stage embryos, retroviral infection of embryos, and genetic manipulation of embryonic stem (ES) cells. As a general rule, each of these techniques is used for a different purpose: microinjection is used where the major goal is to study expression of new genetic information; retroviral infection (of embryos or ES cells) is used for studies of cell lineage and for random insertional mutagenesis; and ES cells are used for site-directed mutagenesis by homologous recombination.

Microinjection is the most commonly used procedure, mainly because of its reliability. The technique requires a concerted training period (often 3–6 months), but once the protocol is mastered, transgenic mice can be efficiently generated with almost any DNA construction. In most cases, transgenic DNAs introduced by microinjection have been found to be efficiently and reproducibly expressed in transgenic mice. Microinjection is the preferred strategy when the objective is to obtain expression of new genetic information (see Chapter 2).

Whereas microinjections are almost always performed on one-cell stage embryos, retroviral infection can be done successfully at various stages of embryonic development, ranging from preimplantation to midgestation. As a result, retroviruses have served as useful markers for cell lineage studies (Soriano and Jaenisch, 1986). For random insertional mutagenesis studies, retroviruses have the valuable attribute of integrating into the host genome without deletions or rearrangements of the chromosomal DNA. One of the best characterized insertional mutations in transgenic mice is the inactivation of the $\alpha 1$ (I) collagen gene by retroviral integration in the *Mov*13 family (Jaenisch *et al.*, 1983; Schnieke *et al.*, 1983; Harbers *et al.*, 1984; Hartung *et al.*, 1986). Retroviruses can also infect ES cells, allowing them to be used for promoter trap and gene trap experiments where there is a preselection for potentially interesting sites of integration (e.g., Friedrich and Soriano, 1991). In contrast to the case with retroviruses, integration of microinjected DNA is often accompanied by major rearrangements or deletions in the host genomic DNA (e.g., Singh *et al.*, 1991; Covarrubias *et al.*, 1986, 1987; reviewed by Meisler, 1992). Such chromosomal changes can complicate the search for the coding sequences of the inactivated gene. Retroviruses have also received attention because of their potential uses for gene therapy (see Chapter 5).

Integration of microinjected DNAs and of retroviruses typically does not occur by homologous recombination, so these procedures are not practical for site-directed mutagenesis. In ES cells, the frequency of integration by homologous recombination is high enough that the targeted integration events can be identified *in vitro* before the cells are used to generate transgenic mice. In those cases where a gene has been cloned, but the phenotype caused by mutation of the gene is not known, ES cells offer a strategy to generate the desired mutants. One technical hurdle to the use of ES cells is mastery of the tissue culture system. The use of ES cells for targeted mutagenesis is described in detail in Chapter 4. Table 1 contains a summary of various features of the techniques for generating transgenic mice.

B. Strains of Mice: Inbred, Outbred, and Hybrid

Inbred strains of mice are defined as strains that have been maintained by successive brother to sister matings over more than 20 generations (Green, 1975). Repetitive inbreeding removes genetic heterogeneity, so that mice of an inbred strain are considered to be genetically identical to each other. There are hundreds of different inbred strains of mice (Festing, 1992). Some of the more commonly used laboratory strains, along with their pigmentation, include the following: A (albino), BALB/c (albino), CBA (agouti), C3H (agouti), C57BL/6 (black), C57BL/10 (black), C57BR (brown), C58 (black), DBA (dilute brown), FVB (albino), NZB (black), NZW (white), SJL (albino), SWR (albino), and 129 (usually albino or chinchilla). The

TABLE 1
Techniques for Generating Transgenic Mice

	Microinjection	Retroviral infection of embryos	Embryonic stem cells
DNA vector	Any cloned DNA, preferably linear with vector sequences removed	Recombinant or wild-type retroviruses	Cloned DNA or retroviruses
Introduction of DNA	Microinjection into pronucleus	Infection after removal of zona pellucida	Electroporation or retroviral infection
Embryonic stage	One-cell stage	Four-cell stage or later	Totipotent ES cells
Embryo transfers	Oviduct	Uterus	Into blastocoel, then into uterus
Genotype of founder mice	Usually nonmosaic	Mosaic	Chimeric
Screening of newborns	Dot blots, Southern blots, or PCR	Southern blots or PCR	Visual coat color markers plus PCR or Southern blots
Copy number of integrated DNA	1–200	1	Can be varied by selection of method for introducing DNA
Percentage of potential founders that are transgenic	10–30%	5–40%	Up to 100%
Expression of the new DNA	Usually	Poor	Enhancer trap, gene trap
Integration	Random, nonhomologous, multicopy, single site	Apparently random using retroviral long terminal repeats (LTRs)	Random plus targeted, depending on method of introducing DNA
Germ line transmission by founders	Usually	Usually	Occasionally a problem
Advantages	a. Straightforward procedure b. Successful expression with many different constructs	Single-copy integration using retroviral LTRs	a. Homologous recombination b. Selection for transformants *in vitro* c. Multiple independent insertions using retroviruses
Disadvantages	a. Physical damage of embryos during microinjections b. Multiple copy integration c. Lack of insertion by homologous recombination	a. Low-level expression b. High titers can be difficult to achieve	Difficult tissue culture system

full name for an inbred strain includes an abbreviation to designate the source of the mice. The abbreviation is placed after a / that follows the name of the inbred strain. For instance A/J mice would be strain A mice from The Jackson Laboratory, whereas FVB/N mice would be FVB mice from the National Institutes of Health. For those strains with a / in the standard name, the abbreviation for the supplier is added to the end of the name (e.g., C57BL/6J mice from The Jackson Laboratory). For more detailed information about naming inbred strains of mice, see Festing (1992).

Fur pigmentation in inbred laboratory mice is controlled primarily by four different genetic loci: agouti (*a*), brown (*b*), albino (*c*), and dilute (*d*) (reviewed by Silvers, 1979). The albino locus encodes tyrosinase, the first enzyme in the pathway to melanin synthesis. When mice are homozygous (*c/c*) for mutations that inactivate the tyrosinase gene, the mice are albino regardless of the genotype at the other loci. All of the common albino strains of mice have the same point mutation in the tyrosinase gene (Yokoyama *et al.*, 1990), indicating that these strains are all derived from a common ancestor. When mice are homozygous or heterozygous for a nonmutated tyrosinase gene (i.e., *C/C* or *C/c*), then the color of pigmentation is determined by the condition of the other genes. If the mice have a wild-type agouti allele (either *A/A* or *A/a*), then the fur will contain both black and yellow bands of pigment and will be agouti (see Fig. 1). If both copies of agouti are mutated (a/a), then the hair becomes uniformly pigmented, and the mice are either black (*B/B* or *B/b*) or brown (*b/b*) (Fig. 1). Mice that are homozygous for a mutation in the dilute gene (i.e., *d/d*) show a decreased intensity of pigmentation (not shown).

Hybrid mice are generated by mating mice from two different inbred strains. The mice from such a mating are termed F1 hybrid mice. They are genetically identical to one another but different from either inbred parent. When F1 mice are mated to one another, the offspring are referred to as F2 hybrids. F2 hybrid mice will be genetically different from one another (owing to meiotic recombination and random sorting of the chromosomes) and will contain different mixtures of the genetic variations that were present in the original inbred progenitors. Hybrid mice exhibit a phenomenon termed hybrid vigor. They show enhanced fertility, they respond better to superovulation regimens, and hybrid embryos can be grown efficiently from the one-cell to blastocyst stage *in vitro*. Outbred strains of mice are propagated by nonstandardized matings and therefore retain substantial genetic variability.

Inbred, outbred, and hybrid mice are used for transgenic research. Table 2 contains a short, subjective list of some of the frequently used strains. The inbred strains most commonly used for transgenic research include C57BL/6 (often referred to as "black 6"), FVB, and 129/Sv/Ev mice. The C57BL/6 strain has been used for laboratory research for many years, so that many known mouse mutations are available on the C57BL/6 background. Young C57BL/6 females superovulate well and C57BL/6 blastocysts have been found to be excellent recipients for genetically engineered ES cells (see Chapter 4). FVB embryos are often used for

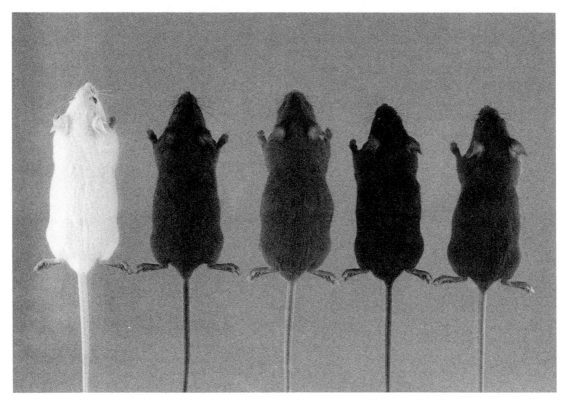

Figure 1. Pigmentation in laboratory mice. Some of the most common pigmentation phenotypes for laboratory mice are pictured. From left to right the mice are albino (genotype c/c), agouti (*A/a B/B C/c*), brown agouti (*A/a b/b C/c*), black (*a/a B/B C/c*), and brown (*a/a b/b C/c*). In agouti mice, the hairs of the fur show a subapical band of yellow melanin (pheomelanin) surrounded by regions of black melanin (eumelanin) (not visible in the photograph). The pheomelanin band is missing in the black and brown mice that are mutated at the agouti locus.

microinjections because they have large distinctive pronuclei (Fig. 2) that are easy targets for microinjection (Taketo *et al.*, 1991). Most of the ES cell lines have been derived from agouti 129/Sv mice. The most commonly used hybrid mice for transgenic research are B6SJL F1, derived by mating the inbred strains C57BL/6 and SJL. Brinster and colleagues (1985) have documented the advantages of the B6SJL mice. Outbred Swiss albino strains such as ICR or Swiss-Webster are often used as recipients for embryo transplantations.

Outbred mice are generally less expensive than inbred or hybrid mice. For example, 6-week-old ICR female mice currently cost less than $2.00 each (Harlan Sprague Dawley, Indianapolis, IN, 1991 catalog), whereas 4- to 5-week-old C57BL/6, FVB, or B6SJL females are approximately $6.00–7.00 each (The Jack-

TABLE 2
Mouse Strains Used for Transgenic Research

Strain	Typical uses
Inbred mice	
C57BL/6	Pronuclear microinjection, recipient blastocysts for embryonic stem (ES) cells
FVB	Pronuclear microinjection
129/Sv	Generate ES cells
Outbred mice	
ICR or CD-1	Pseudopregnant recipient females
Hybrid mice	
B6SJL F1	Pronuclear microinjection
B6D2 F1	Vasectomized males

son Laboratory, Bar Harbor, ME, 1992 catalog; Taconic, Germantown, NY, 1993 catalog). In general, outbred and hybrid mice have better fertility and larger litter sizes than inbred mice. However, the inbred FVB mice have fertility and embryo culture characteristics that nearly match the hybrid and outbred strains (Taketo *et al.*, 1991). The biggest advantage of using inbred mice is the consistency of the

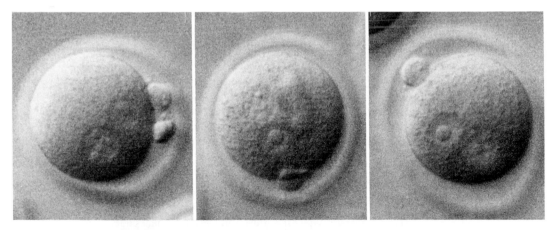

Figure 2. One-cell stage FVB/N embryos. In each embryo, two pronuclei are readily visible. Each pronucleus contains one or more nucleoli (prominent circular organelles within each pronucleus). The two polar bodies are visible for the embryo at left. The FVB/N embryos were cultured in the top of a plastic petri dish (Falcon 1006, Becton Dickinson Labware, Lincoln Park, NJ) and viewed using Hoffman optics. Because the FVB/N pronuclei are large and distinctive, they are easy to inject. Moreover, the fact that FVB/N mice are inbred can simplify the subsequent interpretation of experimental results. [This figure was originally published in Taketo *et al.*, (1991).] (Magnification, 400×.)

genetic background. This can be a pertinent consideration for studies with transgenic mice. When a single DNA construct is used to generate transgenic mice, variability in expression or phenotype between different transgenic mice cannot be attributed to a variable genetic background if an inbred strain of mice is used.

C. Mouse Husbandry: Caging, Mating, Pregnancy, Record Keeping, etc.

1. Animal Welfare

All research with transgenic animals should be carried out as humanely as possible and in accordance with all federal and institutional policies for research with laboratory animals. In the United States, investigators and institutions are expected to comply with the Animal Welfare Act and the Guide for the Care and Use of Laboratory Animals (documents available from the Office for Protection from Research Risks, NIH, Bethesda, MD 20892). As part of this policy, each research institution is required to appoint an Institutional Animal Care and Use Committee, which is required to review and approve animal research proposals and protocols. Before initiating any research with transgenic mice, investigators must document their plans for humane care and research in order to obtain approval from the institutional review committee. In addition, investigators should always confer with the chief veterinarian for the facility where the animals will be housed. The veterinarian will be able to provide information about a variety of important issues, including space allocation for the transgenic mice, acceptable vendors for purchase of mice, responsibilities in the provision of daily and weekly care for the animals, animal husbandry charges, standard housing conditions for the animals, acceptable anesthetics, protocols for animal procedure rooms, treatment of sick animals, disposal of dead animals, and so forth. The chief veterinarian should be consulted concerning all of the husbandry recommendations made in this chapter to be certain that they conform to the standards of the animal facility.

2. Standard Housing Conditions

Laboratory mice are typically housed in polycarbonate cages that are equipped to provide food and water for the mice (see Fig. 3) (Lab Products, Inc., Maywood, NJ). Most facilities use two (or more) standard size cages: a small "shoe box" size of approximately $12 \times 8 \times 5$ in. that can house four or five adult mice, or a nursing female with pups, and a larger cage ($19 \times 10 \times 6$ in.) that can house approximately twice as many mice.

Food for the mice is generally provided *ad libitum*. The food is placed either in a metal container that is positioned inside the cage, or alternatively in the metal

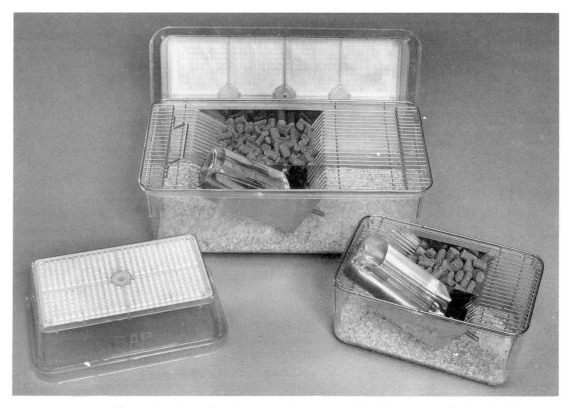

Figure 3. Mouse housing cages. Prototype small and large cages for housing laboratory mice are shown (Lab Products, Inc.). Microisolator tops for each cage size are also pictured. Cage card holders are not shown.

lid on top of the cage. Rodent chow is available in a number of different recipes from suppliers (e.g., Purina Mills, Inc., Richmond, IN). Most animal housing facilities purchase the rodent chow and provide the food as part of the standard animal husbandry services.

Water is also provided *ad libitum,* typically from water bottles equipped with rubber stoppers and sipper tubes (Fig. 3). The water is often acidified to a pH of 2.5 by addition of HCl in order to slow the growth of bacteria in the water. Water bottles should be cleaned and replaced once per week.

Bedding for the mice is traditionally either wood shavings or wood chips and is typically provided by the animal husbandry service. Bedding should be replaced at least once per week. Pregnant females near delivery can be given a cotton block (e.g., Nestlets from Ancare Corporation, North Bellmore, NY) or paper towels to use to build a nest for the pups. Cages with nests should be checked regularly to be certain that the nests do not contact the sipper tubes on the water bottles. Such contact allows the water to drain from the water bottle, resulting in flooding of the animal cage. Flooded cages can lead to death of young mice by hypothermia or drowning. In general, we do not change the bedding in a cage with a female that is 18 days or more pregnant or is nursing pups that are less than 48 hr old. It is best to leave females relatively undisturbed around the time of delivery.

Special microisolator caging systems have been designed to help prevent the spread of pathogens within a mouse colony. As one example, individual cages can be equipped with special filter tops (Fig. 3) that prevent contamination by airborne debris from other cages in the room. When microisolator tops are used, the bedding should be changed twice a week for cages with more than one mouse, because the cage tops restrict evaporation and air circulation within the cages. An alternative isolation strategy is to use microisolator racks (Lab Products, Inc.). These racks are designed with enclosed shelves and specialized ventilation systems. The shelves are equipped with retractable glass doors that can be opened to allow access to the cages. Microisolator racks are expensive, so they are generally used only for small colonies of mice, or for mice that are housed in special locations.

Housing conditions for mice vary according to the stringency of the procedures used to protect the mice from murine pathogens. The two most common conditions are referred to as "conventional" and "specific pathogen free (spf)" (Lang, 1983; Otis and Foster, 1983). For spf facilities, special precautions are employed to protect the mice from murine pathogens, particularly murine viruses. Personnel working with spf mice are required to follow specific guidelines regarding personal hygiene and clothing when they enter the animal husbandry area. In addition, all experimental supplies and surgical equipment must be decontaminated before they are brought into an spf area. Each spf facility will have its own standard operating procedure, which should be adhered to conscientiously. Conventional housing facilities maintain less stringent hygiene standards, typically allowing mice to be returned to the facility after removal for surgical or experimental procedures.

3. Fertile Females and Embryos for Microinjection

Inbred, outbred, or hybrid one-cell stage embryos can be used for microinjections. To generate inbred embryos, both mating partners must be of the same inbred strain (e.g., FVB × FVB), whereas hybrid embryos can be generated by mating mice of two different inbred strains. Hybrid embryos can also be generated by inbred × hybrid or by hybrid × hybrid matings. If either of the parents in the mating are outbred, the embryos are considered to be outbred.

The two most commonly used inbred strains are C57BL/6 and FVB. Although young C57BL/6 females (3–4 weeks old) superovulate well, often producing 30–40 embryos/female, there are some drawbacks to using the C57BL/6 strain to make transgenic mice, particularly for beginners. C57BL/6 embryos from superovulated females show a high percentage of unfertilized and abnormal embryos

(10–40%), and the pronuclei can be small and difficult to identify. A later drawback is the fact that adult C57BL/6 females are not particularly fertile, with average litter sizes of only 5–6 mice. C57BL/6 females are often poor mothers, so that newborn mortality can occur with a frustratingly high frequency. Both of these characteristics can make it difficult to generate extra offspring within a transgenic family in order to do further research.

In contrast, FVB/N embryos have large, well-defined pronuclei, making them easier to inject (Taketo *et al.,* 1991). Superovulated FVB/N females yield an average of 15–25 embryos/female, they mate with high efficiencies (75–90%), and they produce a high percentage (usually >90%) of fertilized healthy embryos. Moreover, FVB females are excellent mothers, they typically do not cannibalize dead newborns, and the average litter size is 9–10 pups. It should be noted that the FVB strain is not well known outside of transgenic mouse research. The FVB strain, particularly at the immunological loci, has not been as fully characterized as the C57BL/6 strain.

The most commonly used embryos for microinjection are not inbred but are hybrid F2 embryos generated from matings of the inbred strains C57BL/6 and SJL. B6SJL F2 mice have been used for many years by Brinster and colleagues, who found that the hybrid embryos gave a substantially higher frequency of transgenic offspring than inbred C57BL/6 mice (Brinster *et al.,* 1985). The hybrid females superovulate well, and the embryos show the traditional hybrid vigor.

Embryos from outbred strains can also be used to generate transgenic mice. Because outbred mice are less expensive, the novice microinjector should consider using outbred embryos for the initial stages of training. Outbred females superovulate well, and one-cell stage embryos can be maintained in culture to the blastocyst stage in order to monitor percent survival. In contrast to outbred (and hybrid) embryos, embryos from inbred strains allowed to develop *in vitro* often stop development at the 2-cell stage (referred to as the 2-cell block). Embryos from a few inbred strains, such as FVB, will develop efficiently from the one-cell stage to the blastocyst stage in embryo culture medium, such as M16 (Whittingham, 1971; Hogan *et al.,* 1986) or BMOC (Gibco, Grand Island, NY).

Female mice can be induced to superovulate by treatment with hormones. The two main advantages of superovulation are the increase in number of embryos and the synchronization of estrus. Without superovulation, female mice generally release 6–10 oocytes during each estrus. With superovulation the number of oocytes can be increased to as many as 40 per female (Rafferty, 1970; Hogan *et al.,* 1986). Superovulation often works best with younger females that have not yet started their own ovulatory cycle. For many strains of mice, females that are 3–5 weeks of age give the best yield of oocytes on superovulation. The young females will mate successfully, so that this strategy is the standard protocol to obtain embryos for microinjection. The regimen for hormone administration is described in Chapter 2.

In the absence of superovulation, female mice begin to ovulate and become fertile around 5–6 weeks of age. Adult female mice have an estrus cycle of 4–6 days, they are receptive to mating only during estrus, and they cycle regularly until 7–9 months of age for most strains. Female mice can be housed together in the same cage before mating, during pregnancy, and after newborns have been weaned. In general, pregnant females are placed in separate cages 1 or 2 days before their scheduled delivery. This helps prevent overcrowding, allows unambiguous identification of which pups belong to which mother, and eliminates the possibility that the pups from one mother will be attacked by another female in the cage.

In most research facilities, mice are housed in rooms with no external lighting. A timer-controlled lighting system is used to provide a consistent daily lights-on, lights-off cycle, which is typically 14 hr of lighting, then 10 hr of darkness (for example, lights-on at 6:00 A.M., lights-off at 8:00 P.M.).

4. Fertile (Stud) Males

Male mice typically become fertile at 7–8 weeks of age. When males are housed together from weaning age, multiple males can generally be kept together in the same cage indefinitely. However, experienced males should not be housed together, since they will fight with one another, often until the most aggressive male has killed the other males in the cage. When males are ready to be used for matings, they should be placed in individual cages. In general it helps to give each male his own cage at least 24 hr before setting up a mating in order to give the male time to establish the cage as his territory. Matings are initiated by adding one or more females to each male cage.

We typically use each fertile male for only one or two matings per week. A more frequent schedule leads to lower mating percentages and a substantially decreased percent fertilization. Vasectomized males can be mated more frequently, since their ability to produce adequate sperm is not relevant. However, vasectomized males only occasionally mate more than twice a week, even if given the opportunity.

The fertility of the stud males typically begins to decline when the mice reach 9–10 months of age, at which time they should be replaced with younger males. Male mice remain fertile longer than female mice. In most strains of mice, experienced males remain fertile up to 14–18 months of age, and inexperienced males can often be successfully mated up to 10–12 months of age. Males that look sick or that have stopped mating should be replaced.

The genetic background of the fertile males should be matched to that of the superovulated females. If the females are outbred, then the males can be outbred. If the females are inbred, then the males should be from the same strain in order to maintain the inbred background. Hybrid females are generally mated to either F1 or F2 hybrid males of the same genetic background.

We maintain a cage card with a mating record for each stud male. The mating record indicates whether a plug was found for each date when the male was set up for mating with a superovulated female. Males that do not produce plugs for four consecutive mating opportunities as well as males that mate less than 50% of the time are replaced. Overly aggressive males will fight with and injure females that are added to their cage. Males that show such behavior more than once are euthanized.

5. Vasectomized Males

The genetic background of the vasectomized males is usually not critical, since these males should not produce any direct descendents. To help recognize the occasional occurrence of an inadequate vasectomy (i.e., a fertile vasectomized male), coat color markers can be used. The vasectomized males can be selected so that inappropriate offspring would have a different coat color than the offspring from injected embryos (e.g., albino versus pigmented).

Vasectomized males should mate consistently and over a reasonable time span. Our favorite males are hybrid B6D2 F1 mice (generated by mating C57BL/6 to DBA/2 mice). The mice are inexpensive and readily available from commercial breeders. The males mate consistently up to at least 1 year of age, and they leave readily identifiable plugs. The males are docile and easy to work with. Males can be purchased already vasectomized (e.g., from Taconic).

6. Recipient Females

Once embryos have been genetically manipulated *in vitro,* they need to be transferred to pseudopregnant recipient females. Because the purpose of the recipient females is to carry the embryos to term and to nurse the newborns to weaning age (3 weeks), it is important to use females that are good mothers. The females should also be relatively docile so that they will allow inspection of the pups, and will accept pups from another mother. Outbred ICR females are inexpensive and make good recipient females.

One strategy to obtain pseudopregnant females is to set up random matings, by placing one or two females per cage with vasectomized males. In a nonsynchronized population, 10–20% of the females will be in estrus on an average day, so that the total number of females to set up for matings should be 5 to 10 times the number that will be needed for the embryo transfers. Mated females are identified by inspection for copulation plugs (Rafferty, 1970). If embryo manipulations are done on consecutive days during the week, this system has the advantage that a large number of females can be set up on the first day, and on subsequent days only previously unmated females need to be checked for plugs.

An alternative strategy to obtain pseudopregnant recipient females is to set up new matings each day using females that are in estrus. Such females can be identified by inspection of the vagina [for swelling, redness, and a rippled folding (wrinkling) of the vaginal wall; see Hogan *et al.* (1986)]. When the females are prescreened for estrus, 50% or more of the females should be successfully mated overnight. The advantage of prescreening is that fewer females need to be checked for plugs each morning. The disadvantage is that new females need to be set up for matings every evening.

A third strategy to obtain pseudopregnant females is to perform superovulations to induce estrus in the recipient females, then to mate the females to vasectomized males. In our experience, naturally mated females have a substantially higher rate of pregnancy after embryo transfer than superovulated females, so we use superovulation only as a last resort.

Pseudopregnant females that are not used for embryo transfers will typically return to an estrous cycle 8–11 days after mating. The females can be saved and remated if desired.

7. Pregnancy

The gestational period in the laboratory mouse is 19–20 days. Females that are more than 11–12 days pregnant with a normal size litter of 7–10 pups can be recognized by visual inspection for abdominal enlargement.

As stated earlier, pregnant recipient females are generally given individual clean cages 1 or 2 days before delivery. If the females are provided with nesting materials such as paper towels or crushed cotton squares, they will generally build a nest in which to place the newborns after delivery. Females that do not build a nest often turn out to be poor mothers. Pregnant females that are housed with a male can be left with the male during delivery if additional offspring are desired. Female mice go into estrus postpartum and are receptive to mating the night after delivery. As a result, female mice can be nursing one litter and pregnant with a second litter at the same time. The gestational time for the second litter will often be extended to 21–25 days, as a consequence of delayed implantation (Hogan *et al.,* 1986). The delayed implantation allows the first set of pups to be weaned at 21 days of age before the second set of newborns arrive.

Female mice often cannibalize fetuses that are dead at birth, so if one wishes to look for prenatal or perinatal lethality of transgenic embryos, then a cesarean section (C-section) should be performed prior to the scheduled delivery. In those cases where one wishes to deliver live pups by C-section, the delivery should be done 24 hr before the scheduled normal delivery, that is, at 18.5 days postfertilization. C-sections are also advisable for females that are pregnant with only 1–3 fetuses, since these fetuses can become oversized, impeding normal delivery. We have not had much success performing C-sections where we keep the pregnant female alive to nurse the newborn pups. The surgical trauma appears to disrupt nursing behavior and milk production. Therefore, it is essential to have a "foster" mother to nurse the C-section pups if they are to be kept alive. The foster mother should have delivered her own litter within the previous 48 hr, and ideally the natural pups of the foster mother should have a different coat color from the C-section pups so that they can be visually distinguished. In addition, the foster mother should have pups that are well fed, with milk visible in their stomachs.

The protocol used for the recovery of midgestation embryos (Hogan *et al.*, 1986) can also be used for C-section delivery of live fetuses. The first step is to sacrifice the pregnant female by cervical dislocation, then to open up the peritoneal cavity and externalize both horns of the uterus. By visual inspection, one can count the number of full-grown fetuses and also identify instances of postimplantation embryonic lethality. Embryos that die shortly after implantation can be recognized by the presence of degenerated (brownish) decidua. (Decidua are sites of cellular proliferation within the uterus induced by fetal implantation.) Fetuses that die at later stages can be identified by their small size and inappropriate color.

To deliver the live fetuses, use a pair of scissors to make an incision at the vaginal end of each uterine horn, and advance the scissors along the antimesometrial side of the uterus up to the oviduct in order to expose the interior of each uterus. With a little practice, this procedure can be done so that each fetus remains encased within its own fetal yolk sac. To deliver each fetus, the individual yolk sacs can be grasped with a pair of watchmaker's forceps and opened by incision with scissors. The edges of the yolk sac can then be peeled back around the fetus, so that the umbilical cord can be located and severed with the scissors. Each fetus can be cleaned up by rolling it back and forth on a paper towel, then transferred to a paper towel on top of a 37°C slide warmer. The entire litter of newborns should be delivered as quickly as possible. Once the pups have been delivered, gently stimulate each fetus every 15–30 sec to induce breathing. Forceps can be used to gently pinch the skin behind the neck, or to spread the legs apart, or to simply roll the newborns on the paper towel. Healthy fetuses will begin to make sporadic inhalations within 1–2 min of delivery, particularly on stimulation. Over the next 1–2 min the fetuses should begin to breathe more regularly. Once breathing becomes stable, the fetuses should become a healthy pinkish color as the blood becomes oxygenated. Fetuses that survive the C-section delivery should be allowed to stabilize, then transferred to a foster mother.

Our strategy for this transfer is to begin by removing the foster female from her cage and placing her on the cage top. Next a sufficient number of the natural pups are removed so that when the C-section newborns are added to the cage the total litter size will be 5–8 pups. The foster mother is more likely to accept the new pups if some of her own pups are left in the cage. The C-section newborns are transferred to the recipient cage and placed with the remaining newborns of the foster female. The entire collection of pups is then buried under approximately 5 mm of the bedding that is already present in the cage of the foster female. This is done to transfer the odors of the cage to the C-section pups before the foster female is returned to the cage. After the appropriate information is entered on the cage card, the foster female is returned inside her cage. One additional recommendation is to treat the nose of the foster female with an alcohol swab just prior to placing her back in her cage. The alcohol treatment will cause a short-term loss of olfactory discrimination. The behavior of the foster female should be monitored for the first few minutes after she is placed back in the cage. In most instances, the foster female will move around the inside of the cage sniffing at both the cage top and the bedding, as well as digging in the bedding at various locations in the cage. The pups in the cage will emerge from the bedding within 2–3 min if the female does not directly dig them out. When the foster female accepts the new pups, the female will begin to groom herself and the new pups. On the other hand, when the female does not accept the new pups, the usual response is to pick up the pups at the nape of the neck and to carry them rapidly around the inside of the cage as if looking for some way to dispose of them. If the newborns are not accepted by the first foster female, then it is typically necessary to repeat the procedure with a new foster mother.

If the natural pups cannot be distinguished from the C-section pups by coat color, then it may be helpful to physically mark one set of pups for later identification. One strategy is to use scissors to snip off a small toe on one of the feet. A small amount of bleeding accompanies toe removal, but the bleeding quickly stops and does not cause the foster female to reject the marked pups. The marked pups can be identified at weaning age by simple visual inspection for missing toes.

Milk production in lactating females is stimulated by nursing, and very small litters often do not provide adequate stimulation to maintain milk production. If the recipient females from 1 day's worth of injections all have small litters, it is often beneficial to consolidate the pups so that the litters average 6–7 pups.

8. Newborn Mice

New litters of mice should be monitored within 12–24 hr after birth to check for newborns that have died, and also to be certain that the newborns are being properly cared for and nursed. At the time of delivery, the mother will eat the placenta and fetal yolk sac associated with each newborn mouse. In addition, the mother will typically groom the newborns by licking in order to stimulate breathing. Within 2–4 hr after birth, properly fed newborns will have milk in their stomachs, which can be seen through their skin. If the mice are not being fed 12 hr after birth, or if they are found unattended and scattered around the cage, they should be transferred to a foster mother as described in the previous section.

Mice are typically old enough to wean by 21 to 23 days after birth. If the mother is not pregnant with another litter of pups, we often leave the pups in with the mother until 25 days of age. If the mother is pregnant with another litter, the first litter should be removed promptly at 21 days of age. If the older pups are not removed, newborn mice in the next litter will be trampled and will not survive. Adolescent mice should weigh 9–10 g or more at weaning. Mice that are smaller than this should be left with the mother, since they often have trouble surviving on their own. At weaning, the mice are sorted by sex and placed into new cages.

9. Mating of Transgenic Mice

Founder transgenic mice can be mated to nontransgenic partners to generate additional transgenic mice. Transgenic offspring of such matings will be hemizygous for the transmitted integration site(s). Interbreeding of mice from different transgenic families that carry the same construct is generally not done. Matings (intercrosses) between hemizygous mice within a given family can be set up in order to generate mice that are homozygous for a transgenic insert. In order to identify homozygous transgenic mice, quantitative Southern or dot-blot hybridizations are generally required. Homozygosity can be confirmed by mating (crossing) the mice to nontransgenic partners. All of the offspring will be transgenic if one of the parents was homozygous. Homozygous mice are often examined in detail for the possibility of insertional mutations. In some families, viable homozygotes may be absent due to a mutation that causes embryonic or perinatal lethality. When the homozygotes of both sexes are viable and fertile, they can be used for subsequent matings (incrossing) to maintain the transgenic family without the need for DNA screenings at each generation.

10. Record Keeping

There is no standardized record keeping system for transgenic research of which I am aware. A good record keeping system should be able to provide updated information about every mouse in the research colony. For each mouse it is important to know whether the mouse is transgenic, what DNA vector the mouse carries, who the parents and the offspring of the mouse are, when the mouse was born, and how the mouse has been characterized to date. I describe below some features of the record keeping systems that we use.

a. Identification Numbers for Each Mouse   Mice that belong to any one of the following three categories are assigned individual, unique identification numbers: recipient females, potential founder mice, or offspring within a transgenic family. Each category is allocated a series of numbers which can be preceded by a letter (e.g., R for recipient females, F for potential founders, and T for transgenic families). Within each category, new mice are assigned sequentially increasing numbers. (One person in the laboratory is responsible for maintaining log books that are used to allocate the identification numbers.) In conjunction, we use a homologous set of ear tags that have the identification numbers engraved on them (National Brand and Tag Company, Newport, KY). When a potential transgenic mouse reaches weaning age and is anesthetized to obtain tissue for DNA isolation, the appropriate ear tag is affixed. Ear tags are applied to recipient females at the time of embryo transplantation. Once the ear tags have been applied, individual mice are uniquely identified and can be readily located within the colony. For studies of embryonic development in transgenic families, the embryos are assigned identification numbers, but from a series for which no corresponding ear tags are purchased.

A common alternative system for keeping records on transgenic mice is to keep track of mice within individual transgenic families. Each mating pair is assigned an identification number, and offspring are identified by ear punching (Hogan *et al.*, 1986). Ear punching can be combined with toe clipping to mark up to 10,000 different mice. This system is less expensive than using ear tags, but mistakes in mouse husbandry are easier to correct if the mice have ear tags.

b. Transgenic Family Identification Numbers   Investigators traditionally assign some type of family identification number (or letter) to each transgenic founder mouse and to all of its offspring. Most laboratories follow their own conventions in the assignment of these identifiers. However, the Committee on Transgenic Nomenclature of the Institute of Laboratory Animal Resources (ILAR) has adopted specific guidelines for naming transgenic families of mice. These guidelines are given in the Appendix at the end of this chapter. At a minimum, a laboratory-assigned number and laboratory identifier (three letters) should be given to each transgenic family. In this fashion each transgenic family becomes uniquely identified. The nomenclature rules should be adhered to by all transgenic research laboratories.

c. Cage Cards   Mouse cages are generally equipped with cage card holders for 3×5 in. index cards. Each card should carry information about the mice inside the cage. We use two different formats for the information: one for cages with mice that are all of the same sex and a different one for mating pairs or nursing females (Fig. 4). For unisex cages the following information is written on each cage card: sex of the mice, number of mice, identification numbers for each of the mice, date of birth, date of weaning (or date of receipt if the mice were received from a vendor), identification numbers for the parents, DNA vector, identification number of the transgenic family, information about recognizable phenotypes for any of the mice, and information about the results of screening the mice for the transgenic DNA (see Fig. 4). When a cage contains a mating pair, comparable information is entered on the cage card: identification numbers for the mating mice, date of birth of the mating pair, identification numbers of the parents of the mating pair, DNA vector, identification number of the transgenic family, date of birth and number of pups on delivery (including the number of stillborn mice), identification numbers assigned to the newborn mice, and comments about phenotypes of the newborn mice (see Fig. 4).

To maintain some organization within the mouse housing area, mice that have similar genetic backgrounds are housed in proximity to each other. When more than one room is available to house the mice, each room can be allocated to different sets of mice. For example, one room could be specifically for the matings to gen-

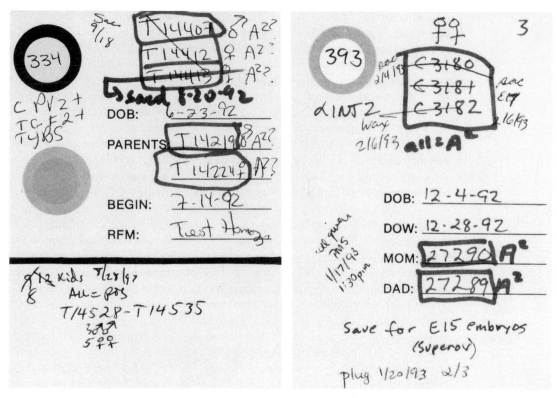

Figure 4. Cage cards. Information about mice within a cage is typically entered on a 3 × 5 in. index card that is held in a cage card holder on each cage. Prototype cage cards are shown for a mating pair (left) and a cage that contains only mice of the same sex (right). The identification numbers and sexes are entered at the top of each card, along with information about the date of birth (DOB), date of weaning (DOW), and parents. For mating pairs, the start date for the mating is entered (BEGIN), along with information about the reason for mating (RFM). When the offspring are born, the relevant information is recorded, including number of pups and date of birth. New identification numbers are assigned to the offspring and entered on the cage card. At weaning, the sexes of the offspring are recorded. To simplify the recognition of specific sets of related transgenic mice, each construction that is used for microinjection is assigned a code of one small colored dot within a different colored larger dot. These dots are adhesive and are placed in the upper left-hand corner of each card. Each founder mouse is assigned a family identification number, and the family number is written within the smaller dot (334 for the left card, 393 for the right card). The names of the transgenic constructions are written on the cards under the dots. When a mouse is sacrificed, a line is drawn through the corresponding identification number.

erate embryos for injection and pseudopregnant females; a separate room could be used to house transgenic families. To simplify the recognition of various sets of mice we have found that different color index cards are helpful. For example, we use one color card to identify FVB/N mice, a different color for outbred ICR mice, a third color for vasectomized B6D2 F1 males, and a fourth color for all of the cages that contain potential transgenic mice. In addition, we use a color-coding scheme to identify families of transgenic mice that were all derived from injections with the same DNA vector. Colored dots are added to each transgenic cage card (see Fig. 4). By placing a small dot of one color inside a larger dot of a different color, unique color patterns are generated and assigned to each microinjected DNA. When a founder transgenic mouse is identified, it is assigned a transgenic family identification number, and that number is written on the small dot on each cage of that family. We have found that this cage card identification system greatly simplifies the search for specific cages of mice within an animal housing room. When the mice in a specific cage have all been removed, the cage card is saved and placed in a file (organized by family number) so that it can be retrieved if needed.

d. Pedigree Charts   For each family of transgenic mice, a continuously updated pedigree chart is maintained (Fig. 5). Males are designated by squares, females by circles, and mice that die or are sacrificed without sexing are indicated by diamonds. Transgenic mice are indicated by placing a diagonal line within their symbol. Putative homozygous mice are given an X within their symbol. For families where the founder mouse has more than one site of integration, the different sites are designated A, B, C, etc. (which is different from the nomenclature rules in the Appendix), and the relevant site is written below the symbol for each transgenic mouse. Matings are indicated by entering a short descending line below the symbol for each mouse of the mating pair, then by drawing a horizontal mating line to connect the two mice. Offspring are indicated by drawing a vertical line running down from the mating pair line to the next tier of the pedigree chart, followed by a horizontal line on which to indicate information about the offspring (see Fig. 5). Such information will include the date of birth, the identification numbers, and the sexes of the mice, once determined. Mice that are deceased are indicated by drawing an X through the stem line that attaches to their sex symbol (see Fig. 5). The pedigree charts are left in the mouse housing rooms and are updated whenever new information is available. These continuously updated pedigree charts provide a simple summary of the status of each transgenic family.

e. Newborn Records   Separate hard copy records about each litter of mice can also be maintained. Desirable information might include number of mice in the litter, identification number and sex of each mouse, parents, date of birth, date of weaning, transgenic vector, transgenic family identification number, phenotype of each mouse, results of genomic screening for each mouse, results of all assays for transgene expression, mating record for each mouse, and information about the date of death, cause of death, and status of any tissues that were saved for each animal. Investigators setting up a new transgenic laboratory will need to decide what type(s) of record keeping system to establish, and then to establish the discipline necessary to maintain the records.

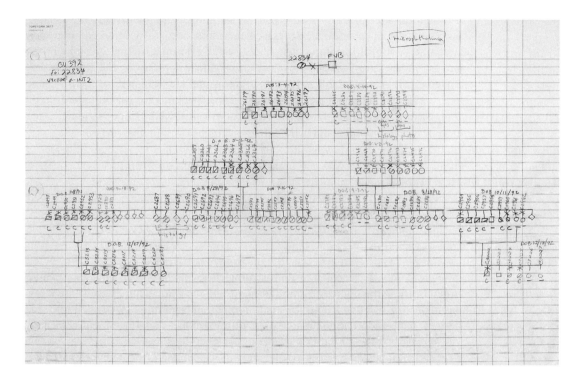

Figure 5. Pedigree chart. Pedigree charts are kept in the room where the mice are housed and are updated whenever new information becomes available. The charts provide a quick overview of the status and history of each transgenic family. Males are indicated by squares, females by circles, and mice that die (or are sacrificed) before sexing are indicated by diamonds. Identification numbers and dates of birth are written above the sex information for each mouse. Transgenic mice are indicated by placing a / within the sex symbol, and information about the phenotype or screening results is entered below the symbol (e.g., c for cataracts). Homozygous mice can be signified by placing an X inside the sex symbol. When mice are sacrificed, we indicate this by drawing an X through the vertical line that attaches the sex symbol to the horizontal litter line.

f. Computerized Record Keeping In some laboratories, it may be preferable to employ a computerized record keeping system, although such systems have not yet achieved widespread use. Ideally, a data management program could be designed so that the relevant information for each cage of mice could be entered into the computer, and the computer would print out an appropriately labeled cage card. The computer could automatically assign identification numbers and could maintain the pedigree charts and newborn records for all of the transgenic families in the colony. Data would need to be entered into the computer for each recipient female at the time of embryo transfer, for each litter at the time of birth and also at the time of weaning, and for each mating pair. In addition, data concerning the results of assays on the mice and dates of sacrifice would need to be entered. For such a system to be useful, the computer should be fully portable so that it can be brought into the mouse housing rooms whenever husbandry is performed. I am unaware of any commercially available software programs that are designed for record keeping in a transgenic mouse colony. Custom designed programs for mouse record keeping do exist (e.g., Silver, 1987; or Dr. Richard Woychik, Oak Ridge National Laboratory, personal communication, 1993).

III. TROUBLESHOOTING

A. Assaying for Successful Pronuclear Microinjection

One problem with learning to do microinjections is that it can be a long wait between the time the microinjections are done and the time that the results are known, particularly if one waits until the microinjected embryos have developed into weaning age mice before screening. By the time tail DNA is isolated and screened, it will be nearly 2 months after the time of the microinjections. If a novice is making a consistent mistake, a considerable amount of time and effort will be wasted.

There are a number of constructions that are particularly useful when learning to do microinjections. At the outset, either the metallothionein–β-galactosidase (MT–βgal) construct (Stevens *et al.,* 1989) or alcohol dehydrogenase under control of the Rous sarcoma virus promoter (RSV–ADH) (Nielsen and Pedersen, 1991) can be used to determine whether the microinjections are properly introducing genetic information into the pronucleus. After correct microinjection, either MT–βgal or RSV–ADH can be expressed in embryos at the 2- to 4-cell stage. As a result, embryos can be microinjected, incubated at 37°C for 24–48 hr, then stained for β-galactosidase or ADH activity (Stevens *et al.,* 1989; Neilsen and Pedersen, 1991). When the microinjections are done properly, a significant proportion of the embryos will show histochemical staining.

B. Visual Identification of Transgenic Mice

1. Elastase–*ras*

The promoter for the elastase gene has been shown to be active in pancreatic acinar cells in transgenic mice (Swift *et al.,* 1984; Ornitz *et al.,* 1985). Transgenic mice that carry the elastase promoter linked to the *ras* oncogene develop pancreatic tumors (Quaife *et al.*, 1987). These tumors are dominant and dramatic and can typically be recognized by simple visual inspection of the perinatal fetus. In a non-transgenic fetus, the whitish pancreas can be identified through the slightly transparent skin. Transgenic fetuses exhibit substantially enlarged pancreata (Quaife *et al.*, 1987; Pinkert, 1990). Pancreatic neoplasia can be confirmed by surgery to allow visual inspection of the viscera. For the novice microinjector, elastase–*ras* permits rapid identification of successful generation of transgenic mice within 3 weeks after injection of the DNA (Pinkert, 1990). One of the advantages of this construct is the fact that it can be used with embryos of any genotype (in contrast to the tyrosinase minigene, which is described in the next section). One drawback to the elastase–*ras* construct is the fact that the pancreatic tumors are generally lethal at an early age for the transgenic mice (Quaife *et al.*, 1987).

2. Tyrosinase Minigene

Classic albino strains of mice have a mutation in the gene encoding tyrosinase, the first enzyme in the pathway to melanin synthesis (Yokoyama *et al.*, 1990). Microinjection of a tyrosinase minigene into embryos of an albino strain of mice can result in gene cure of the albino defect and the synthesis of pigment (Tanaka *et al.*, 1990; Beermann *et al.*, 1990; Yokoyama *et al.*, 1990). The conversion from albinism to pigmentation is easy to recognize (Fig. 6), so the tyrosinase minigene offers a number of advantages for microinjection training. The microinjections can be done using albino inbred strains such as FVB/N or BALB/c. Alternatively, inexpensive outbred albino strains such as ICR can be used. Pigmented mice have dark eyes that can be identified by simple visual inspection at birth. In fact, the pigment epithelial cells of the retina begin to synthesize melanin by day 12 of embryonic development (Theiler, 1989), so that transgenic mice can typically be identified by visual inspection of the fetuses 2 weeks after microinjection (Fig. 7). Another advantage of the tyrosinase minigene is the fact that it is not detrimental to the health of the transgenic mice. The tyrosinase minigene is not useful in strains of mice that are already pigmented.

C. Superovulation Problems

1. Poor or No Superovulation

After an overnight mating opportunity, 70% or more of superovulated females should have mated, on average. If the mating percentage is consistently below 50%, either the males or the females may be guilty. A good first step to take in diagnosing the problem is to check for oocytes in the oviducts of the females with no visible plug.

Figure 6. Mice transgenic for a tyrosinase minigene. A nontransgenic FVB/N albino mouse is shown at left. The other three mice represent independent sites of integration (i.e., independent founder mice) for the tyrosine minigene *TyBS* (Yokoyama *et al.,* 1990). The mouse at right is mosaic. [This figure was originally published in a Dutch book entitled *De DNA-Makers, Architecten van het leven* (H. Schellekens, ed.), p. 121. Natuur and Techniek, Maastrich, Netherlands, 1993.]

If the nonplugged females have ovulated a normal number of oocytes (10–20/ side) and if the oocytes are mostly fertilized, then the problem is not in the super-ovulation regimen, but in the identification of copulation plugs. It may help to check for plugs earlier in the morning, before they have a chance to fall out. If the non-plugged females have ovulated, but the eggs are unfertilized, the fault may lie with the males rather than the females. Check to be sure that the males are not too old or mating too often. If the males are not the problem, then it may be the mating conditions. Check to be sure that the lights are set for the proper lights-on, lights-off cycle, and that the lights are truly turning off at night. Also check for problems with loud noises or other disturbances, and be certain that the temperature in the animal room is not elevated (80°F is too hot).

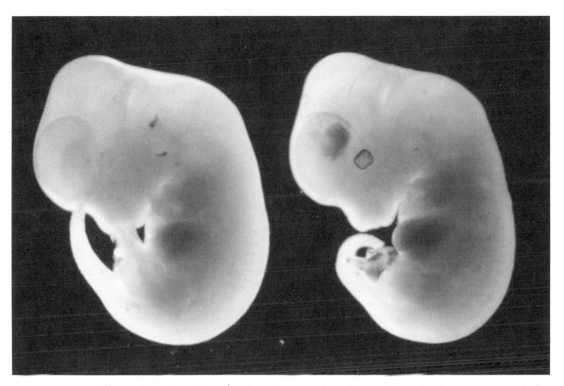

Figure 7. Visual identification of transgenic embryos. Most mice that are transgenic for the tyrosinase minigene can be identified by visual inspection for ocular pigmentation anytime after embryonic day 11 (E11). A nontransgenic E12 embryo is shown at left, whereas a littermate embryo transgenic for the tyrosinase minigene is shown at right. (Magnification, 10×.)

When the nonplugged females have not ovulated, the most likely explanation is a problem with the hormones, particularly the follicle stimulating hormone (FSH), which is usually provided by administration of pregnant mare's serum gonadotropin. If the stock solutions of hormones (50 units/ml) are more than 2 months old, make up new solutions and be sure to store them at 4°C. If the solutions were just recently prepared, check to be certain that they were prepared at the correct concentration. Another possible cause of problems is superovulation of female mice immediately after arrival in a new housing room. A minimum of 48 hr is often required for acclimation.

If the mating percentage is fine, but there are very few embryos, it might be useful to isolate the embryos earlier in the day. In some cases, the cumulus cells disaggregate early and the embryos proceed down the oviduct, making them more difficult to identify and to isolate. The most conclusive test of poor superovulation is to save the females to see whether any of them become pregnant, and whether they have large or small litters. If problems persist, try another strain of mice, or obtain the mice from a different vendor.

2. Unfertilized Eggs

If the females have copulation plugs, but the embryos are unfertilized, the problem lies with the males. In most cases, poor male fertility is caused by overuse of the males or by old age. Fertile males will typically show high fertilization rates if they are mated just once a week up to 9–10 months of age.

3. Malformed Eggs

The frequency of malformed eggs is generally a function of the strain of mice and the age of the females at superovulation. In our experience, 10–25% of the eggs from superovulated 4- to 5-week-old C57BL/6 females are immature or malformed, whereas analogous FVB females typically yield less then 5% defective eggs. The younger the females are at the time of superovulation, the higher the percentage of misshapen oocytes. If the percentage of defective eggs is over 25%, then the concentrations of the hormones used for superovulation should be rechecked. Poor health or abnormal stress of the superovulated females can also result in a high percentage of malformed eggs.

When almost all of the embryos appear unhealthy or abnormal, even the fertilized embryos, the most common cause is a problem with the osmolarity or pH of the mouse embryo medium. Alternatively, the CO_2 concentration in the 37°C incubator may be incorrect (it should be 5%, v/v), or the embryos may have been outside of the incubator for too long during embryo isolation or manipulation. (See Chapter 2 for information about embryo culture media and culture conditions.)

D. Colony Problems

1. Males Not Mating

Males not mating is a rare problem, most often caused by overbreeding of the males. If the females are in estrus, healthy adult males will mate at least 75% of the time.

2. No Recipient Females in Estrus

When female mice are housed together, they will often begin to cycle in synchrony. When this happens, there will be days with multiple recipient females in estrus and other days with very few recipients in estrus. One way to circumvent this problem is to superovulate recipient females when needed. The drawback to this strategy, as stated earlier, is that superovulated females seem to have a poor pregnancy rate when used as recipients. An alternative strategy is to maintain a larger stock of potential recipient females in the colony.

3. High Rates of Abortion in Recipients

Recipients that have as few as one healthy fetus will normally maintain pregnancy to full term. Pregnant females rarely miscarry or deliver before term ($<1\%$ of the time). Frequent miscarriage indicates that the females are either sick or unduly stressed.

4. Small Litter Sizes in Recipient Females

When recipient females consistently have small litters (only one or two pups), the problem is unlikely to be simply embryonic lethality of transgenic fetuses, and it is more likely to be a technical problem. The mouse embryo medium might be at fault, the microinjection procedure might be flawed, the concentration (or quality) of the microinjected DNA(s) might be incorrect, or there may be problems with the embryo transfer procedure. To identify the problem, a first step is to do embryo transfers using uninjected embryos. Approximately 75–100% of uninjected embryos should yield healthy newborns. If the percentage is substantially less than that, there is a serious problem with either the culture conditions or the embryo manipulation procedures. A second troubleshooting procedure is to culture injected and uninjected embryos overnight. Most ($>90\%$) of the uninjected embryos should reach the 2-cell stage after 24 hr in culture. If the embryos do not divide, it suggests that there are serious deficiencies in the culture conditions. For embryos that survive microinjection, a majority (50–60%) should divide to the 2-cell stage after overnight culture. If not, either the microinjection procedure is causing irreparable damage, or the DNA solution is lethal. To test the microinjection procedure, do the injections with DNA buffer (10 mM Tris, 0.1 mM EDTA) alone. If the embryos survive mock injections but not injections of DNA, then the DNA should be repurified. The DNA must be free of contamination by phenol, chloroform, ethanol, etc., and the concentration should not exceed 5μg/ml.

Another strategy to assess small litter sizes is to perform C-sections on recipient females at 19 days of gestation. When females are pregnant with one or more healthy fetuses, it will be possible to determine the total number of embryos that progressed to the implantation stage or beyond. Embryonic implantation induces proliferation of the wall of the uterus and formation of a deciduum. Even if the embryo ceases development shortly after implantation, the deciduum is not resorbed until after pregnancy is complete. Embryos that develop to midgestation or later can be recognized and their stage of development estimated. Contaminants in the microinjected DNA often cause early postimplantation lethality, resulting in multiple decidua per pregnant female. Recombinant DNAs that cause dominant embryonic lethality are very rare, so the absence of transgenic mice for a specific DNA construction is typically due to some factor other than prenatal lethality of transgenic fetuses.

5. Recipient Females Do Not Become Pregnant

If the recipient females do not become pregnant, do some test embryo transfers using uninjected embryos. If pregnancies still do not occur, then embryo transfers should be practiced until the embryos can be consistently transferred to the proper region of the oviduct and the recipient females consistently become pregnant. Lack of pregnancy can also be caused by health impairment, inappropriately prepared anesthetic, or stresses in the animal housing area. Much more likely factors are poor embryo culture conditions or improper reimplantation techniques. If uninjected embryos give pregnant females, but injected embryos do not, then there is a problem with the embryo manipulations (either the protocol or the DNA).

IV. TRANSGENIC PHENOMENOLOGY

Multiple independent transgenic families have been generated in my laboratory by microinjection of the tyrosinase minigene. They provide a visual demonstration of many of the common phenomena seen in transgenic mice.

A. Transgene Expression: Effects of Integration Site and Copy Number

Microinjected DNA usually integrates at only one site, or a very limited number of different sites, in individual embryos. The number of copies that integrate is highly variable, ranging from just one copy to as many as 200. When multiple copies integrate, they are almost always found linked in a tandem head-to-tail array at the site of integration (Brinster *et al.*, 1981). The molecular events responsible for this pattern of integration are not well understood.

One other common observation is that there is often considerable variability in the level of transgene expression from one independent transgenic family to another. For example, with the tyrosinase minigene, pigmentation intensities have been found to range from nearly normal agouti to gray to brownish to tan to pigmented only in the ears (Figs. 8 and 9). This variability is generally attributed to

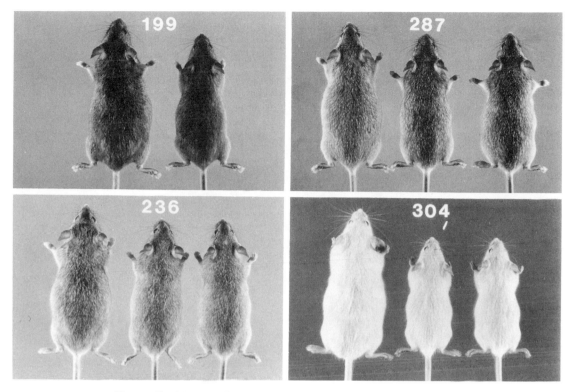

Figure 8. Variation in pigmentation. Different sites of integration of the tyrosinase mini-gene produce different levels of pigmentation, ranging from normal agouti (199) to gray (287) to brownish gray (236) to light tan (304). The color and intensity of pigmentation are consistent within each of the families. For each family, the mouse at left is a parent and representative pigmented offspring are shown at right.

influences of the chromosomal sequences flanking the different sites of integration. For the tyrosinase minigene there is a modest, but not consistent, correlation between the copy number of the transgene and its level of expression (see Table 3). Agouti pigmentation is present in some families that carry only two copies of the transgene, whereas other families that are pigmented only in their ears have much higher copy numbers (Table 3). The brown and tan families have consistently low copy numbers.

For some genes (such as the globins) regulatory sequences have been identified that will give levels of expression of transgenic DNA that are copy number dependent and site-of-integration independent (Grosveld *et al.*, 1987; Ryan *et al.*, 1989). These regulatory sequences often contain DNase superhypersensitive regions. Position-independent levels of expression can also be obtained by coinjection

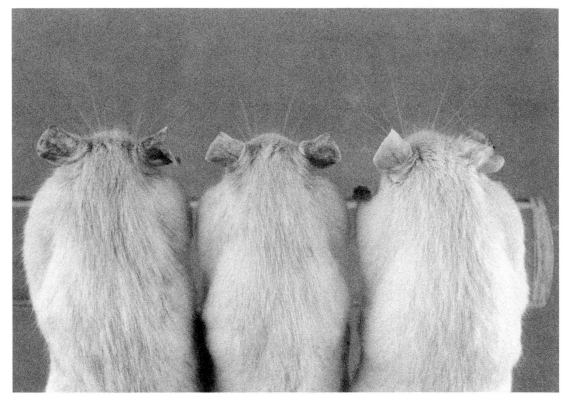

Figure 9. Pigmentation in the ears. In some of the tyrosinase transgenic families, pigmentation is readily visible only in the ears of the transgenic mice (left and center). A nontransgenic albino mouse is shown at right for comparison. (The right ear of the nontransgenic mouse has a scar where an ear tag was once located.)

of a matrix-attachment region (McKnight *et al.*, 1992). DNA purification for microinjection and removal of cloning vector sequences are discussed in Chapter 2.

B. Genetic Mosaicism in Founder Mice

DNA synthesis in mouse embryos begins 12 to 14 hr after fertilization, so that replication of the genome is generally in progress by the time that microinjections are performed. If the microinjected DNA integrates into an unreplicated location in the genome, then it can be duplicated prior to cell division, and every cell in the developing embryo will receive a copy of the integrated transgenic DNA. However, if the integration site has replicated prior to integration, then the embryo will be a genetic mosaic composed of both transgenic and nontransgenic cells. When integration occurs after just one round of DNA replication, then the founder mouse will be

TABLE 3
Comparison of Transgene Copy Number and Pigmentation Intensity for 78 Independent Integration Sites of the Tyrosinase Minigene

Pigmentation[a]	Number of families	Copy number (range)[b]	Average copy number
Agouti	14	2–40	8.0
Gray	18	2–40	7.7
Himalayan	6	4–30	16.0
Brown	13	1–8	3.8
Tan	9	1–4	2.0
Ears only	5	2–30	9.6
Dark eyes only	2	2	2.0
Mottled	5	4–16	9.6
Variable	6	4–16	10.0

[a] Pigmentation intensities were classified subjectively. The agouti mice had nearly normal pigmentation intensity, the gray mice resembled a dilute black pigmentation, himalayan mice were light gray with dark ears, brown mice showed a dilute brown phenotype, tan mice were very light brown, and certain families were pigmented exclusively in their ears or eyes. The mottled mice were partially pigmented, and the variable families showed variable intensities of pigmentation for a single integration site.

[b] Copy number was determined by Southern hybridizations. The intensity of the hybridization bands for the transgenic DNA was compared to the intensity for the endogenous tyrosinase gene.

approximately 50% mosaic, although the percentage may vary depending on the rates of cell division and the relative contributions of the transgenic and nontransgenic cells to the various tissues of the developing mouse.

Nonmosaic transgenic mice with one site of integration should transmit the transgenic DNA in Mendelian fashion to about 50% of their offspring, whereas mosaic mice generally show a frequency of transmission of 25% or less. (Note: Transgenic founder mice that have more than one site of integration can produce litters where 75% or more of the offspring are transgenic, although the percent transmission for any one site of integration is expected to average 50% or less. See Section IV,E for further discussion of multiple integration sites.) In some cases the transgenic DNA will integrate after the 2-cell stage, resulting in founder mice that are transgenic in substantially less than 50% of their cells. In most cases, mosaicism is recognized when the copy number of the transgenic insert is higher in the transgenic offspring than in the founder, and when the percentage transmission is substantially below 50%. With the tyrosinase minigene, mosaic mice can be identified visually, since they show a mottled pigmentation pattern (Fig. 10) that is reminiscent of chimeric mice (see Chapter 4). When the tyrosinase founder mice are mottled because of mosaicism, their transgenic offspring show a uniform pattern of pigmentation (Fig. 10). The tyrosinase minigene allows visual identification of mosaic founder mice that have only a small number of pigmented cells, and it also allows rapid identification of their rare transgenic offspring. Approximately 35% of the founder mice from our tyrosinase minigene injections were found to be genetic

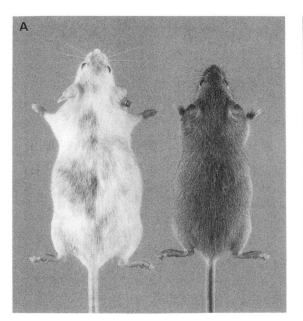
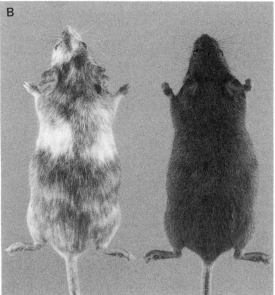
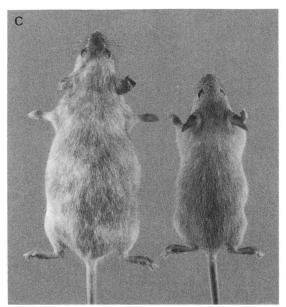

Figure 10. Mosaic founder mice. The tyrosinase minigene allows visual identification of mosaic founder mice, since the mosaic mice (at left in A, B, and C) show a heterogeneous, mottled pattern of pigmentation, whereas their transgenic offspring (at right) show uniform pigmentation.

mosaics (data not shown). For five of the mottled founder mice, the transgenic offspring still exhibited a mosaic pattern of pigmentation (e.g., Fig. 11), indicating that in these mice the mottled pigmentation was due to a mosaic pattern of transgene expression rather than to genetic mosaicism. These mice provide visual evidence that specific sites of integration can not only bias the level of transgene expression, but may also influence the cell-by-cell pattern of expression.

C. *Intrafamily Variation in Expression*

In some transgenic families, the level of transgene expression may vary from mouse to mouse even though the mice are inbred and have the same site of integration

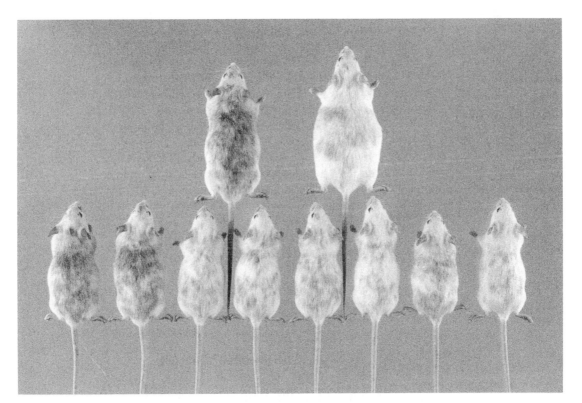

Figure 11. Mosaic expression. Some of the tyrosinase transgenic families show a persistent pattern of heterogeneous pigmentation even though the transgenic mice are genetically nonmosaic. The two mice in the top row are homozygous for the tyrosinase minigene and are the parents of the eight mice below (family OVE159). The transgenic mice in this family show variation in the intensity and specific pattern of pigmentation, but all of the mice are mottled.

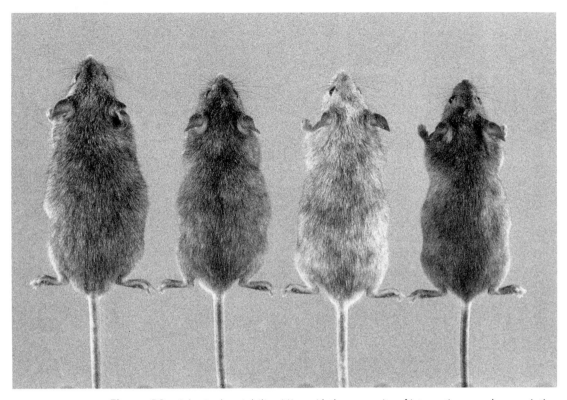

Figure 12.   Inherited variability. Mice with the same site of integration can show variation in the intensity of pigmentation. In the family that is shown (OVE195), the parent at left produced the three offspring at right. Southern hybridizations showed an identical pattern of bands for all of the mice, indicating a single stable site of integration (data not shown). When the offspring mice were mated, the same variation in pigmentation was seen in the next generation. The lightly pigmented mouse had light and dark offspring, as did the more darkly pigmented mice.

(Fig. 12). In our collection of tyrosinase transgenic families, over 90% of the integration sites gave stable patterns of expression. In these cases, all of the heterozygous transgenic mice in a given family had an identical, or nearly identical, color, intensity, and pattern of pigmentation. The pigmentation did not vary from generation to generation, and the males showed the same pigmentation as the females. There were six transgenic families in which the intensity and color of pigmentation were not consistent (e.g., Fig. 12). Southern hybridizations were done to look for multiple sites of integration in these families, but the hybridization patterns were stable and consistent within each family (data not shown). The mice in these six families show an innate variability in their level of transgene expression and demonstrate that mice of apparently identical genotype can exhibit variable phenotypes. The variation in expression is not correlated with the sex of the mice or of the transgenic parent, since the variability is seen between siblings.

The level of expression of transgenic DNA can also be influenced by genomic imprinting (e.g., Chaillet *et al.,* 1991; Reik *et al.,* 1990). No examples of genomic imprinting were seen in the families transgenic for the tyrosinase minigene.

D. Integration into a Sex Chromosome

Integration into either of the sex chromosomes can be identified by mating a (nonfounder) transgenic male to a nontransgenic female. If the transgenic DNA has integrated into a nonpseudoautosomal region of the X chromosome, then all the female offspring will be transgenic, whereas all the male offspring will be nontransgenic. In other words, the transgenic DNA will be transmitted along with the X chromosome to only the female offspring. The transmission pattern will be reversed for integration into the Y chromosome. Among the tyrosinase families, two integrations were found in the X chromosome and one in the Y chromosome. In the Y chromosome family, all the males and only the males were pigmented. The pigmentation was limited exclusively to the ears. In both instances where the tyrosinase minigene integrated into the X chromosome, the transgene was subject to X chromosome inactivation (e.g., Fig. 13). Heterozygous female mice showed a mottled pattern of pigmentation, whereas heterozygous males and homozygous females were uniformly pigmented.

E. Multiple Sites of Integration in Founder Mice

Founder transgenic mice occasionally have transgenic DNA integrated at more than one site in the genome. This phenomenon is typically recognized when the offspring are analyzed and found to contain two or more different copy numbers or hybridization patterns for the transgenic DNA. [Multiple sites of integration are difficult to identify by polymerase chain reaction (PCR) screening.] In general, different sites of integration are assumed to represent independent integration events. The copy numbers are variable, the levels of expression can be different, and transmission is generally random. However, transgenic families have occasionally been discovered to have different insertions in the same region of the genome (Xiang *et al.,* 1990; P. A. Overbeek, unpublished, 1990). The molecular basis for related integration events is not clear.

The tyrosinase transgenic mice allow easy identification of transgenic founder mice with multiple sites of integration, since the founder mice produce offspring

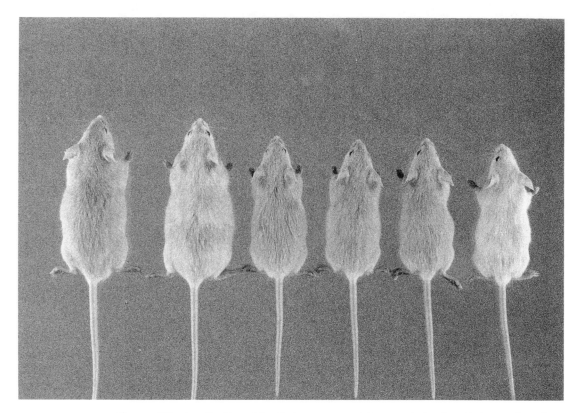

Figure 13. Inactivation of the X chromosome. A transgenic family with integration of the tyrosine minigene into the X chromosome is shown. The two mice at left are a heterozygous male and female, respectively, and are parents to the four mice at right (two males and two females). One of the females, second from the right, is homozygous for the tyrosinase minigene, whereas the other (far right) is heterozygous and mottled.

with two or more consistent colors or intensities of pigmentation (Fig. 14). Approximately 15% of the tyrosinase founder mice had either two or three different integration sites that were recognized by visual inspection of the offspring and confirmed by Southern hybridizations (not shown).

F. Homozygous versus Hemizygous

One of the major advantages of using the tyrosinase minigene to generate transgenic mice is the fact that homozygous mice in most families can be identified by simple visual inspection, since the homozygous mice have darker coat colors (Fig. 15), reflecting the increased gene dosage. In some families it has not been possible to identify homozygotes by simple visual inspection. However, homozygotes can still

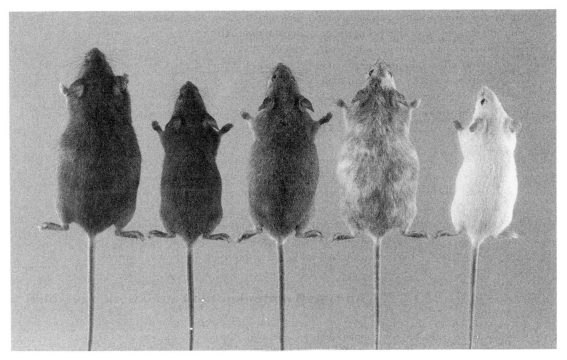

Figure 14. Founder mouse with multiple integration sites. The founder mouse for family OVE376 (at left) produced offspring with four different types of pigmentation. The middle two of the four offspring actually have the same site of integration, which is in the X chromosome. The male (left) and female (right) patterns of expression are shown.

be identified by matings to nontransgenic partners, since homozygous mice produce 100% transgenic offspring. Breeding studies to generate homozygous mice have allowed the recognition of insertional mutations that cause cleft palate (OVE270) or situs inversus (*inv* mutation, family OVE210) (Yokoyama *et al.*, 1993). In addition, five other tyrosinase families have insertional mutations that result in embryonic or neonatal lethality for the homozygous mice. These families have still been easy to maintain, since the transgenic mice in each generation can be visually identified.

G. Coinjections

The tyrosinase minigene can be coinjected with other constructs of interest. The coinjected constructs typically cointegrate into the genome, where they can be independently expressed, thereby allowing visual identification of the transgenic mice in the first and all subsequent generations (see Overbeek *et al.*, 1991).

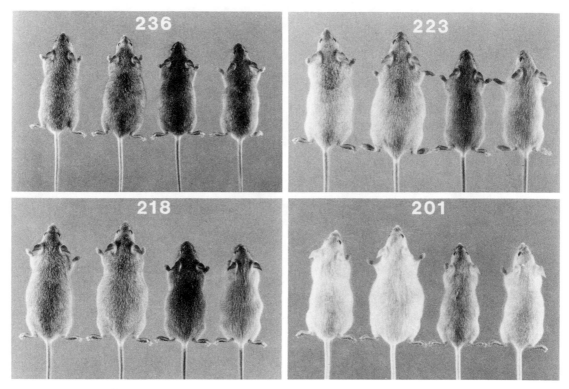

Figure 15. Visual identification of homozygotes. For most of the tyrosinase transgenic families, homozygous transgenic mice can be recognized by simple inspection for darker coat color, reflecting the 2-fold higher gene dosage. In each family shown, two heterozygous parents are shown at left, and two weanling age offspring are shown at right, one of which is homozygous (closest to the parents) while the other is heterozygous.

H. Genetic Instability

In almost all instances, transgenic DNA is stably maintained once it has integrated into the genome. Even when there are multiple copies of the DNA integrated in a tandem head-to-tail array, the transgenic DNA is transmitted stably from one generation to the next without genomic rearrangements and without deletions. However, examples of transgenic families with genetic instability under selective pressure have been identified (e.g., Sandgren *et al.*, 1991; Wilkie *et al.*, 1991). The tyrosinase transgenic families have all displayed stable Southern hybridization patterns, and those families with stable pigmentation have maintained a uniform pigmentation intensity over more than 10 generations of mating. Because alterations of the integration site would be expected to change both tyrosinase expression and pigmentation, these results imply that genetic instability is very rare for these transgenic inserts.

V. SUMMARY

In this chapter, I have tried to provide a simplified overview of factors that often need to be taken into consideration by laboratories that are just beginning to do research with transgenic mice. I have delineated the three general strategies for generating transgenic mice and discussed the situations where each strategy is most commonly used. An introduction to different strains of mice and to husbandry techniques and troubleshooting protocols has been provided. In Section III I have reviewed some of the characteristics of transgenic mice generated by microinjection. In general, microinjected DNA appears to integrate randomly into the mouse genome. In most cases the DNA integrates stably as a tandem head-to-tail repeat containing from 1 to 200 copies of the injected DNA. Multiple factors can influence the pattern of expression of the transgenic DNA. The tyrosinase minigene can be used to allow visual identification of transgenic mice. Interestingly, certain tyrosinase transgenic mice show that, even for a stable integration at a single site in the genome, there can be variations in the level and pattern of transgene expression.

APPENDIX: STANDARDIZED NOMENCLATURE FOR TRANSGENIC ANIMALS[1]

Rules for Naming Transgenic Families

Transgenic animals should be named according to the following conventions.

Symbol. A transgene symbol consists of four parts, all in Roman type, as follows:

$$TgX(YYYYYY)\#\#\#\#\#Zzz$$

where TgX is the mode, (YYYYYY) the insert designation, $\#\#\#\#\#$ the laboratory-assigned number, and Zzz the laboratory code.

Mode. The first part of the symbol always consists of the letters Tg (for "transgene") and a letter designating the mode of insertion of the DNA: N for nonhomo-

---

1. Adapted with permission from ILAR NEWS, Volume 34, Number 4, 1992. Courtesy of the National Academy Press, Washington, D.C.

logous insertion, R for insertion via infection with a retroviral vector, and H for homologous recombination. The purpose of this designation is to identify it as a symbol for a transgene and to distinguish among three fundamentally different organizations of the introduced sequence relative to the host genome, not simply to indicate the method of insertion or nature of the vector. For example, mice derived by infection of embryos with murine leukemia virus (MuLV) vectors will be designated TgR, and mice derived by microinjection or electroporation of MuLV DNA into zygotes will be designated TgN; mice derived from ES cells by introduction of DNA followed by recombination with the homologous genomic sequence will be designated TgH, whereas mice derived by insertions of the same sequence by non-homologous crossing-over events will be designated TgN.

Insert designation. The second part of the symbol indicates the salient features of the transgene as determined by the investigator. It is always in parentheses and consists of no more than eight characters, either letters (capitals or lowercase) or a combination of letters and numbers. Italics, superscripts, subscripts, internal spaces, and punctuation should not be used. The choice of the insert designation is up to the investigator, but the following guidelines should be used.

• Short symbols (six or fewer characters) are preferred. The total number of characters in the insert designation plus the laboratory-assigned number may not exceed 11 (see below); therefore, if seven or eight characters are used, the number of digits in the laboratory assigned number will be limited to four or three, respectively.

• The insert designation should identify the inserted sequence and indicate important features. If the insertion uses sequences from a named gene, it is preferable that the insert designation contain the standard symbol for that gene. If the gene symbol would exceed the spaces available, its beginning letters should be used. Hyphens should be omitted when normally hyphenated gene symbols are used. For example, insl should be used in the symbols of transgenes that contain either coding or regulatory sequences from the mouse insulin gene (*Ins-1*) as an important part of the insert designation. Resources are available to identify standard gene symbols (see pp. 111–112). Symbols that are identical with other named genes in the same species should be avoided. For example, the use of Ins to designate "insertion" would be incorrect.

• For consistency, a series of transgenic animals produced with the same construct might be given the same insert designation. However, that is not required; some lines might manifest unique and important characteristics (e.g., insertional mutations) that would warrant a unique insert designation. If two different symbols are used for the same construct in different transgenic lines, the published descriptions should clearly identify the construct as being the same in both lines. Two different gene constructs used for transgenic animal production, either within a laboratory or in separate laboratories, should not be identified by identical insert designations. Designations can be checked through the available resources (see pp. 111–112).

• A standard abbreviation can be used as part of the insert designation (see below for an example). If a standard abbreviation is used, it should be placed at the end of the insert. These now include the following:

An   Anonymous sequence
Ge   Genomic clone
Im   Insertional mutation
Nc   Noncoding sequence
Rp   Reporter sequence
Sn   Synthetic sequence
Et   Enhancer trap construct
Pt   Promoter trap construct

This list will be expanded as needed and maintained by appropriate international nomenclature committees.

• The insert designation should identify the inserted sequence, not its location or phenotype.

Laboratory-assigned number and laboratory code. The laboratory-assigned number is a unique number that is assigned by the laboratory to each stable transmitted insertion when germ line transmission is confirmed. As many as five characters (numbers as high as 99,999) may be used; however, the total number of characters in the insert designation plus the laboratory-assigned number may not exceed 11. No two lines generated within one laboratory should have the same assigned number. Unique numbers should be given even to separate lines with the same insert integrated at different positions. The number can have some intralaboratory meaning or simply be a number in a series of transgenes produced by the laboratory. The laboratory code is uniquely assigned to each laboratory that produces transgenic animals. A laboratory that has already been assigned such a code for other genetically defined mice and rats or for DNA loci should use that code. The registry of these codes is maintained by the Institute of Laboratory Animal Resources (ILAR).

Examples. The complete designation identifies the inserted DNA, provides a symbol for ease of communication, and supplies a unique identifier to distinguish it from all other insertions. Each insertion retains the same symbol even if it is placed on a different genetic background. Specific lines of animals carrying the insertion should be additionally distinguished by a stock designator preceding the transgene symbol. In general, this designator will follow the established conventions for the naming of strains or stocks of the particular animal used. If the background is a mixture of several strains, stocks, or both, the transgene symbol should be used without a strain or stock name. The following examples are typical:

• C57BL/6J-TgN(CD8Ge)23Jwg. The human **CD8** genomic clone (**Ge**) inserted into C57BL/6 mice from The Jackson Laboratory (J); the **23**rd mouse screened in a series of microinjections in the laboratory of Jon W. Gordon (**Jwg**).

- Crl:ICR-TgN(SVDhfr)432Jwg. The **SV**40 early promoter driving a mouse dihydrofolate reductase (***Dhfr***) gene; **4** kilobase plasmid; the **32**nd animal screened in the laboratory of Jon W. Gordon (**Jwg**). The ICR outbred mice were obtained from Charles River Laboratories (Crl).

- TgN(GPDHIm)lBir. The human glycerol phosphate dehydrogenase (**GPDH**) gene inserted into zygotes retrieved from (C57BL/6J × SJL/J)F1 females; the insertion caused an insertional mutation (**Im**) and was the **1**st transgenic mouse named by Edward H. Birkenmeier (**Bir**). No strain designation is provided because each zygote derived from such an F1 hybrid mouse has a different complement of alleles derived from the original inbred parental strains.

- 129/J-TgH(SV40Tk)65Rpw (hypothetical). An **SV**40 thymidine kinase (**Tk**) transgene targeted by homologous recombination to a specific but anonymous locus using embryonic stem cells derived from mouse strain 129/J. This was the **65**th mouse of this series produced by Richard P. Woychik (**Rpw**).

Abbreviations

Transgene symbols can be abbreviated by omitting the insert. For example, the full symbol TgN(GPDHIm)1Bir would be abbreviated TgN1Bir. The full symbol should be used the first time the transgene is mentioned in a publication; thereafter, the abbreviation may be used.

Insertional Mutations and Phenotypes

The symbol should not be used to identify the specific insertional mutation or phenotype caused directly or indirectly by the transgene. If an insertional mutation produces an observable phenotype, the locus so identified must be named according to standard procedures for the species involved [Lyon, M. F. (1989). Rules and guidelines for gene nomenclature. In *"Genetic Variants and Strains of the Laboratory Mouse"* (M. F. Lyon and A. G. Searle, eds.), 2nd Ed., pp. 2–11. Oxford Univ. Press, London]. The allele of the locus identified by the insertion can then be identified by the abbreviated transgene symbol (see above) according to the conventions adopted for the species.

Examples

- $ho^{TgN447Jwg}$. The insertion of a transgene into the hotfoot locus (*ho*).

- $xxx^{TgN21Jwg}$. The insertion of a transgene that leads to a recessive mutation in a previously unidentified gene. A gene symbol for *xxx* must be obtained from a species-genome database or a member of a nomenclature committee (see next section).

Resources Available for Assistance with Transgenic Nomenclature

Before naming a transgene, an investigator should obtain a laboratory code from the ILAR at the address given in the list below. An investigator who has already been assigned such a code for other genetically defined mice and rats or for DNA loci should use the same code. The transgene should be named as stated in the rules. Assistance in selecting transgene symbols is available from several organizations (see below). Lists of named genes for mice and rats are published periodically in *Mouse Genome* (Oxford University Press, Journal Subscriptions Department, Pinkhill House, Southfield Road, Eynsham, Oxford OX8 1JJ, UK) and *Rat News Letter* (Dr. Viktor Stolc, editor, *Rat News Letter,* 2542 Harlo Drive, Allison Park, Pittsburgh, PA 15101). The list of mouse genes is also maintained in GBASE, a genomic database for the mouse maintained by Dr. Don P. Doolittle, Dr. Alan L. Hillyard, Ms. Lois J. Maltais, Dr. Muriel T. Davisson, Dr. Thomas H. Roderick, and Mr. John N. Guidi at The Jackson Laboratory (see below). Human gene symbols are recorded in the Genome Data Base (GDB), which is maintained at the Johns Hopkins University (see below).

Institute of Laboratory Animal Resources (ILAR). Assigns laboratory codes; assists in naming transgenes; provides rules for naming transgenes. Contact: Dr. Dorothy D. Greenhouse, ILAR, National Research Council, 2101 Constitution Avenue, NW, Washington, DC 20418. Tel: 1-202-334-2590; Fax: 1-202-334-1687; Bitnet: DGREENHO@NAS.

The Jackson Laboratory. Assists in naming transgenes; provides rules for standardized nomenclature for mice; provides lists of named mouse genes. Contact: Dr. Muriel T. Davisson, The Jackson Laboratory, Bar Harbor, ME 04609. Tel: 1-207-288-3371; Fax: 1-207-288-8982.

Medical Research Council Radiobiology Unit. Assists in naming transgenes; provides lists of named mouse genes. Contact: Dr. Josephine Peters, MRC Radiobiology Unit, Chilton, Didcot, Oxford OX11 ORD, UK. Tel: 44-235-834-393; Fax: 44-235-834-918.

Genome Data Base (GDB). Records, stores, and provides information on mapped human genes and clones. Contact: GDB, Welch Medical Library, The Johns Hopkins University, 1830 East Monument Street, Baltimore MD 21205; Tel: 1-301-955-9705; Fax: 1-301-955-0054. For assistance in naming human genes, the contact is Dr. Phyllis J. McAlpine, GDB Nomenclature Editor, University of Manitoba, Department of Human Genetics, 250 Old Basic Sciences Building, 770 Bannatyne Avenue, Winnipeg, Manitoba, Canada R3E 0W3. Tel: 1-204-788-6393; Fax: 1-204-786-8712; Bitnet: GENMAP@UOFMCC.

The Transgenic Animal Data Base (TADB). The Transgenic Animal Data Base is intended to be a comprehensive, on-line, computerized record of all lines of transgenic animals and animals with targeted mutations that have been generated worldwide. Transgenic animals of little interest to one researcher might be of enormous interest to others. This situation is addressed through the database by making available to the scientific community extensive information about transgenic constructs, including methods, expression, and phenotypes.

Scientists provide data on their own lines of transgenic animals. The TADB office faxes a set of specific questions to each scientist, who answers the questions on a floppy disk with a word-processing program. The floppy disk is returned to the TADB office, where the data are formatted and transferred to the on-line computer. The result is that pertinent information, published or unpublished, on each line of animals is organized in the database.

The TADB is accessible internationally via a toll-free number through the Tymnet telecommunications network. Users can get information at no cost from the TADB office. Contact: Ms. Karin Schneider, TADB Coordinator, Oak Ridge National Laboratory, P.O. Box 2008, MS 6050, Oak Ridge, TN 37831-6050; Tel: 1-615-574-7776; Fax: 1-615-574-9888; Bitnet: TUG@ORNLSTC; Internet: OWENSET@IRAVAX.HSR.ORNL.GOV.

ACKNOWLEDGMENTS

I thank Gerri Hanten for many years of mouse embryo microinjections, Mick Kovac for management of the colony of transgenic mice, Lindsey Lampp for photographing the mice, and Tammy Reid for typing the manuscript. I also thank Dr. Glenn Merlino for his helpful comments on the manuscript. Research using the tyrosinase minigene was supported by National Institutes of Health Grant HD 25340.

REFERENCES

Beermann, F., Ruppert, S., Hummler, E., Bosch, F. X., Müller, G., Rüther, U., and Schütz, G. (1990). Rescue of the albino phenotype by introduction of a functional tyrosinase gene into mice. *EMBO J.* **9**, 2819–2826.

Brinster, R. L., Chen, H. Y., and Trumbauer, M. (1981). Somatic expression of herpes thymidine kinase in mice following injection of a fusion gene into eggs. *Cell (Cambridge, Mass.)* **27**, 223–231.

Brinster, R. L., Chen, H. Y., Trumbauer, M. E., Yagle, M. K., and Palmiter R. D. (1985). Factors affecting the efficiency of introducing foreign DNA into mice by microinjecting eggs. *Proc. Natl. Acad. Sci. U.S.A.* **82**, 4438–4442.

Chaillet, J. R., Vogt, T. F., Beier, D. R., and Leder, P. (1991). Parental-specific methylation of an imprinted transgene is established during gametogenesis and progressively changes during embryogenesis. *Cell (Cambridge, Mass.)* **66**, 77–83.

Covarrubias, L., Nishida, Y., and Mintz, B. (1986). Early postimplantation embryo lethality due to DNA rearrangements in a transgenic mouse strain. *Proc. Natl. Acad. Sci. U.S.A.* **83**, 6020–6024.

Covarrubias, L., Nishida, Y., Terao, M., D'Eustachio, P., and Mintz, B. (1987). Cellular DNA rearrangements and early developmental arrest caused by DNA insertion in transgenic mouse embryos. *Mol. Cell. Biol.* **7**, 2243–2247.

Dickie, M. M. (1975). Keeping records. *In* "Biology of the Laboratory Mouse" (E. L. Green, ed.), 2nd Ed., pp. 23–27. Dover, New York.

Festing, M. F. W. (1990). Choice of an experimental animal and rodent genetics. *In* "Postimplantation Mammalian Embryos: A Practical Approach" (A. J. Copp and D. L. Cockroft, eds.), pp. 205–219. Oxford Univ. Press, Oxford.

Festing, M. F. W. (1992). Origins and characteristics of inbred strains of mice, 14th listing. *Mouse Genome* **90**, 231–352.

Friedrich, G., and Soriano, P. (1991). Promoter traps in embryonic stem cells: A genetic screen to identify and mutate developmental genes in mice. *Genes Dev.* **5**, 1513–1523.

Green, E. L. (1975). Breeding systems. *In* "Biology of the Laboratory Mouse" (E. L. Green, ed.), 2nd Ed., pp. 11–22. Dover, New York.

Grosveld, F., van Assendelft, G. B., Greaves, D. R., and Kollias, G. (1987). Position-independent, high-level expression of the human β-globin gene in transgenic mice. *Cell (Cambridge, Mass.)* **51**, 975–985.

Harbers, K., Kuehn, M., Delius, H., and Jaenisch, R. (1984). Insertion of retrovirus into the first intron of αl(I) collagen gene leads to embryonic lethal mutation in mice. *Proc. Natl. Acad. Sci. U.S.A.* **81**, 1504–1508.

Hartung, S., Jaenisch, R., and Breindl, M. (1986). Retrovirus insertion inactivates mouse αl(I) collagen gene by blocking initiation of transcription. *Nature (London)* **320**, 365–367.

Hetherington, C. M. (1987). Mouse husbandry. *In* "Mammalian Development: A Practical Approach" (M. Monk, ed.), pp. 1–12. IRL Press, Oxford.

Hogan, B., Costantini, F., and Lacy E. (1986). "Manipulating the Mouse Embryo: A Laboratory Manual." Cold Spring Harbor Laboratory, Cold Spring Harbor, New York.

Jaenisch, R., Harbers, K., Schnieke, A., Chumakov, I., Jähner, D., Löhler, J., Grotkopp, D., and Hoffmann, E. (1983). Germline integration of Moloney murine leukemia virus at the *Mov13* locus leads to recessive lethal mutation and early embryonic death. *Cell (Cambridge, Mass.)* **32**, 209–216.

Lang, C. M. (1983). Design and management of research facilities. *In* "The Mouse in Biomedical Research" (H. L. Foster, J. D. Small, and J. G. Fox, eds.), Vol. 3, pp. 38–50. Academic Press, New York.

Les, E. P. (1975). Husbandry. *In* "Biology of the Laboratory Mouse" (E. L. Green, ed.), 2nd Ed., pp. 29–37. Dover, New York.

McKnight, R. A., Shamay, A., Sankaran, L., Wall, R. J., and Henninghausen, L. (1992). Matrix-attachment regions can impart position-independent regulation of a tissue-specific gene in transgenic mice. *Proc. Natl. Acad. Sci. U.S.A.* **89**, 6943–6947.

Meisler, M. H. (1992). Insertional mutation of "classical" and novel genes in transgenic mice. *Trends Genet.* **8**, 341–344.

Nielsen, L. L., and Pedersen, R. A. (1991). *Drosophila* alcohol dehydrogenase: A novel reporter gene for use in mammalian embryos. *J. Exp. Zool.* **257**, 128–133.

Ornitz, D. M., Palmiter, R. D., Hammer, R. E., Brinster, R. L., Swift, G. H., and MacDonald, R. J. (1985). Specific expression of an elastase–human growth hormone fusion gene in pancreatic acinar cells of transgenic mice. *Nature (London)* **313**, 600–602.

Otis, A. P., and Foster, H. L. (1983). Management and design of breeding facilities. *In* "The Mouse in Biomedical Research" (H. L. Foster, J. D. Small, and J. G. Fox, eds.), Vol. 3, pp. 17–35. Academic Press, New York.

Overbeek, P. A., Aguilar-Cordova, E., Hanten, G., Schaffner, D. L., Patel, P., Lebovitz, R. M., and Lieberman, M. W. (1991). Coinjection strategy for visual identification of transgenic mice. *Transgenic Res.* **1**, 31–37.

Pinkert, C. A. (1990). A rapid procedure to evaluate foreign DNA transfer in mammals. *BioTechniques* **9**, 38–39.

Quaife, C. J., Pinkert, C. A., Ornitz, D. M., Palmiter, R. D., and Brinster, R. L. (1987). Pancreatic neoplasia induced by *ras* expression in acinar cells of transgenic mice. *Cell (Cambridge, Mass.)* **48**, 1023–1034.

Rafferty, K. A., Jr. (1970). "Methods in Experimental Embryology of the Mouse." Johns Hopkins Press, Baltimore, Maryland.

Reik, W., Howlett, S. K., and Surani, M. A. (1990). Imprinting by DNA methylation: From transgenes to endogenous gene sequences. *Development (Cambridge, UK)* (Suppl.), 99–106.

Ryan, T. M., Behringer, R. R., Martin, N. C., Townes, T. M., Palmiter, R. D., and Brinster, R. L. (1989). A single erythroid-specific DNase I superhypersensitive site activates high levels of human β-globin gene expression in transgenic mice. *Genes Dev.* **3**, 314–323.

Sandgren, E. P., Palmiter, R. D., Heckel, J. L., Daugherty, C. C., Brinster, R. L., and Degen, J. L. (1991). Complete hepatic regeneration after somatic deletion of an albumin–plasminogen activator transgene. *Cell (Cambridge, Mass.)* **66**, 245–256.

Schnieke, A., Harbers, K., and Jaenisch, R. (1983). Embryonic lethal mutation in mice induced by retroviral insertion into the αl(I) collagen gene. *Nature (London)* **304**, 315–320.

Silver, L. M. (1987). Mouse news: Princeton. *Mouse News Lett.* **78**, 71–72.

Silvers, W. K. (1979). "The Coat Colors of Mice: A Model for Mammalian Gene Action and Interaction." Springer-Verlag, New York.

Singh, G., Supp, D. M., Schreiner, C., McNeish, J., Merker, H.-J., Copeland, N. G., Jenkins, N. A., Potter, S. S., and Scott, W. (1991). *legless* insertional mutation: Morphological, molecular, and genetic characterization. *Genes Dev.* **5**, 2245–2255.

Soriano, P., and Jaenisch, R. (1986). Retroviruses as probes for mammalian development: Allocation of cells to the somatic and germ cell lineages. *Cell (Cambridge, Mass.)* **46**, 19–29.

Stevens, M. E., Meneses, J. J., and Pedersen, R. A. (1989). Expression of a mouse metallothionein–*Escherichia coli* β-galactosidase fusion gene (MT–βgal) in early mouse embryos. *Exp. Cell Res.* **183**, 319–325.

Swift, G. H., Hammer, R. E., MacDonald, R. J., and Brinster, R. L. (1984). Tissue-specific expression of the rat pancreatic elastase I gene in transgenic mice. *Cell (Cambridge, Mass.)* **38**, 639–646.

Taketo, M., Schroeder, A. C., Mobraaten, L. E., Gunning, K. B., Hanten, G., Fox, R. R., Roderick, T. H., Stewart, C. L., Lilly, F., Hansen, C. T., and Overbeek, P. A. (1991). FVB/N: An inbred mouse strain preferable for transgenic analyses. *Proc. Natl. Acad. Sci. U.S.A.* **88**, 2065–2069.

Tanaka, S., Yamamoto, H., Takeuchi, S., and Takeuchi T. (1990). Melanization in albino mice transformed by introducing cloned mouse tyrosinase gene. *Development (Cambridge, UK)* **108**, 223–227.

Theiler, K. (1989). "The House Mouse: Atlas of Embryonic Development." Springer-Verlag, New York.

Whittingham, D. G. (1971). Culture of mouse ova. *J. Reprod. Fertil.* **14**, 7–21.

Wilkie, T. M., Braun, R. E., Ehrman, W. J., Palmiter, R. D., and Hammer, R. E. (1991). Germ-line intrachromosomal recombination restores fertility in transgenic MyK-103 male mice. *Genes Dev.* **5**, 38–48.

Xiang, X., Benson, K. F., and Chada, K. (1990). Mini-mouse: Disruption of the pygmy locus in a transgenic insertional mutant. *Science* **247**, 967–969.

Yokoyama, T., Silversides, D. W., Waymire, K. G., Kwon, B. S., Takeuchi, T., and Overbeek, P. A. (1990). Conserved cysteine to serine mutation in tyrosinase is responsible for the classical albino mutation in laboratory mice. *Nucleic Acids Res.* **18**, 7293–7298.

Yokoyama, T., Copeland, N. G., Jenkins, N. A., Montgomery, C. A., Elder, F. F. B., and Overbeek, P. A. (1993). Reversal of left–right asymmetry: A situs inversus mutation. *Science* **260**, 679–682.

Gene Transfer in Embryonic Stem Cells

Thomas Doetschman

Department of Molecular Genetics, Biochemistry, and Microbiology

University of Cincinnati College of Medicine

Cincinnati, Ohio 45267

I. INTRODUCTION TO EMBRYONIC STEM CELLS

A wealth of information about embryogenesis has been generated from the study of cultured stem cells of the embryo, a source of early postimplantation embryonic material. Stem cells of the embryo are generally classified into two major groups, namely, embryonal carcinoma (EC) and embryonic stem (ES) cells. The former are derived from the stem cells of teratocarcinomas, the latter from the stem cells (inner cell mass) of blastocysts.

A significant portion of the information obtained from studies on EC and ES cells has been derived from gene transfer experiments. Because the last few years have seen an exponential increase in gene transfer studies utilizing ES cells, and because several reviews have covered gene transfer in EC cells (Nicolas and Berg, 1983; Wagner and Stewart, 1986), this chapter concentrates on gene transfer in ES cells.

A. Establishment of Embryonic Stem Cells

Embryonic stem cell lines have been established from the inner cell mass cells of mouse blastocysts using a variety of conditions. They have been established from implantationally delayed blastocysts in standard embryonal carcinoma (EC) culture medium (Evans and Kaufman, 1981; Kaufman, *et al.*, 1983; Wobus *et al.*, 1984), from nondelayed blastocysts grown in culture medium conditioned by EC cells (Martin, 1981; Martin and Lock, 1983), from nondelayed blastocysts cultured in standard EC culture medium (Robertson *et al.*, 1983a; Axelrod, 1984; Doetschman *et al.*, 1985), and from morulae cultured in standard medium (Eistetter, 1989). To prevent differentiation during establishment of the cell lines, ES cells must be grown either on feeder layers of the embryonic fibroblast cell line STO (Evans and Kaufman, 1981; Martin, 1981), on primary murine embryonic fibroblasts (Wobus *et al.*,

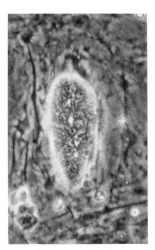

Figure 1. Undifferentiated ES cells. D3 cells growing on a primary embryonic fibroblast feeder layer. Note the refractile edges of the ES cell colony and the lack of apparent cell borders between the ES cells, both characteristics of undifferentiated ES cells. Magnification: 200×.

1984; Doetschman *et al.*, 1985), with Buffalo rat liver (BRL) cell-conditioned medium in combination with feeder cells (Handyside *et al.*, 1989), or in the absence of feeder cells but in the presence of leukemia inhibitory factor (LIF) (Pease *et al.*, 1990; Nichols *et al.*, 1990). Figure 1 shows a colony of undifferentiated ES cells growing on a primary embryonic fibroblast feeder layer.

Establishment of ES cells lines does not appear to depend on strain specificity (reviewed in Robertson *et al.*, 1983b). In addition to delayed and normal blastocysts, ES lines have been established from blastocysts of haploid and diploid parthenogenetically activated embryos (Kaufman *et al.*, 1983; Robertson *et al.*, 1983a), from diploid androgenetic embryos (Mann *et al.*, 1990), as well as from embryos with homozygous lethal mutations (Magnuson *et al.*, 1982; Martin *et al.*, 1987; Conlon *et al.*, 1991). Although ES cell lines have been reported for other species, including hamsters (Doetschman *et al.*, 1988a), pigs (Notarianni *et al.*, 1990; Piedrahita *et al.*, 1990; Strojek *et al.*, 1990), and sheep (Piedrahita *et al.*, 1990), no blastocyst injection or gene transfer studies have yet been reported for the ES cells from any of those species.

B. Differentiation Potential of Embryonic Stem Cells

When grown in the absence of feeder layer cells, BRL-conditioned medium, or LIF, ES cells will spontaneously differentiate into embryonic structures which, depending on the environment, will vary in the degree of similarity to those of the embryo. In the culture dish they will form embryoid bodies (EB) containing embryonic endoderm and ectoderm, yolk sac, nerve cells, epithelial cells, tubular structures, and cartilage (Martin and Lock, 1983; Wobus *et al.*, 1988), as well as heart muscle and blood islets (Doetschman *et al.*, 1985). EB with similarities to the egg cylinder (day 6) and yolk sac (day 9) stages of embryonic development are shown in Fig. 2. When injected subcutaneously into syngeneic or immunosuppressed hosts, ES cells will form teratocarcinomas containing cartilage, secretory epithelium, glandular structures, keratinizing epithelium, melanin pigmentation, and muscle (Martin, 1981; Martin and Lock, 1983; Kaufman *et al.*, 1983; Mann *et al.*, 1990). Figure 3 shows lung epithelium and early bone formation in a tumor section. When injected back into blastocysts, ES cells will colonize some extraembryonic and all embryonic tissues (Beddington and Robertson, 1989; Suemori *et al.*, 1990; Lallemand and Brûlet, 1990; Nagy *et al.*, 1990) and all of the tissues tested in the adult (Gossler *et al.*, 1986; Bradley and Robertson, 1986), including the germ line (Bradley *et al.*, 1984). Figure 4 illustrates a mouse family in which the agouti offspring demonstrate that the chimeric father is a germ-line chimera. ES cells cultured continuously for as long as 5 months have been shown to colonize the germ line (Suda *et al.*, 1987).

The capacity of ES cells to undergo differentiation in these three different en-

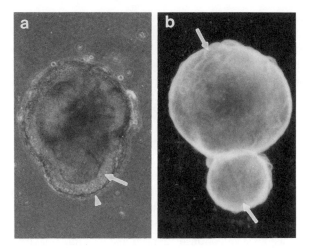

Figure 2. Embryoid bodies formed from differentiating ES cells. D3 cells were cultured on petri dishes in the absence of feeder cells, conditions under which ES cells spontaneously differentiate (a) Complex EB formed after 8 days of differentiation culture. Note the outer layer of endoderm (arrowhead) with an underlying layer of embryonic ectoderm (arrow). Phase-contrast optics. Magnification: $100\times$ (b) Cystic EB formed after 14 days of differentiation culture. Arrows indicate locations where red blood islets could be found. Dark-field optics. Magnification: $15\times$.

vironments makes them useful for investigating the effects of genetic modifications of either the "gain of function" or "loss of function" types. Here we review gene transfer studies involving transfection of undifferentiated ES cells with DNA and analysis of transgene expression in undifferentiated ES cells, EB, teratocarcinomas, or chimeric animals. A summary of the routes of investigation that can be pursued by gene transfer in ES cells is schematized in Fig. 5.

II. GENE TRANSFER AND EXPRESSION

Genetic material has been transferred into ES cells by several commonly used procedures, some of which have clear advantages over others, depending on the application. Infection by retroviral vectors specifically designed for expression in ES cells has provided the highest efficiency of gene transfer. For each infection, approximately 25% of the cells contain integrated vector (Robertson *et al.*, 1986). Multiple integrations have been achieved using multiple infections, and the resulting cell lines have been useful for rapid generation of insertional mutations (Robertson *et al.*, 1986; Conlon *et al.*, 1991). Retroviral vectors have also been used to infect ES cells for gain of function studies in chimeric animals (Stewart *et al.*, 1985; Williams *et al.*, 1988).

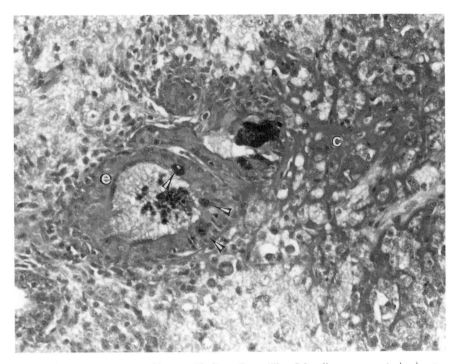

Figure 3. Tumor section of lung epithelium. Ten million D3 cells were injected subcutaneously into an immunodeficient mouse. After 6 weeks the tumor was removed, sectioned, and periodic acid–Schiff (PAS) stained. Note bronchiolar-like epithelium containing ciliated cells (e) and mucous cells (arrowheads). Also note well-developed cartilage or early bone (no mineralization) formation (c). Magnification: $250\times$.

The high efficiency, single-site integrations afforded by retroviral vectors would be ideal for gene targeting strategies. Unfortunately, these vectors have a propensity to integrate at random sites (for discussion, see review by Camerini-Otero and Kucherlapati, 1990) and therefore require further refinement. Although DNA microinjection has been used for gene targeting (Zimmer and Gruss, 1989), it is not clear whether the method is any more efficient than electroporation, and there may be technical difficulties in keeping the cells undifferentiated during the procedure. Calcium phosphate/DNA coprecipitation (Gossler *et al.*, 1986; Lovell-Badge *et al.*, 1985; Shinar *et al.*, 1989; Szyf *et al.*, 1990; Lindenbaum and Grosveld, 1990) and protoplast fusion (Takahashi *et al.*, 1988) have also been used to transfect ES cells with efficiencies of 10^{-5} to 10^{-6} and 1 to 5×10^{-5}, respectively. These methods usually produce multiple copies of the transgene at the insertion site, which is not a disadvantage unless specific genetic modifications are required. In experiments where precise genetic modifications are desired, as is the case for gene targeting by means of homologous recombination, electroporation is the method of choice be-

Figure 4. Family of a germ line chimera. D3 cells (male cells) of a strain 129 mouse (agouti, black) were injected into blastocysts of strain C57BL/6 mice (nonagouti, black). One male chimera (agouti and black) was bred to a CF1 (albino, nonagouti) female. In this particular litter all progeny were agouti, indicating that their paternal genome was from the ES cells and not from the injected blastocyst.

cause the DNA transfers are usually, though not always, single-copy integration events (Boggs *et al.*, 1986; Reid *et al.*, 1991).

A. *Transgenes in Undifferentiated Embryonic Stem Cells*

Experiments in which genes have been expressed in undifferentiated ES cells can be categorized by whether they are designed to identify regulatory sequences involved in ES cell host-range specificity. Because viruses have very strict host-range specificities with respect to EC and ES cells, they have often been used as the starting point for understanding control of differentiation during early embryogenesis. In the process, the viral regulatory sequences have been useful for expressing exogenous genes in ES cells.

Expression constructs designed for EC cells which use the herpes simplex virus (HSV) thymidine kinase promoter alone (Wagner *et al.*, 1985), or in combination with the SV40 regulatory region (Rubenstein *et al.*, 1984), work equally well in ES cells (Gossler *et al.*, 1986). When coupled with the thymidine kinase promoter (Thomas and Capecchi, 1987), specific mutations in the polyoma virus enhancer known to allow expression in EC cells (Linney and Donerly, 1983) are also effective in ES cells. Other regulatory sequences used to obtain high levels of expression in undifferentiated ES cells are the human β-actin promoter (Joyner *et al.*, 1989), the polymerase II (Pol2) promoter (Soriano *et al.*, 1991b), and the phos-

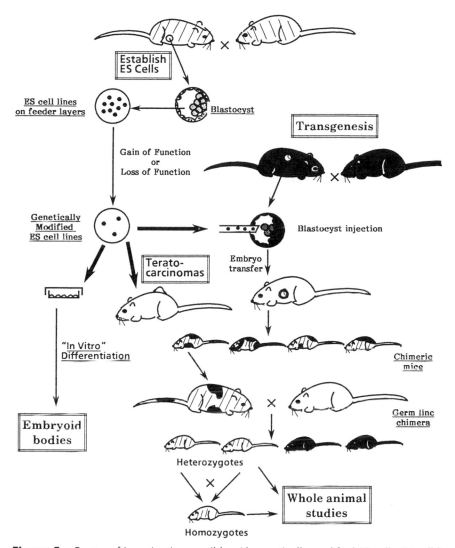

Figure 5. Routes of investigation possible with genetically modified ES cells. ES cell lines are established from the inner cell mass of blastocyst stage embryos. ES cells are genetically modified by addition of a gene or by gene targeting. Genetically modified ES cells can then be investigated either in EB *in vitro,* in tumors, or in chimeric or germ line chimeric mice.

phoglycerate kinase I (PGK) promoter (Soriano *et al.,* 1991b; Mombaerts *et al.,* 1991; Tybulewicz *et al.,* 1991). Expression has also been enhanced by replacing nonmammalian poly(A) addition signal sequences with those from the human hypoxanthine-guanine phosphoribosyl transferase (*Hprt*) gene (Thomas and Capecchi, 1987), the bovine growth hormone gene (Soriano *et al.,* 1991b; Hasty *et al.,*

1991), the Pol2 gene (Soriano *et al.*, 1991b), and the PGK gene (Soriano *et al.*, 1991b; Mombaerts *et al.*, 1991; Tybulewicz *et al.*, 1991; Mortensen *et al.*, 1991). In addition, transcriptional enhancement has been obtained by placing rat β-actin regulatory sequences downstream of the poly(A) addition signal of the marker gene (McMahon and Bradley, 1990). The topic of efficiency of transgene expression is revisited in more detail in a later section on detection of gene targeting events (Section III).

With respect to retroviral expression vectors, MPSVneo, a *neo* gene driven by myeloproliferative sarcoma virus long terminal repeat (LTR) sequences (Stocking *et al.*, 1985), was used to achieve high titers for infection of ES cells (Evans *et al.*, 1985; Robertson *et al.*, 1986). Further investigations into the myeloproliferative sarcoma virus have led to the identification of viral sequences that restrict expression to ES cells (Grez *et al.*, 1990). High levels of expression of a selectable marker gene have also been obtained using the Rous sarcoma virus (RSV) promoter combined with the prokaryotic Tn5 promoter (Le Mouellic *et al.*, 1990) and with four repeats of the GTIIC + GTI sequence motifs of the SV40 enhancer (Lufkin *et al.*, 1991). One study used a comparison between ES cells and 3T3 cells to demonstrate that a viral enhancer affects an internal promoter independently of expression from the viral promoter (Soriano *et al.*, 1991a).

Based on the assumption that genes expressed in ES cells may have important regulatory and functional significance for early developmental processes, gene and enhancer traps (Gossler *et al.*, 1989) and promoter traps (Lallemand and Brûlet, 1990; Macleod *et al.*, 1991) are being used to identify developmentally interesting genes. Some of these have been combined with a *lacZ* tag (Gossler *et al.*, 1989; Lallemand and Brûlet, 1990) for following the expression pattern associated with the trapped sequences. In another type of experiment, ES cells are being used to investigate sequences responsible for maintaining the hypomethylated state of a *thy-1* transgene (Szyf *et al.*, 1990). ES cells were well suited for this study because of their relatively high background level of methylation.

B. Transgenes in Embryoid Bodies

EB provide an *in vitro* model system for early stages of postimplantation embryogenesis. In several studies differentiation-permissive culture conditions have been used to test the tissue specificity of promoters on transgene expression and function, and to analyze the effects of an insertional mutation on early stages of embryogenesis. To find a ubiquitous promoter that has potential for marking ES cells in chimeric mice, an ES cell line transfected with the *lacZ* gene driven by a Rous sarcoma virus LTR/β-actin promoter combination was first screened for widespread expression in EB and tumors before injection into blastocysts to make chimeric animals. Expression in many tissues of the EB and tumors correlated with wide-ranging expression in the chimeras (Suemori *et al.*, 1990).

In one gene targeting experiment, the effects of *pim-1* oncogene deficiency generated by the consecutive knockout of both chromosomal genes in ES cells has been followed in differentiation culture. Surprisingly, although the oncogene is expressed at high levels in ES cells, its ablation had no apparent effect on EB development (te Riele *et al.*, 1990). In another gene targeting experiment, one copy of the *int-2* gene was mutated by the insertion of a *lacZ* marker gene. Expression of β-galactosidase in EB was as expected from *int-2* gene expression studies (Mansour *et al.*, 1990). A recessive lethal mutation generated by multiple infection of ES cells with a retroviral vector was analyzed by producing an ES cell line from homozygous recessive embryos. The differentiation potential of these cells was studied *in vitro* as well as in tumors and in chimeric animals. The gastrulation stage lethal defect had no apparent restriction in developmental potential in culture or in tumors, and the lethal mutation was circumvented in chimeric animals made from the cells (Conlon *et al.*, 1991). Finally, EB are a suitable embryo model system for using transgenes to study the regulation of globin isoform switching during mouse embryogenesis (Lindenbaum and Grosveld, 1990). This study indicated that ε-globin gene expression is not dependent on the presence of other globin genes or the dominant control region of the β-globin locus. Together these studies suggest that the *in vitro* differentiation of ES cells into EB can be a useful approach for investigating the embryonic effects of transgene expression and insertion, both random and targeted.

C. Transgenes in Teratocarcinomas

ES cells have the potential to form teratocarcinomas when injected subcutaneously into syngeneic or nude mice. Even though there is less organization to the embryonic structures, the tumor environment allows ES cells to differentiate into a broader array of tissues for investigation. This advantage was utilized in the study described above in which tumors as well as EB were used to prescreen for promoters that might allow ubiquitous expression of the ES cell phenotype in chimeric animals (Suemori *et al.*, 1990). In one experiment the tissue specificity of *ras, myc,* and large T-antigen gene expression was determined in tumor tissues as well as in EB and chimeric mice (Suda *et al.*, 1988). In all cases, in contrast to *ras* and *myc* expression, T-antigen expression was suppressed as ES cells that had been prescreened for high levels of expression were allowed to form tumors. In a study described previously (Conlon *et al.*, 1991), an ES cell line homozygous for a recessive embryonic lethal mutation was also studied in tumors. As in the EB, there was no indication of developmental restriction. In a third study, muscle-specific transgene expression in ES cell-derived tumors was abrogated by the cointroduction of specific plasmic sequences that led to methylation of the transgene (Shinar *et al.*, 1989). In the absence of the plasmid sequences, inhibition of expression of the muscle-specific gene in nonmuscle tissues occurred without methylation of the transgene.

D. Transgenes in Chimeric Mice

Embryonic stem cells have the potential to colonize the germ line, allowing transmission of the transgene or its associated genetic modification to the progeny of the chimera. However, expression of the transgene in the chimeric animal itself has been the object of several investigations. The first transgene experiments in ES cell-derived chimeric animals demonstrated that genes introduced into ES cells by retroviral infection could be stably expressed in differentiated tissues of chimeric animals produced by blastocyst injection (Stewart *et al.*, 1985). At the same time, chimeric mice were used to demonstrate tissue-specific expression of the human type II collagen transgene (Lovell-Badge *et al.*, 1987). Similarly, deletion analysis on the promoter of the chicken δ-crystallin transgene in ES cells led to the identification of sequences responsible for lens-specific expression in chimeric mice (Takahashi *et al.*, 1988).

An important tool for cell lineage and fate mapping studies of genetically manipulated ES cells in chimeric animals would be an easily identifiable marker gene, or tag, driven by a ubiquitous promoter so that the genetically altered cells could be followed during embryonic and fetal development of the chimeric animal. In one attempt to find such a promoter, a RSV LTR/chicken β-actin promoter driving expression of the *lacZ* gene has been used (Suemori *et al.*, 1990). Although it is expressed in most cell lineages up to gastrulation, but has a more limited expression at later stages of development, this construct was useful for describing the process of vascularization in the early mouse embryo (Kadokawa *et al.*, 1990). Rather than using a ubiquitous promoter to follow cell fate in a chimeric environment, several studies have utilized a *lacZ* reporter gene to follow expression of a transgene. The expression of genes associated with sequences trapped by gene and enhancer traps (Gossler *et al.*, 1989) or by promoter traps (Lallemand and Brûlet, 1990) has been followed in chimeric animals to help determine whether the genes will be of developmental interest. In gene targeting studies, ES cells expressing a *lacZ* reporter gene placed by homologous recombination under control of the promoter of the target gene have been used to follow the expression of the *hox-3.1* (Le Mouellic *et al.*, 1990) and *int-2* (Mansour *et al.*, 1990) target genes in chimeric animals. These techniques may be useful for correlating, in a chimeric context, the presence of manipulated cells with the developmental defect.

Based on the assumption that overexpression of a transgene in an animal produced by DNA microinjection may have too severe a phenotype to permit analysis of its effects, the poloyoma middle T-antigen gene (Williams *et al.*, 1988) was introduced into ES cells by retroviral infection. The chimeras produced from these cells developed hemangiomas during embryonic development. Similarly, to circumvent the inability to analyze an embryonic lethality that occurred as the result of an insertional mutation, ES cells derived from blastocysts homozygous for the mutation were used to make chimeric embryos which survived the lethal event (Conlon *et al.,* 1991). These reports demonstrate the value of the chimeric environment for rescuing an embryo from a lethal mutation so that the effects of the genetic defect can still be investigated in the context of the whole animal.

III. GENE TARGETING IN EMBRYONIC STEM CELLS

The vast majority of work being done on ES cells involves preplanned genetic modifications introduced by homologous recombination between exogenous (targeting) DNA and an endogenous (target) gene (reviewed by Baribault and Kemler, 1989; Capecchi, 1989a, b; Frohman and Martin, 1989; Evans, 1989; Camerini-Otero and Kucherlapati, 1990; Mansour, 1990; Robertson, 1991). In a few gene targeting experiments, effects of the genetic modification have been analyzed in EB *in vitro* (te Riele *et al.,* 1990; Mansour *et al.,* 1990; Mortensen *et al.,* 1991). In others, effects of the targeted locus either *in vitro* (Mansour *et al.,* 1990) or in chimeric animals (Mansour *et al.,* 1990; Le Mouellic *et al.,* 1990) has been followed. In most gene targeting studies, however, analysis has been carried out in the offspring of germ line chimeras (Joyner *et al.,* 1989; Koller *et al.,* 1989; Schwarzberg *et al.,* 1989; Thompson *et al.,* 1989; Koller *et al.,* 1990; Zijlstra *et al.,* 1990; DeChiara *et al.,* 1990; Thomas and Capecchi, 1990; McMahon and Bradley, 1990; Stanton *et al.,* 1990; Yagi *et al.,* 1990; Joyner *et al.,* 1991; Pevny *et al.,* 1991; Soriano *et al.,* 1991b; Kitamura *et al.,* 1991; Chisaka and Capecchi, 1991; Fung-Leung *et al.,* 1991; Mucenski *et al.,* 1991; Tybulewicz *et al.,* 1991; Schorle *et al.,* 1991; Cosgrove *et al.,* 1991; Rahemtulla *et al.,* 1991; Grusby *et al.,* 1991; Lufkin *et al.,* 1991). Gene targeting experiments in ES cells and data pertaining to targeting schemes and targeting efficiencies are listed in Table 1.

A. *Improving Targeting Efficiency*

Targeting efficiencies, determined as the number of targeted cells per treated cell, range from 1.7×10^{-5} to $\sim 1 \times 10^{-9}$. The reasons for the 10^4-fold difference are not clear. In an early study the *Hprt* gene was targeted using varying lengths of homologous sequence, leading to the observation that targeting frequency is directly dependent on length of homology (Thomas and Capecchi, 1987). However, when all of the published targeting studies are examined, no obvious correlation between length of homology and targeting frequency is found (Fig. 6). It is clear that factors besides length of homology are also involved in determining targeting efficiency.

Knowledge of the rate-limiting aspects of homologous recombination in mammalian cells would be useful for designing more efficient targeting constructs. Because targeting efficiency is not appreciably changed by the number of targeting

TABLE 1
Gene Targeting: Strategies and Efficiencies

(a) Gene	(b) Strategy	(c) Homology (kb)	(d) Strain compatibility	(e) Cell line (+ = germ line)	(f) Targeted cells per transfectants	(g) Enrichment[a] factor	(h) Targeted cells per enriched cells	(i) Overall targeting efficiency
$hprt^{KT/MC}$ [b]	$E_{Py}P_{tk}neo{>}A^-/6TG$ [c]	4.0	−	ND[d]	4.9×10^{-5}	NA[e]	NA	$\sim2.1 \times 10^{-8}$
		5.4			1.4×10^{-4}		NA	$\sim4.6 \times 10^{-8}$
		9.1			1.1×10^{-3}		NA	$\sim2.1 \times 10^{-7}$
	$E_{Py}P_{tk}neo{>}A^-/6TG$ [f]	3.7		ND	5.3×10^{-5}	NA	NA	$\sim1.9 \times 10^{-8}$
		9.3			8.9×10^{-4}		NA	$\sim2.5 \times 10^{-7}$
$hprt^{TD/OS}$	HAT	2–4	+	E14TG2a + BK/OS	1.8×10^{-1}	NA	NA	1.8×10^{-6}
$hprt^{TD/OS}$	$P^-neo{>}A_{SV}/6TG/PCR$	1.3	+	D3	ND	ND	NA	8.0×10^{-7}
$hprt^{SM/MC}$	PNS: $E_{Py}P_{tk}neo{>}A^-/E_{Py}tk{>}$ [g]	9.1	ND	ND	2.5×10^{-4}	3000	7.5×10^{-1}	$\sim1.6 \times 10^{-6}$
$int\text{-}2$	PNS: $E_{Py}P_{tk}neo{>}A^-/E_{Py}tk{>}$	10	ND	ND	2.5×10^{-5}	2000	5×10^{-2}	$\sim1.4 \times 10^{-7}$
$hprt^{ST/DM}$	HAT	2–4	−	E14TG2a +	1.2×10^{-2}	NA	NA	$\sim1.5 \times 10^{-7}$
		1–3		E14TG2a +	2.4×10^{-2}	NA	NA	$\sim1.5 \times 10^{-7}$
$hox1.1^{AZ/PG}$	Microinjection	1.6	ND	D3	3.3×10^{-2}	NA	NA	4.2×10^{-3}
$En\text{-}2^{AJ/JR\ 89}$	$P_{h\beta A}neo{>}A^-/PCR$ (1.0)[h]	3.7	−	D3 + AJ/JR 91	3.8×10^{-3}	2	7.6×10^{-3}	6.7×10^{-8}
$c\text{-}fos^{RU/BS}$	PNS: $E_{Py}P_{tk}neo{>}A_{tk}/E_{Py}tk{>}$	4	ND	CC1.2	NA	1250	NA	$\sim2.5 \times 10^{-8}$ i
Adipsin	PNS: $E_{Py}P_{tk}neo{>}A_{tk}/E_{Py}tk{>}$	3	ND	CC1.2	4×10^{-5}	2270	9.1×10^{-2}	2×10^{-8}
aP2	PNS: $E_{Py}P_{tk}neo{>}A_{tk}/E_{Py}tk{>}$	4	ND	CC1.2	3×10^{-4}	282	8.5×10^{-2}	1.5×10^{-7}
$\beta_2 - M^{BK/OS\ 89}$	$E_{Py}P_{tk}neo{>}A_{tk}/PCR$	6.2	−	E14TG2a + BK/OS 90	8.3×10^{-3}	NA	NA	4.5×10^{-8}
$c\text{-}abl^{PS/SG\ 90}$	$P^-neo{>}A_{SV}$ (1.4)[h]	~7	−	CCE + PS/ER 89;PS/SG 91	$\sim9 \times 10^{-4}$	75	$\sim7 \times 10^{-2}$	2×10^{-7}
$\beta_2\text{-}M^{MZ/RJ\ 89}$	$tk^-/E_{Py}P_{tk}neo{>}A^-/PCR$ [j]	10	−	D3 + MZ/RJ 89;90	4.0×10^{-2}	4	1.6×10^{-1}	4.0×10^{-7}
$n\text{-}myc^{JC/FA}$ [k]	$P^-neo{>}A_{SV}$ (2.0)[h]	3.3	−	CCE	3.6×10^{-3}	~120	$\sim4 \times 10^{-1}$	3.2×10^{-8}
	$P^-neo{>}A^-$ (0.28)[h]	4.8		CCE	4.8×10^{-3}	~52	$\sim2.5 \times 10^{-1}$	4.3×10^{-8}
$IGF\text{-}II^{TD/DR}$	PNS: $E_{Py}P_{tk}neo{>}A_{tk}/E_{Py}tk{>}$ (0.25)[h]	10	ND	CCE +	$\sim4.8 \times 10^{-4}$	47	2.3×10^{-2}	3.7×10^{-9}
$hox\text{-}3.1^{HL/PB}$	$P^-lacZ{>}A_{SV}/E_{Py}P_{RSV+Tn5}neo{>}A_{SV}/$ PCR	8.3	−	CCE	$\sim1.1 \times 10^{-3}$	NA	NA	1.3×10^{-7}
	$P^-lacZ{>}A_{SV}/E_{Py}P_{RSV+Tn5}neo{>}A^-$ A+T/ PCR/ (7.2)[m] (0.13)[h]	8.3		CCE	$\sim4.0 \times 10^{-3}$	2.4	$\sim1 \times 10^{-2}$	2.7×10^{-7}
$PolII^{CS/AB}$	ama	5.8	−	D3	3.4×10^{-2}	NA	NA	7.1×10^{-7}
$int\text{-}1^{KT/MC}$ [n]	PNS: $E_{Py}tk{>}/E_{Py}P_{tk}neo{>}A_{tk}/E_{Py}tk{>}$	13.5	−	CC1 +	2×10^{-7} i	12,500	2.5×10^{-3} i	$\sim1 \times 10^{-9}$ i

Gene	Construct							
wnt-1[AM/AB] [n]	PNS: $E_{Py}tk^</E_{Py}P_{tk}neo> A_{tk}E_{r\beta A}/E_{Py}tk^<$	8.4	−	AB1+	1.4×10^{-4}	26	3.6×10^{-3}	2.7×10^{-8}
int-2[SM/MC]	PNS: $E_{Py}tk>/P^-lacZ> A_{SV}/$ $E_{Py}P_{tk}neo> A_{tk}$ (4.4)[m]	10	−	ND	8.9×10^{-5}	1,135	1×10^{-1}	9.0×10^{-7}
hprt	6TG (1.0)[m]	9.1	−	ND	NA	NA	NA	$\sim 5 \times 10^{-6}$
	(3.4)	9.1		ND	NA	NA	NA	$\sim 4.5 \times 10^{-6}$
	(4.3)	ND		ND	NA	NA	NA	$\sim 2.1 \times 10^{-6}$
	(8.0)	9.1		ND	NA	NA	NA	$\sim 4 \times 10^{-6}$
	(12)	ND		ND	NA	NA	NA	$\sim 3.8 \times 10^{-6}$
n-myc[BS/LP]	$P^-neo> A_{tk}$/PCR	5.5	−	D3+	ND	20–100	NA	1.4×10^{-7}
c-fyn[TY/SA]	PNS: $P^-neo> A_{tk}/E_{Py}P_{tk}neo> A^-$	6.3	ND	E14+	2×10^{-2}	~ 25	$\sim 5 \times 10^{-1}$	3.2×10^{-8}
pim-1[HU/AB]	Double targeting[o] $P^-neo> A_{tk}E_{Py}$ (0.3)[h]	6.1	−	E14	$\sim 4.5 \times 10^{-2}$	~ 19	$\sim 9 \times 10^{-1}$	6.8×10^{-6}
	$P^-hyg> A^-$ (0.3)[h]	6.1		E14	$\sim 9 \times 10^{-3}$	~ 69	$\sim 6 \times 10^{-1}$	3.2×10^{-7}
gata-1[LP/FC]	PNS: $E_{Py}<k/E_{Py}P_{tk}neo> A_{tk}$	4.8	ND	CCE+	3.3×10^{-3}	23	7.6×10^{-2}	3×10^{-7}
c-src[PS/A3]	$P_{Pol2}neo> A_{BGH}$/PCR	10.5	ND	CCE+	6.8×10^{-3}	NA	NA	2.0×10^{-7}
hprt[VV/OS]	In-out[p] In step	5.0	+	E14TG2a	ND	NA	NA	2.8×10^{-6}
	Out step	NA			NA	NA	NA	8.0×10^{-7} [q]
Hox2.6[PU/AB]	Hit and run[p] Hit step: PNS: $E_{Py}tk>/$ $P_{PGK}neo> A_{BGH}$	3.1	+	AB1	3.2×10^{-2}	ND	NA	ND
	Run step	NA			NA	NA	NA	3.8×10^{-3} [r]
hprt	Hit step: PNS: $E_{Py}P_{tk}neo> A_{tk}$	6.8	−	AB1	7.2×10^{-3}	ND	NA	ND
	Run step	NA			NA	NA	NA	4.3×10^{-6} [r]
TCR$_\beta$[PM/ST]	PNS: $tk>/P_{PGK}neo< A_{PGK}/E_{Py}tk>$ (15)[h] or $tk>/P_{PGK}neo< A_{PGK}$ (15)[h]	8.5	−	E14	4×10^{-3}	2	8×10^{-3}	ND
μ chain[DK/KR]	PNS: $E_{Py}neo> A_{tk}/E_{Py}tk>$	9.0	ND	D3+	3.2×10^{-3}	8	2.6×10^{-2}	1.8×10^{-7}
hox-1.5[OC/MC]	PNS: $tk>/E_{Py}P_{tk}neo> A_{tk}/tk>$ [s]	11.5	ND	CC1.2+	ND	ND	NA	ND
hprt[LR/OS]	Cotransfection: $E_{Py}P_{tk}neo> A_{tk}$	7.4	+	E14TG2a	ND	100	NA	4.4×10^{-6}
CD8[WF/TM]	$E_{Py}P_{tk}neo< A_{tk}$/PCR	2.2	−	D3+	5×10^{-3}	NA	NA	1×10^{-7}
c-myb[MM/SP]	$E_{Py}P_{tk}neo> A_{tk}$/PCR	9.9	ND	D3+	2.7×10^{-2}	NA	NA	1.3×10^{-7}
MCK[JD/BW] [k]	Double targeting[j] $E_{Py}P_{tk}neo> A_{tk}$ (0.5)[h]	8.5	−	ES5	NA	NA	NA	NA
	$P_{tk}hyg> A_{tk}E_{Py}$ [f]	9.0		ES5-2697	3.2×10^{-2}	NA	NA	2×10^{-7}
c-abl[VT/RM]	PNS: $P_{PGK}tk< A_{PGK}/P_{PGK}neo< A_{PGK}$ (0.3)[h]	6.6	ND	CCE+	3×10^{-2}	5.5	1.5×10^{-1}	ND

(continued)

TABLE 1
(continued)

(a) Gene	(b) Strategy	(c) Homology (kb)	(d) Strain compatibility	(e) Cell line (+ = germ line)	(f) Targeted cells per transfectants	(g) Enrichment factor	(h) Targeted cells per enriched cells	(i) Overall targeting efficiency
Double targeting								
α_{12}[RM/JS]	PNS: $P_{PGK}tk^>/P_{PGK}neo^> A_{PGK}$	6.8	−	CCE	2.8×10^{-2}	4.4	2.8×10^{-2}	1.7×10^{-5}
	PNS: $P_{PGK}hyg^>A_{PGK}/P_{PGK}tk^>$	6.8			1.1×10^{-2}	3.4	1.1×10^{-2}	2.3×10^{-6}
IL-2[HS/IH]	$E_{Py}P_{tk}neo^<A_{tk}$/PCR	6.0	ND	E14+	5×10^{-3}	NA	5×10^{-3}	ND
MHCII A_β^b [DC/DM]	$E_{Py}tk^</E_{Py}P_{tk}neo^<A_{tk}/E_{Py}tk^<$	~6	−	D3+	ND	ND	2.6×10^{-2}	ND
CDC4 [AR/TM]	$E_{Py}P_{tk}neo^>A_{tk}$/PCR	2.8	ND	D3+	3.3×10^{-3}	NA	NA	5×10^{-8}
MHC II A_β^b [MG/LG]	PNS: $E_{Py}tk^>/E_{Py}P_{tk}neo^>A_{tk}$	5.4	ND	D3+	9.2×10^{-3}	5	4.6×10^{-2}	3.3×10^{-7}
Hox1.6[TL/PC]	PNS: $E_{GTIIC+GTII}P_{RSV}neo^>A^-$/PCR	8.0	ND	D3+	ND	ND	4.9×10^{-2}	2.3×10^{-6}

[a] Applicable to any scheme that should decrease the number of random integrants selected or surviving.

[b] Superscript after the name of the gene indicates the initials of the first and last authors.

[c] neo, Selectable marker gene; E, enhancer; P, promoter; P⁻, promoterless marker gene; A, poly (A) addition signal; A⁻, marker gene lacking poly (A) addition signal; subscripts, source of E, P, or A; superscripts (>) or (<), orientation of marker gene with respect to target gene. The order of genes, enhancers, and promoters is same as in targeting construct. If no promoter or poly (A) addition signal is indicated, these elements are present but endogenous to the marker gene.

[d] ND, No data.

[e] NA, Not applicable.

[f] "Insertion" or "O" type targeting vector. All other vectors used were "replacement" or "omega" type targeting vectors.

[g] PNS, Positive/negative selection scheme.

[h] The amount of endogenous sequence removed and replaced (if greater than 0.1 kb) by the nonhomologous sequences of the targeting vector.

[i] Because the number of targeted cells achieved was only one, this number is not statistically significant.

[j] The negative tk selection system was not used to isolate targeted cells even though the tk gene was in the construct.

[k] Data were used only from experiments in which targeted colonies were observed.

[l] Instead of poly (A) addition signal on the neo gene, an A + T-rich mRNA degradation sequence has been placed downstream of the neo gene inside a Hox-3.1 intron to reduce neo transcript levels after nonhomologous integration.

[m] Kilobases of heterologous sequence disrupting the homologous sequence. With the exception of Mansour et al. (1990), this parameter is listed only for experiments in which the inserted sequences were greater than 4 kb. Insertions produced from "insertion" or "O" type targeting vectors are not included. As the specific purpose of studies by Mansour et al. (1990) was to determine whether the size of disrupting sequence affects homologous recombination frequencies, insertions of less than 4 kb are also listed.

[n] wnt-1 and int-1 are synonymous.

[o] Consecutive targeting of each allele, first with a promoterless neo gene and then with a promoterless hygromycin gene as positive selectors.

[p] In–out and hit and run are synonymous strategies.

[q] Reversion frequency = revertants/starting cells or colonies.

[r] Reversion frequency = revertants/cell generation.

[s] The two tk genes were from a different source (HSV I and II) in order to reduce potential recombination between them.

[t] Consecutive targeting of the same allele, first with neo and then with the hygromycin gene as positive selectors.

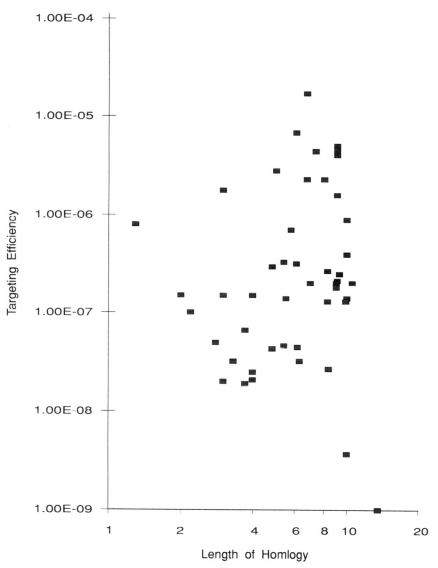

Figure 6. Targeting efficiency as a function of length of homology. The abscissa indicates kilobases of homology between targeting and target DNA. The ordinate indicates the overall targeting efficiency. The data are from Table 1, columns c and i.

molecules introduced into a cell (Thomas *et al.*, 1986) or the number of target loci per cell (Zheng and Wilson, 1990), it appears that the cellular machinery for homologous recombination may be rate limiting. This suggests that a sequence in the targeting vector represented in multiple genomic loci would be less likely to recombine homologously with a specific target locus. This suggests that choosing targeting fragments containing few mouse repetitive sequences may improve efficiencies.

In one study the effect of cotransfecting the targeting sequence with other DNA molecules was found to increase the number of cotransfectants that were targeted, but the degree to which cotransfection increased the ease of isolating targeted cells was limited (Reid *et al.*, 1991). In another study, an AT-rich region was hypothesized to improve the ratio of homologous to nonhomologous recombination by allowing for easier strand melting during the initiation of homologous recombination (Le Mouellic *et al.*, 1990).

The orientation of the selectable marker gene has been the "same" as that of the target gene in all but a few cases (Mombaerts *et al.*, 1991; Fung-Leung *et al.*, 1991; Mucenski *et al.*, 1991; Tybulewicz *et al.*, 1991; Schorle *et al.*, 1991; Cosgrove *et al.*, 1991). In none of these cases, however, was it reported that the "same" orientation did not work. In only one case was a particular orientation found to be necessary, and that was because the *neo* gene was not expressed in ES cells when oriented "opposite" to the β_2-microglobulin targeting sequences (Koller and Smithies, 1989).

A survey of gene targeting experiments makes it clear that some sequence polymorphism built into the targeting DNA does not appreciably affect the targeting frequency. For example, a 4-bp insertion (Doetschman *et al.*, 1987; Stanton *et al.*, 1990), 20-bp insertion (Zimmer and Gruss, 1989), 51-bp insertion (Le Mouellic *et al.*, 1990), 27- and 14-bp insertions (Hasty *et al.*, 1991), 16-bp insertion (Schwartberg *et al.*, 1990), and point mutations (Steeg *et al.*, 1990) do not affect targeting efficiency. A small 4-bp deletion also did not seem to affect targeting frequency (Koller and Smithies, 1989). In addition, replacements by the positive selector gene of up to 2 kb (Joyner *et al.*, 1989; DeChiara *et al.*, 1990; te Riele *et al.*, 1990; Charron *et al.*, 1991b; van Deursen *et al.*, 1991; Tybulewicz *et al.*, 1991), and even 15 kb (Mombaerts *et al.*, 1991; Lufkin *et al.*, 1991), as well as insertions of the positive selector gene of up to 12 kb (Mansour *et al.*, 1990) have caused no significant change in targeting frequency. These results speak to the question of the strain compatibility of DNA between the targeting sequence and target cell. Clearly, small strain sequence polymorphisms alone will not necessarily decrease targeting efficiencies. Table 1 (column d) shows that there is no significant difference in targeting efficiencies between experiments in which there was ($+$) or was not ($-$) strain compatibility. In one targeting experiment, a known restriction site polymorphism in the β_2-microglobulin gene did not appear to impede targeting (Zijlstra *et al.*, 1989).

However, to the extent that specific DNA sequences or conformations of which we are yet unaware can interfere with homologous recombination, strain differences between target and targeting DNA could play an important role in either increasing or decreasing targeting efficiencies. In the extreme case of the renin gene where different strains have either one or two renin loci, it would be impossible to target a gene that is not present in the ES cells. Consequently, to avoid any unnecessary difficulties in targeting or in diagnosing the targeted cells, it would be safest to use strain-compatible DNA for gene targeting.

B. Improving Detection of Targeted Cells

Another approach to facilitate the isolation of targeted cells is to decrease detection of random integrants rather than increase targeting efficiency. The most commonly used scheme is termed positive–negative selection (PNS). In this scheme it is assumed that nonhomologous integration events will retain the negative selector gene and confer sensitivity to gancyclovir, whereas homologous events will remove the negative selector gene and confer resistance. In most experiments, the *neo* gene is the positive selector and the HSV *tk* gene is the negative selector (Mansour *et al.*, 1988, 1990; Johnson *et al.*, 1989; DeChiara *et al.*, 1990; Thomas and Capecchi, 1990; McMahon and Bradley, 1990; Pevny *et al.*, 1991; Mombaerts *et al.*, 1991; Kitamura *et al.*, 1991; Chisaka and Capecchi, 1991; Tybulewicz *et al.*, 1991; Mortensen *et al.*, 1991). In one experiment, in order to avoid the potentially mutagenic effects of gancyclovir, a diphtheria toxin-A gene was used as a negative selector instead of the HSV *tk* gene. The toxin gene was missing a poly(A) addition signal to reduce transient expression of the toxin which might lead to killing of the targeted cells. (Yagi *et al.*, 1990).

The enrichment factors for the *neo/tk* experiments range widely from 2 to 12,500 (Table 1). The reason for this variation is not clear. One experiment has shown that about half of the neomycin-resistant cells are not gancyclovir sensitive (Zijlstra *et al.*, 1989), and another experiment has shown that *tk* expression causes nonspecific cytotoxicity in about half of the transfectants (Tybulewicz *et al.*, 1991). These factors, however, account for only a small portion of the variation. The diphtheria toxin-A gene yielded an enrichment of 25-fold.

In other enrichment schemes, a promoterless positive selector gene has been used (Doetschman *et al.*, 1988b; Schwartzberg *et al.*, 1990; Charron *et al.*, 1990; Stanton *et al.*, 1990; Yagi *et al.*, 1990; te Riele *et al.*, 1990). This scheme is based on the idea that the selector gene will be expressed only if inserted in the correct orientation relative to the promoter of an endogenous gene, thereby reducing substantially the number of integrants capable of expressing the selector gene. The use of this targeting scheme has yielded enrichments ranging from 19 to 120. Similarly, another scheme relies on the integration site providing the poly(A) addition signal for the positive selector gene. Here, an integration in intergenic regions or in the wrong orientation within a gene would prevent its expression. Such experiments have yielded an enrichment of from 2- to 5-fold (Thomas and Capecchi, 1987; Mansour *et al.*, 1988; Joyner *et al.*, 1989; Zijlstra *et al.*, 1990; Charron *et al.*, 1990; Le Mouellic *et al.*, 1990; te Riele *et al.*, 1990; Lufkin *et al.*, 1991). In a few of these studies both a promoterless and a poly(A) addition signal-less positive selector gene were used in the same targeting construct (Charron *et al.*, 1990; te Riele *et al.*, 1990).

Another approach for facilitating the isolation of targeted cells is to increase the detectability of homologous events. The most commonly used method for improving the detection of targeted cells among nontargeted transfectants is polymerase chain reaction (PCR) amplification (Joyner *et al.*, 1989; Koller and Smithies, 1989; Zijlstra *et al.*, 1989; Le Mouellic *et al.*, 1990; Stanton *et al.*, 1990; Soriano *et al.*, 1991b; Fung-Leung *et al.*, 1991; Mucenski *et al.*, 1991; Schorle *et al.*, 1991; Rahemtulla *et al.*, 1991; Lufkin *et al.*, 1991) in which one PCR primer is specific for the targeting sequence and another for the target sequence. In this way cells pooled from several transfected colonies, or from individual colonies, will yield an amplified fragment specific for one of the recombination junctions only if one of the colonies represented in the pool is targeted. The fidelity of the targeting event can also be determined from direct sequencing of the PCR-amplified product (Doetschman *et al.*, 1988b).

Based on the assumption that some promoters for positive selector genes do not work well in certain target loci, promoters other than the commonly used *tk* promoter have been used. These include the β-actin (Joyner *et al.*, 1989), RSV (Le Mouellic *et al.*, 1990; Lufkin *et al.*, 1991), Pol2 (Soriano *et al.*, 1991b), and PGK promoters (Soriano *et al.*, 1991b; Mombaerts *et al.*, 1991; Tybulewicz *et al.*, 1991). Similarly, based on evidence that poly(A) addition signals from mammalian genes such as the *Hprt* gene (Thomas and Capecchi, 1987) increase transfection efficiencies in ES cells, the bovine growth hormone (Soriano *et al.*, 1991b; Hasty *et al.*, 1991), Pol2 (Soriano *et al.*, 1991b), and PGK (Soriano *et al.*, 1991b; Mombaerts *et al.*, 1991; Tybulewicz *et al.*, 1991; Mortensen *et al.*, 1991) poly(A) addition signals have been used in targeting experiments to increase *neo* gene expression to selectable levels.

C. Targeting Subtle Mutations into Genes

Most of the gene targeting experiments published to date have been gene ablations in which large mutations have been introduced into a gene, effectively generating a complete loss of activity of the gene product. Two experiments, however, have demonstrated a targeting scheme for introducing subtle mutations into a gene locus. In these experiments an insertional or O-type recombination vector has been applied in conjunction with a positive/negative selection scheme using either a mini *Hprt* gene (Reid *et al.*, 1990) in an HPRT-deficient ES cell line (Valancius and Smithies, 1991) or the *neo/tk* combination in a wild-type cell line (Hasty *et al.*, 1991). The authors have termed this approach an "in–out" or "hit and run" targeting scheme, respectively. The approach is as follows: a targeted insertion leads to a duplication of some sequences so that in a small number of targeted cells (10^{-3} to 10^{-7}) there occurs an intrachromosomal recombination between the duplicated sequences, resulting in loss of the intervening sequences which contain the marker genes. Negative selection for loss of the marker genes isolates the cells undergoing the intrachromosomal recombination event. If a subtle mutation is present in the targeting sequences that lie outside of the duplication, this mutation can be incorporated into the targeted locus while at the same time all other mutations are removed. These and similar schemes designed to make fine mutations in ES cells will be very useful for studying functional domains and structure/function questions at the whole animal level.

D. Tagging the Targeted Locus

Two targeting experiments have used the *lacZ* gene to tag either the *hox-3.1* (Le Mouellic *et al.*, 1990) or the *int-2* (Mansour *et al.*, 1990) genes. In both studies the *lacZ* gene has been placed in the reading frame with the target gene so as to produce a fusion protein with β-galactosidase activity. Included in the targeting constructs was the selectable marker gene *neo* with its own promoter. This scheme was designed to determine the developmental and tissue-specific expression pattern of the target gene by histological methods in chimeric mice without having to produce germ-line chimeras.

IV. METHODS

The methods given here are those used in my laboratory and are not to be considered the only approaches that have been successfully used. Other approaches are often used and can be found in the individual papers cited.

A. Embryonic Stem Cell Culture

1. Fetal Calf Serum

Fetal calf serum must be screened for ability to support ES cell growth and to minimize differentiation of ES cells when grown on feeder cells (obtain 50-ml aliquots of serum from several companies, test them, and buy the best).

2. Culture Medium

The culture medium for ES cells is 15% (v/v) fetal calf serum (heat-inactivated at 55°C for 30 min) in high glucose-containing Dulbecco's modified Eagles medium

(DMEM) (with glutamine and bicarbonate, no HEPES). Culture medium also contains 0.1 mM β-mercaptoethanol, and cells are grown in the presence of 10% (v/v) CO_2.

3. Passaging Embryonic Stem Cells

Every 2 to 3 days passage ES cells according to standard procedures using the same trypsin/EDTA solution used for embryonic fibroblast preparation. It is best to rinse the cultures 3 times in phosphate-buffered saline (PBS) before adding trypsin/EDTA. Swish the trypsin/EDTA around a bit and remove it. Place the culture in the incubator for 2–4 min, then resuspend the cells in complete ES cell culture medium and plate 1/10 to 1/20 of the cells onto a new feeder layer of cells. Try to keep about 10 to 20 ES cell colonies per field ($10\times$ objective). Passage *before* ES cell colonies express differentiated endoderm cells at their periphery.

4. Freezing and Thawing

Standard slow freeze/rapid thaw techniques work fine. For freeze medium, make complete ES cell culture medium 10% with dimethyl sulfoxide (DMSO) (use DMSO with $OD_{275} < 0.3$).

B. Differentiation Culture

The following procedure is essentially the same as that described in detail elsewhere (Doetschman *et al.*, 1985).

1. Sedimentation to Remove Feeder Cells

Trypsinize cells in the late afternoon and allow to sediment 30 min in the original dish (fibroblasts sediment preferentially). Remove unattached cells to a new tissue culture dish and let stand an additional 1 hr. Remove nonattached cells (nearly all of these cells are ES cells because most of the fibroblasts have attached). Count the ES cells that remained in suspension and plate on bacterial dishes rather than tissue culture dishes at about 500,000 cells per 100-mm dish in standard ES culture medium (15% FCS).

2. Suspension Culture

Change the medium every 2 days of culture until embryonic ectoderm appears sometime between 4 and 6 days of culture (see Doetschman *et al.*, 1985, for description of embryonic ectoderm). When ectoderm appears, switch to medium with 20% FCS (other ingredients as before). From this point on, change the medium every 2 days but add 1/2 volume of medium on the in-between days. This is necessary to prevent exhaustion of the medium and allows for much better differentiation. EB with beating hearts should be observed at 8–12 days. Shortly thereafter, the EB should become cystic, with some bodies containing blood islands. (Note: In some bacterial dishes the embryoid bodies still attach to the substrate. When this happens passage the embryoid bodies to fresh plates.)

C. *Electroporation of Embryonic Stem Cells*

The electroporation procedure is essentially the same as that described in detail elsewhere (Doetschman *et al.*, 1988b). Electroporate 2 days after passage and 4 hr after a medium change. This ensures that the cells are growing as rapidly as possible during the transfection.

1. Buffer

The buffer for electroporation is normal culture medium (15% FCS in high-glucose DMEM with 10^{-4} M β-mercaptoethanol). This medium does not hinder the transfection or targeting efficiency and prevents unnecessary stress on the cells, which are under enough shock as it is.

2. Electroporation Parameters

Electroporation at room temperature or on ice will work fine. Use from 10^7 to 10^8 cells per milliliter of culture medium. Use no more than 5 nM of targeting DNA fragment to minimize integration of concatemers. Use 800 V/cm and 200 μF; this will yield about 50% kill. Replate electroporated cells 2 min after treatment. Handle cells gently and do not centrifuge.

D. *Preparation of Primary Mouse Embryonic Fibroblasts*

1. Sterilely remove embryos that are 14–17 days old (day of plug = day 1). Remove the liver and heart, rinse in PBS to remove as much blood as possible, and tease the embryos apart with a pair of forceps.

2. Add, per embryo, 3–5 ml of 0.05% trypsin/EDTA solution [0.05% trypsin (Sigma, St. Louis, MO, type XI) in 8 g/liter NaCl, 0.4 g/liter KCl, 1 g/liter

glucose, 0.4 g/liter NaHCO$_3$, 0.22 g/liter Na$_2$EDTA (0.6 mM), pH 7.0] and let stand at 4°C overnight. This allows trypsin, which has nearly no activity at 4°C, to diffuse into the tissue.

3. The next morning, without disturbing the pellet of tissue, aspirate as much of the trypsin solution as possible and bring to 37°C for 30 min.

4. Add tissue culture medium (10% FCS in high-glucose DMEM and 10^{-4} M β-mercapotoethanol) and pipette vigorously to break up the tissue.

5. Gravity sediment for 1–2 min to let remaining clumps fall to the bottom of the tube.

6. Count the cells remaining in suspension and plate in four 100-mm dishes per starting embryo.

7. When confluent, freeze using procedures described above at 1 cryotube/dish; alternatively, passage as desired.

8. Passage 1:5. If passaged at too low a density, embryonic fibroblasts will not do well, especially after they have been passaged a few times.

When ready to use the fibroblasts as feeders, allow the plates to become heavily confluent, treat with mitomycin C (10 μg/ml) for 1 hr, wash well (3 times in DMEM), and culture in 15% FCS medium. After mitomycin C treatment [irradiation (300 rads) should work just as well], the cells can then be used as feeders. Immediately before use, however, the feeder cells should receive a medium change in order to ensure that no trace amount of mitomycin C remains, as it could be mutagenic to the ES cells.

The cells can also be passaged once more after mitomycin C treatment or irradiation. However, since they won't grow, they must be plated at a confluent density: about 6 to 7 × 10^5 cells per 50-mm tissue culture dish or 1.5 to 2 × 10^6 cells per 100-mm dish. This method is preferred because the passaging selects against poorly metabolizing or dying cells so that the resulting feeders will be healthier.

The cells will make good feeders after the third passage (the preparation being counted as passage 1) and, depending on the batch, will not become senescent until at least the sixth passage, often later. Consequently, if passaged at a 1:5 split, each of the 25 third-passage plates will yield at least 125 plates of feeders. After senescence they will not do well as feeders.

E. Screening for Targeted Colonies

The underlying principle of the screening procedure is to screen individual colonies by Southern analysis in such a way that one has the results before having to freeze down or passage the colonies. This saves a tremendous amount of time and effort, and the results are much more reliable than those obtained from PCR analysis.

1. Picking Colonies

Pick G418[R] colonies in their entirely using a Pipetman and a white tip (sequence gel loading tip). Transfer each picked colony to one well of a 24-well tray seeded earlier that day with G418[R] primary embryonic feeder cells (one 100-mm plate of feeders to one 24-well tray). Maintain G418 selection on the colonies at this point. Any colonies that were not actually G418 resistant will be eliminated by keeping the selective pressure on.

Because the white tip is so large with respect to the colony, the colony will be transferred in one to three large portions. Consequently, on the day after picking, rinse well (at least 3 times) with PBS and then trypsinize each well again. Resuspend cells thoroughly in complete medium without G418 and replate right back into the same wells. This should give a fine lawn of ES cell colonies on a healthy feeder layer. Without selection the colonies should grow rapidly.

Every 3 to 5 days, as required to prevent differentiation but maximize growth, retrypsinize each well and replate, as before, into the same well. There is no need to add new feeders. Repeat this procedure (1 or 2 times should suffice) until the ES cell colonies cover about two-thirds of the entire surface of the well. At this time the colonies are ready to harvest for DNA. Each well will yield from 20 to 50 μg of DNA, enough for two lanes on a Southern blot.

2. Rapid DNA Preparation

The following procedure is a modified form of that published by Laird *et al.*, (1991). Trypsinize each well and resuspend cells in 1 ml complete medium. Put one drop of suspension in a new well preseeded earlier that day with G418[R] primary embryonic feeder cells. At this point put the cells back on selective pressure. This will ensure against contamination and will also slow the growth of the cells. Put the remainder of the suspension into a microcentrifuge tube, pellet the cells, remove the supernatant, resuspend the cell pellet in 500 μl of lysis buffer without sodium dodecyl sulfate (SDS), then add SDS and proteinase K. Work each sample vigorously up and down in a blue tip about $10\times$ to shear the DNA a bit. Put at 55°C while rocking. During this time rework the samples with a blue tip every hour. Lysis should be complete after 3 to 4 hr.

Precipitate the DNA with 1 volume of 2-propanol: gently mix the 2-propanol into the DNA solution by rocking the microcentrifuge tube back and forth for about 10 min. After the DNA has precipitated, pellet the DNA by centrifugation, rinse well with 70% (v/v) ethanol, dry just long enough to remove the ethanol (do not overdry, e.g., ~30 sec in a SpeedVac concentrator), and resuspend in 400 μl water [DNA goes into solution much faster in water than in Tris–EDTA buffer (TE)], again working each sample with a blue tip. After solution is reached, add EDTA or render the solution TE to prevent degradation. You should now have a solution of nucleic acid, about one-third of which is RNA and two-thirds DNA. RNase treatment during restriction digestion will eliminate the RNA.

Digest the DNA and perform Southern analysis. It is preferable to use nick translation because the concentration of probe DNA is much higher, leading to more rapid hybridization kinetics. Use dextran sulfate and hybridize for only 12 hr. Expose on a phosphorimager screen if available (should take about 12 hr).

The Southern analysis should be done before the one drop of passaged cells need to be handled again. At this point only the potentially positive lines need to be expanded and frozen. All the rest (probably 99% of them) can be pitched. If trouble with the Southern blots has been encountered and the cells need to be handled before the results of Southern analysis are in, simply passage the cells onto new feeders again. This is much faster than freezing individual colonies.

F. Estimated Size of Mouse Colony

This section provides guidelines for estimating the number of mice one would need to maintain a minimum-sized, self-contained mouse colony to support a blastocyst injection operation in a small laboratory. Procedures are listed along with actual estimates for the number of animals needed. The numbers are given as a daily census. If animal charges are in terms of cages rather than animals, note that my mouse colony averages 2.5 mice per cage in cages that hold a maximum of 4 to 5 adult mice.

1. Blastocyst Production for Blastocyst Injection Procedure

To minimize the costs of purchasing mice, a breeding colony capable of producing all the experimental animals needed will be established and maintained. Assuming three blastocyst injection experiments a week (each experiment requiring 16 superovulated C57BL/6 females), 200 4- to 6-week-old females will need to be produced monthly, necessitating a breeding colony of about 40 males and females. We estimated a census of 600 mice to maintain this level of blastocyst production. At the very beginning less than 16 superovulated females will be needed, but, as the proficiency of personnel improves, 16 will probably be needed. A typical blastocyst injection is shown in Fig. 7.

2. Pseudopregnant Females

Two pseudopregnant females per experiment will be used for blastocyst transfer. With increased proficiency of the microinjector, this number will increase.

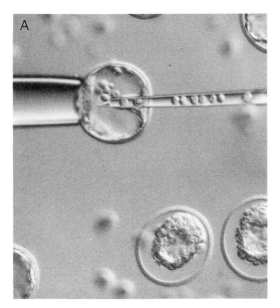 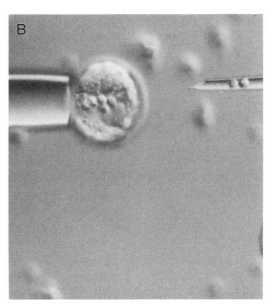

Figure 7. ES cell transfer into mouse blastocysts. (a) During ES cell transfer. ES cells are transferred into the host blastocyst. (b) After transfer. ES cells have been transferred into the blastocyst and reside by the inner cell mass. Also note bevel on the transfer pipette. (Courtesy of H. G. Polites and C. A. Pinkert.)

Twenty vasectomized males and 40 females [C57BL/6 (♀)/CBA (♂) or C57BL/6 (♀)/C3H (♂) F1 animals for both] will be required at any given time to ensure that enough pseudopregnant females are available for two transfers per session. There will be an estimated turnover of about 40 females weekly. The pseudopregnant females that are not used will be recycled with vasectomized males after 2 weeks. We anticipate an average census of 100 females and 20 vasectomized males for the production of pseudopregnant females.

3. Embryos Required for Embryonic Fibroblast Production

We routinely prepare embryonic fibroblasts about once a year. Embryos from four litters will be prepared each time. This will require an average yearly census of about 4 males and 4 females to ensure that you do not lose the strain. We use 129/Sv mice for feeder cells because they usually do quite well as feeders, although we have used other strains with fairly good success. Because we also prepare embryonic fibroblasts from *neo* expressing mice, we also maintain a small colony of transgenic *neo* mice (strain FVB/N). We maintain an average census of about 16 mice for feeder cell production.

4. Establish Embryonic Stem Cell Lines

Initially, a small colony of about 4 males and 16 females, with the females turning over at about 1 per week, should be sufficient for establishing ES cell lines and for maintaining this small colony. This, including the breeding colony to replenish the females, will require about 30 animals.

5. Breeding Test for Germ Line Chimerism

As soon as the blastocyst injections start yielding chimeric animals, the animals will have to be mated with normal C57BL/6 females to test whether the targeted ES cells that were injected had colonized the germ line. The offspring of germ line chimeras will then be used to establish a breeding colony and produce experimental animals from that strain. Although it is difficult to ascertain the number of animals that will be housed for these purposes, we estimate an additional 100 animals for each genetically modified strain.

6. Totals

About 766 mice will be needed to establish a base operation as outlined above. From each germ line chimera an additional 100 mice will be needed for breeding and experimental purposes.

ACKNOWLEDGMENTS

I thank Marcia Shull for improving the clarity of the manuscript and for assembling some of the data for Table 1, Ilona Ormsby for the germ line chimeric mouse, and Jessica Shaw-White, Jim Stringer, and Ann Kier for the tumor data. This work was supported by National Institutes of Health Grants PO1 HL41496 and RO1 HD26471.

REFERENCES

Axelrod, H. R. (1984). Embryonic stem cell lines derived from blastocysts by a simplified technique. *Dev. Biol.* **101**, 225–228.
Baribault, H., and Kemler, R. (1989). Embryonic stem cell culture and gene targeting in transgenic mice. *Mol. Biol. Med.* **6**, 481–492.

Beddington, R. S. P., and Robertson, E. J. (1989). An assessment of the developmental potential of embryonic stem cells in the midgestation mouse embryo. *Development (Cambridge, UK)* **105**, 733–737.

Boggs, S. S., Gregg, R. G., Borenstein, N., and Smithies, O. (1986). Efficient transformation and frequent single-site, single-copy insertion of DNA can be obtained in mouse erythroleukemia cells transformed by electroporation. *Exp. Hematol. (N.Y.)* **14**, 988–994.

Bradley, A., and Robertson, E. (1986). Embryo-derived stem cells: A tool for elucidating the developmental genetics of the mouse. *"Curr. Top. Dev. Biol.* **20**, 357–371.

Bradley, A., Evans, M., Kaufman, M. H., and Robertson, E. (1984). Formation of germ-line chimeras from embryo-derived teratocarcinoma cell lines. *Nature (London)* **309**, 255–256.

Camerini-Otero, R. D., and Kucherlapati, R. (1990). Right on target. *New Biol.* **2**, 337–341.

Capecchi, M. R. (1989a). The new mouse genetics: Altering the genome by gene targeting. *Trends Genet.* **5**, 70–76.

Capecchi, M. R. (1989b). Altering the genome by homologous recombination. *Science* **244**, 1288–1292.

Charron, J., Malynn, B. A., Robertson, E. J., Goff, S. P., and Alt, F. W. (1990). High-frequency disruption of the N-*myc* gene in embryonic stem and pre-B cell lines by homologous recombination. *Mol. Cell. Biol.* **10** 1799–1804.

Chisaka, O., and Capecchi, M. R. (1991). Regionally restricted developmental defects resulting from targeted disruption of the mouse homeobox gene *hox-1.5*. *Nature (London)* **350**, 473–479.

Conlon, F. L., Barth, K. S., and Robertson, E. J. (1991). A novel retrovirally induced embryonic lethal mutation in the mouse: Assessment of the developmental fate of embryonic stem cells homozygous for the 413.d proviral integration. *Development (Cambridge, UK)* **111**, 969–981.

Cosgrove, D., Gray, D., Dierich, A., Kaufman, J., Lemeur, M., Benoist, C., and Mathis, D. (1991). Mice lacking MHC class II molecules. *Cell (Cambridge, Mass.)* **66**, 1051–1066.

DeChiara, T. M., Efstratiadis, A., and Robertson, E. J. (1990). A growth-deficiency phenotype in heterozygous mice carrying an insulin-like growth factor II gene disrupted by targeting. *Nature (London)* **345**, 78–80.

Doetschman, T. C., Eistetter, H., Katz, M., Schmidt, W., and Kemler, R. (1985). The *in vitro* development of blastocyst-derived embryonic stem cell lines: Formation of visceral yolk sac, blood islands and myocardium. *J. Embryol. Exp. Morphol.* **87**, 27–45.

Doetschman, T., Gregg, R. G., Maeda, N., Hooper, M. L., Melton, D. W., Thompson, S., and Smithies, O. (1987). Targeted correction of a mutant HPRT gene in mouse embryonic stem cells. *Nature (London)* **330**, 576–578.

Doetschman, T., Williams, P., and Maeda, N. (1988a). Establishment of hamster blastocyst derived embryonic stem (ES) cells. *Dev. Biol.* **127**, 224–227.

Doetschman, T., Maeda, N., and Smithies, O. (1988b). Targeted mutation of the HPRT gene in mouse embryonic stem cells. *Proc. Natl. Acad. Sci. U.S.A.* **85**, 8583–8587.

Eistetter, H. R. (1989). Pluripotent embryonal stem cell lines can be established from disaggregated mouse morulae. *Dev. Growth Differ.* **31**, 275–282.

Evans, M. J. (1989). Potential for genetic manipulation of mammals. *Mol. Biol. Med.* **6**, 557–565.

Evans, M. J., and Kaufman, M. (1981). Establishment in culture of pluripotential cells from mouse embryos. *Nature (London)* **292**, 154–156.

Evans, M. J., Bradley, A., Kuehn, M. R., and Robertson, E. J. (1985). The ability of EK cells to form chimeras after selection of clones in G418 and some observations on the integration of retroviral vector proviral DNA into EK cells. *Cold Spring Harbor Symp. Quant. Biol.* **50**, 685–689.

Frohman, M. A., and Martin, G. R. (1989). Cut, past, and save: New approaches to altering specific genes in mice. *Cell (Cambridge, Mass.)* **56**, 145–147.

Fung-Leung, W. P., Schilham, M. W., Rahemtulla, A., Kündig, T. M., Vollenweider, M., Potter, J., van Ewijk, W., and Mak, T. W. (1991). CD8 is needed for development of cytotoxic T cells but not helper T cells. *Cell (Cambridge, Mass.)* **65**, 443–449.

Gossler, A., Doetschman, T., Serfling, E., and Kemler, R. (1986). Transgenesis by means of blastocyst-derived embryonic stem cell lines. *Proc. Natl. Acad. Sci. U.S.A.* **83**, 9065–9069.

Gossler, A., Joyner, A. L., Rossant, J. and Skarnes, W. C. (1989). Mouse embryonic stem cells and reporter constructs to detect developmentally regulated genes. *Science* **244**, 463–465.

Grez, M., Akgün, E., Hilberg, F., and Ostertag, W. (1990). Embryonic stem cell virus, a recombinant murine retrovirus with expression in embryonic stem cells. *Proc. Natl. Acad. Sci. U.S.A.* **87**, 9202–9206.

Grusby, J. J., Johnson, R. S., Papaioannou, V. E., and Glimcher, L. H. (1991). Depletion of CD4$^+$ T cells in major histocompatibility complex class II-deficient mice. *Science* **253**, 1417–1420.

Handyside, A. H., O'Neill, G. T., Jones, M., and Hooper, M. L. (1989). Use of BRL-conditioned medium in combination with feeder layers to isolate a diploid embryonic stem cell line. *Roux's Arch. Dev. Biol.* **198**, 48–55.

Hasty, P., Ramírez-Solis, R., Krumlauf, R., and Bradley, A. (1991). Introduction of a subtle mutation into the *Hox-2.6* locus in embryonic stem cells. *Nature (London)* **350**, 243–246.

Johnson, R. S., Sheng, M., Greenberg, M. E., Kolodner, R. D., Papaioannou, V. E., and Spiegelman, B. M. (1989). Targeting of nonexpressed genes in embryonic stem cells via homologous recombination. *Science* **245**, 1234–1236.

Joyner, A. L., Skarnes, W. C., and Rossant, J. (1989). Production of a mutation in mouse *En-2* gene by homologous recombination in embryonic stem cells. *Nature (London)* **338**, 153–156.

Joyner, A. L., Herrup, K., Auerbach, B. A., Davis, C. A., and Rossant, J. (1991). Subtle cerebellar phenotype in mice homozygous for a targeted deletion of the *En-2* homeobox. *Science* **251**, 1239–1243.

Kadokawa, Y., Suemori, H., and Nakatsuji, N. (1990). Cell lineage analyses of epithelia and blood vessels in chimeric mouse embryos by use of an embryonic stem cell line expressing the β-galactosidase gene. *Cell Differ. Dev.* **29**, 187–194.

Kaufman, M. H., Robertson, E. J., Handyside, A. H., and Evans, M. J. (1983). Establishment of pluripotential cell lines from haploid mouse embryos. *J. Embryol. Exp. Morphol.* **73**, 249–261.

Kitamura, D., Roes, J., Kühn, R., and Rajewsky, K. (1991). A B cell-deficient mouse by targeted disruption of the membrane exon of the immunoglobulin mu chain gene. *Nature (London)* **350**, 423–426.

Koller, B. H., and Smithies, O. (1989). Inactivating the β_2-microglobulin locus in mouse embryonic stem cells by homologous recombination. *Proc. Natl. Acad. Sci. U.S.A.* **86**, 8932–8935.

Koller, B. H., Hagemann, L. J., Doetschman, T., Hagaman, J. R., Huang, S., Williams, P. J., First, N. L., Maeda, N., and Smithies, O. (1989). Germ-line transmission of a planned alteration made in a hypoxanthine phosphoribosyltransferase gene by homologous recombination in embryonic stem cells. *Proc. Natl. Acad. Sci. U.S.A.* **86**, 8927–8931.

Koller, B. H., Marrack, P., Kappler, J. W., and Smithies, O. (1990). Normal development of mice deficient in β_2M, MHC class I proteins, and CD8$^+$ T cells. *Science* **248**, 1227–1230.

Laird, P. W., Zijderveld, A., Linders, K., Rudnicki, M. A., Jaenisch, R., and Berns, A. (1991). Simplified mammalian DNA isolation procedure. *Nucleic Acids Res.* **19**, 4293–4294.

Lallemand, Y., and Brûlet P. (1990). An *in situ* assessment of the routes and extents of colonization of the mouse embryo by embryonic stem cells and their descendants. *Development (Cambridge, UK)* **110**, 1241–1248.

Le Mouellic, H., Lallemand, Y., and Brûlet, P. (1990). Targeted replacement of the homeobox gene *Hox-3.1* by the *Escherichia coli lacZ* in mouse chimeric embryos. *Proc. Natl. Acad. Sci. U.S.A.* **87**, 4712–4716.

Lindenbaum, M. H., and Grosveld, F. (1990). An *in vitro* globin gene switching model based on differentiated embryonic stem cells. *Genes Dev.* **4**, 2075–2085.

Linney, E., and Donerly, S. (1983). DNA fragments from F9 PyEC mutants increase expression of heterologous genes in transfected F9 cells. *Cell (Cambridge, Mass.)* **35**, 693–699.

Lovell-Badge, R. H., Bygrave, A. E., Bradley, A., Robertson, E., Evans, M. J., and Cheah, K. S. E. (1985) *Cold Spring Harbor* **50**, 707–711.

Lovell-Badge, R. H., Bygrave, A., Bradley, A., Robertson, E., Tilly, R., and Cheah, K. S. E. (1987). Tissue-specific expression of the human type II collagen gene in mice. *Proc. Natl. Acad. Sci. U.S.A.* **84**, 2803–2807.

Lufkin, T., Dierich, A., LeMeur, M., Mark, M., and Chambon, P. (1991). Disruption of the *Hox-1.6* homeobox gene results in defects in a region corresponding to its rostral domain of expression. *Cell (Cambridge, Mass.)* **66**, 1105–1119.

Macleod, D., Lovell-Badge, R., Jones, S., and Jackson, I. (1991). A promoter trap in embryonic stem (ES) cells selects for integration of DNA into CpG islands. *Nucleic Acids Res.* **19**, 17–23.

McMahon, A. P., and Bradley, A. (1990). The *Wnt-1* (*int-1*) proto-oncogene is required for development of a large region of the mouse brain. *Cell (Cambridge, Mass.)* **62**, 1073–1085.

Magnuson, T., Epstein, C. J., Silver, L. M., and Martin, G. R. (1982). Pluripotent embryonic stem cell lines can be derived from t^{w5}/t^{w5} blastocysts. *Nature (London)* **298**, 750–753.

Mann, J. R., Gadi, I., Harbison, M. L., Abbondanzo, S. J., and Stewart C. L. (1990). Androgenetic mouse embryonic stem cells are pluripotent and cause skeletal defects in chimeras: Implications for genetic imprinting. *Cell (Cambridge, Mass.)* **62**, 251–260.

Mansour, S. L. (1990). Gene targeting in murine embryonic stem cells: Introduction of specific alterations into the mammalian genome. *Genet. Anal.* **7**, 219–227.

Mansour, S. L., Thomas, K. R., and Capecchi, M. R. (1988). Disruption of the proto-oncogene *int-2* in mouse embryo-derived stem cells: A strategy for targeting mutations to non-selectable genes. *Nature (London)* **336**, 348–352.

Mansour, S. L., Thomas, K. R., Deng, C., and Capecchi, M. R. (1990). Introduction of a *lacZ* reporter gene into the mouse *int-2* locus by homologous recombination. *Proc. Natl. Acad. Sci. U.S.A.* **87**, 7688–7692.

Martin, G. (1981). Isolation of pluripotent cell line from early mouse embryos cultured in medium conditioned by teratocarcinoma stem cells. *Proc. Natl. Acad. Sci. U.S.A.* **78**, 7634–7638.

Martin, G. R., and Lock, L. F. (1983). Pluripotent cell lines derived from early mouse embryos cultured in medium conditioned by teratocarcinoma stem cells. *In* "Teratocarcinoma Stem Cells, Cold Spring Harbor Conference on Cell Proliferation" (L. M. Silver, G. R. Martin, and S. Strickland, eds.), Vol. 10, pp. 635–663. Cold Spring Harbor Laboratory, Cold Spring Harbor, New York.

Martin, G. R., Silver, L. M., Fox, H. S., and Joyner, A. L. (1987). Establishment of embryonic stem cell lines from preimplantation mouse embryos homozygous for lethal mutations in the *t*-complex. *Dev. Biol.* **121**, 20–28.

Mombaerts, P., Clarke, A. R., Hooper, M. L., and Tonegawa, S. (1991). Creation of a large genomic deletion at the T-cell antigen receptor β-subunit locus in mouse embryonic stem cells by gene targeting. *Proc. Natl. Acad. Sci. U.S.A.* **88**, 3084–3087.

Mortensen, R. M., Zubiaur, M., Neer, E. J., and Seidman, J. G. (1991). Embryonic stem cells lacking a functional inhibitory G-protein subunit (α_{i2}) produced by gene targeting of both alleles. *Proc. Natl. Acad. Sci. U.S.A.* **88**, 7036–7040.

Mucenski, M. L., McLain, K., Kier, A. B., Swerdlow, S. H., Schreiner, C. M., Miller, T. A., Pietryga, D. W., Scott, W. J., and Potter, S. S. (1991). A functional c-*myb* gene is required for normal murine fetal hepatic hematopoiesis. *Cell (Cambridge, Mass.)* **65**, 677–689.

Nagy, A., Gócza, E., Diaz, E. M., Prideaux, V. R., Iványi, E., Markkula, M., and Rossant, J. (1990). Embryonic stem cells alone are able to support fetal development in the mouse. *Development (Cambridge, UK)* **110**, 815–821.

Nicolas, J. F., and Berg, P. (1983). Regulation of expression of genes transduced into embryonal carcinoma cells. *In* "Teratocarcinoma Stem Cells, Cold Spring Harbor Conference on Cell Proliferation" (L. M. Silver, G. R. Martin, and S. Strickland, eds.), Vol. 10, pp. 469–485. Cold Spring Harbor Laboratory, Cold Spring Harbor, New York.

Nichols, J., Evans, E. P., and Smith A. G. (1990). Establishment of germ-line-competent embryonic stem (ES) cells using differentiation inhibiting activity. *Development (Cambridge, UK)* **110**, 1341–1348.

Notarianni, E., Laurie, S., Moor, R. M., and Evans, M. J. (1990). Maintenance and differentiation in culture of pluripotential embryonic cell lines from pig blastocysts. *J. Reprod. Fertil. Suppl.* **44**, 51–56.

Pease, S., Braghetta, P., Gearing, D., Grail, D., and Williams, L. (1990). Isolation of embryonic stem (ES) cells in media supplemented with recombinant leukemia inhibitory factor (LIF). *Dev. Biol.* **141**, 344–352.

Pevny, L., Simon, M. C., Robertson, E., Klein, W. H., Tsai, S. F., D'Agati, V., Orkin, S. H., and Costantini, F. (1991). Erythroid differentiation in chimaeric mice blocked by a targeted mutation in the gene for transcription factor GATA-1. *Nature (London)* **349**, 257–260.

Piedrahita, J. A., Anderson, G. B., and BonDurant, R. H. (1990). On the isolation of embryonic stem cells: Comparative behavior of murine, porcine and ovine embryos. *Theriogenology* **34**, 879–901.

Rahemtulla, A., Fung-Leung, W. P., Schilham, M. W., Kündig, T. M., Sambhara, S. R., Narendran, A., Arabian, A., Wakeham, A., Paige, C. J., Zinkernagel, R. M., Miller, R. G., and Mak, T. W. (1991). Normal development and function of CD8$^+$ cells but markedly decreased helper cell activity in mice lacking CD4. *Nature (London)* **353**, 180–184.

Reid, L. H., Gregg, R. G., Smithies, O., and Koller, B. H. (1990). Regulatory elements in the introns of the human HPRT gene are necessary for its expression in embryonic stem cells. *Proc. Natl. Acad. Sci. U.S.A.* **87**, 4299–4303.

Reid, L. H., Shesely, E. G., Kim, H. S., and Smithies, O. (1991). Cotransformation and gene targeting in mouse embryonic stem cells. *Mol. Cell. Biol.* **11**, 2769–2777.

Robertson, E. J., Evans, M. J., and Kaufman, M. H. (1983a). X-chromosome instability in pluripotential stem cell lines derived from parthenogenetic embryos. *J. Embryol. Exp. Morphol.* **74**, 297–309.

Robertson, E. J. (1991). Using embryonic stem cells to introduce mutations into the mouse germ line. *Biol. Reprod.* **44**, 238–245.

Robertson, E. J., Kaufman, M. H., Bradley, A., and Evans, M. J. (1983b). Isolation, properties and karyotype analysis of pluripotent (EK) cell lines from normal and parthenogenetic embryos. *In* "Teratocarcinoma Stem Cells, Cold Spring Harbor Conference on Cell Proliferation" (L. M. Silver, G. R. Martin, and S. Strickland, eds.), Vol. 10, pp. 647–663. Cold Spring Harbor Laboratory, Cold Spring Harbor, New York.

Robertson, E., Bradley, A., Kuehn, M., and Evans, M. (1986). Germ-line transmission of genes introduced into cultured pluripotential cells by retroviral vector. *Nature (London)* **323**, 445–448.

Rubenstein, J. L. R., Nicolas, J. F., and Jacob, F. (1984). Construction of a retrovirus capable of transducing and expressing genes in multipotential embryonic cells. *Proc. Natl. Acad. Sci. U.S.A.* **81**, 7137–7140.

Schorle, H., Holtschke, T., Hünig, T., Schimpl, A., and Horak, I. (1991). Development and function of T cells in mice rendered interleukin-2 deficient by gene targeting. *Nature (London)* **352**, 621–624.

Schwarzberg, P. L., Goff, S. P., and Robertson, E. J. (1989). Germ-line transmission of a c-*abl* mutation produced by targeted gene disruption in ES cells. *Science* **246**, 799–802.

Schwartzberg, P. L., Robertson, E. J., and Goff, S. P. (1990). Targeted gene disruption of the endogenous c-*abl* locus by homologous recombination with DNA encoding a selectable fusion protein. *Proc. Natl. Acad. Sci. U.S.A.* **87**, 3210–3214.

Schwartzberg, P. L., Stall, A. M., Hardin, J. D., Bowdish, K. S., Humaran, T., Boast, S., Harbison, M. L., Robertson, E. J., and Goff, S. P. (1991). Mice homozygous for the *abl*^{m1} mutation show poor viability and depletion of selected B and T cell populations. *Cell (Cambridge, Mass.)* **65**, 1165–1175.

Shinar, D., Yoffe, O., Shani, M., and Yaffe, D. (1989). Regulated expression of muscle-specific genes introduced into mouse embryonal stem cells: Inverse correlation with DNA methylation. *Differentiation (Berlin)* **41**, 116–126.

Soriano, P., Friedrich, G., and Lawinger, P. (1991a). Promoter interactions in retrovirus vectors introduced into fibroblasts and embryonic stem cells. *J. Virol.* **65**, 2314–2319.

Soriano, P., Montgomery, C., Geske, R., and Bradley, A. (1991b). Targeted disruption of the c-*src* proto-oncogene leads to osteopetrosis in mice. *Cell (Cambridge, Mass.)* **64**, 693–702.

Stanton, B. R., Reid, S. W., and Parada, L. F. (1990). Germ line transmission of an inactive N-*myc* allele generated by homologous recombination in mouse embryonic stem cells. *Mol. Cell. Biol.* **10**, 6755–6758.

Steeg, C. M., Ellis, J., and Bernstein, A. (1990). Introduction of specific point mutations into RNA polymerase II by gene targeting in mouse embryonic stem cells: Evidence for a DNA mismatch repair mechanism. *Proc. Natl. Acad. Sci. U.S.A.* **87**, 4680–4684.

Stewart, C. L., Vanek, M., and Wagner, E. F. (1985). Expression of foreign genes from retroviral vectors in mouse teratocarcinoma chimaeras. *EMBO J.* **4**, 3701–3709.

Stocking, C., Kollek, R., Bergholz, U., and Ostertag, W. (1985). Long terminal repeat sequences impart hematopoietic transformation properties to the myeloproliferative sarcoma virus. *Proc. Natl. Acad. Sci. U.S.A.* **82**, 5746–5750.

Strojek, R. M., Reed, M. A., Hoover, J. L., and Wagner, T. E. (1990). A method for cultivating morphologically undifferentiated embryonic stem cells from porcine blastocysts. *Theriogenology* **33**, 901–913.

Suda, Y., Suzuki, M., Ikawa, Y., and Aizawa, W. (1987). Mouse embryonic stem cells exhibit indefinite proliferative potential. *J. Cell. Physiol.* **133**, 197–201.

Suda, Y., Hirai, S. I., Suzuki, M., Ikawa, Y., and Aizawa, S. (1988). Active *ras* and *myc* oncogenes can be compatible, but SV40 large T antigen is specifically suppressed with normal differentiation of mouse embryonic stem cells. *Exp. Cell Res.* **178**, 98–113.

Suemori, H., Kadodawa, Y., Goto, K., Araki, I., Kondoh, H., and Nakatsuji, N. (1990). A mouse embryonic stem cell line showing pluripotency of differentiation in early embryos and ubiquitous β-galactosidase expression. *Cell Differ. Dev.* **29**, 181–186.

Szyf, M., Tanigawa, G., and McCarthy, P. L., Jr. (1990). A DNA signal from the *Thy-1* gene defines *de novo* methylation patterns in embryonic stem cells. *Mol. Cell. Biol.* **10**, 4396–4400.

Takahashi, Y., Hanaoka, K., Hayasaka, M., Katoh, K., Kato, Y., Okada, T. S., and Kondoh, H. (1988). Embryonic stem cell-mediated transfer and correct regulation of the chicken δ-crystallin gene in developing mouse embryos. *Development (Cambridge, Mass.)* **102**, 259–269.

te Riele, H., Maandag, E. R., Clarke, A., Hooper, M., and Berns, A. (1990). Consecutive inactivation of both alleles of the *pim-1* proto-oncogene by homologous recombination in embryonic stem cells. *Nature (London)* **348**, 649–651.

Thomas, K. R., and Capecchi, M. R. (1987). Site-directed mutagenesis by gene targeting in mouse embryo-derived stem cells. *Cell (Cambridge, Mass.)* **51**, 503–512.

Thomas, K. R., and Capecchi, M. R. (1990). Targeted disruption of the murine *int-1* proto-oncogene resulting in severe abnormalities in midbrain and cerebellar development. *Nature (London)* **346**, 847–850.

Thomas, K. R., Folger, K. R., and Capecchi, M. R. (1986). High frequency targeting of genes to specific sites in the mammalian genome. *Cell (Cambridge, Mass.)* **44**, 419–428.

Thompson, S., Clarke, A. R., Pos, A. M., Hooper, M. L., and Melton, D. W. (1989). Germ line transmission and expression of a corrected HPRT gene produced by gene targeting in embryonic stem cells. *Cell (Cambridge, Mass.)* **56**, 313–321.

Tybulewicz, V. L. J., Crawford, C. E., Jackson, P. K., Bronson, R. T., and Mulligan, R. C. (1991). Neonatal lethality and lymphopenia in mice with a homozygous disruption of the c-*abl* proto-oncogene. *Cell (Cambridge, Mass.)* **65**, 1153–1163.

Valancius, V., and Smithies, O. (1991). Testing an "in–out" targeting procedure for making subtle genomic modifications in mouse embryonic stem cells. *Mol. Cell. Biol.* **11**, 1402–1408.

van Deursen, J., Lovell-Badge, R., Oerlemans, F., Schepens, J., and Wieringa, B. (1991). Modulation of gene activity by consecutive gene targeting of one creatine kinase M allele in mouse embryonic stem cells. *Nucleic Acids Res.* **19**, 2637–2643.

Wagner, E. F., and Stewart, C. L. (1986). Integration and expression of genes introduced into mouse embryos. *In* "Experimental Approaches to Mammalian Embryonic Development" (J. Rossant and R. A. Pederson, eds.), pp. 509–549. Cambridge Univ. Press, Cambridge.

Wagner, E. F., Vanek, M., and Vennström, B. (1985). Transfer of genes into embryonal carcinoma cells by retrovirus infection: Efficient expression from an internal promoter. *EMBO J.* **4**, 663–666.

Williams, R. L., Courtneidge, S. A., and Wagner, E. F. (1988). Embryonic lethalities and endothelial tumors in chimeric mice expressing polyoma virus middle T oncogene. *Cell (Cambridge, Mass.)* **52**, 121–131.

Wobus, A. M., Holzhausen, H., Jäkel, P., and Schöneich, J. (1984). Characterization of a pluripotent stem cell line derived from a mouse embryo. *Exp. Cell Res.* **152**, 212–219.

Wobus, A. M., Grosse R., and Schöneich, J. (1988). Specific effects of nerve growth factor on the differentiation pattern of mouse embryonic stem cells *in vitro*. *Biomed. Biochim. Acta* **41**, 965–973.

Yagi, T., Ikawa, Y., Yoshida, K., Shigetani, Y., Takeda, N., Mabuchi, I., Yamamoto, T., and Aizawa, S. (1990). Homologous recombination at c-*fyn* locus of mouse embryonic cells with use of diphtheria toxin A-fragment gene in negative selection. *Proc. Natl. Acad. Sci. U.S.A.* **87**, 9918–9922.

Zheng, H., and Wilson, J. H. (1990). Gene targeting in normal and amplified cell lines. *Nature (London)* **344**, 170–173.

Zijlstra, M., Li, E., Sajjadi, F., Subramani, S., and Jaenisch, R. (1989). Germ-line transmission of a disrupted β_2-microglobulin gene produced by homologous recombination in embryonic stem cells. *Nature (London)* **342**, 435–438.

Zijlstra, M., Bix, M., Simister, N. E., Loring, J. M., Raulet, D. H., and Jaenisch, R. (1990). β_2-Microglobulin deficient mice lack CD4$^-$8$^+$ cytolytic T cells. *Nature (London)* **344**, 742–746.

Zimmer, A., and Gruss, P. (1989). Production of chimaeric mice containing embryonic stem (ES) cells carrying a homoeobox *Hox1.1* allele mutated by homologous recombination. *Nature (London)* **338**, 150–153.

5

Retrovirus-Mediated Gene Transfer

Philip A. Wood

Department of Comparative Medicine

Schools of Medicine and Dentistry

The University of Alabama at Birmingham

Birmingham, Alabama 35294

I. INTRODUCTION

Recombinant retrovirus-mediated gene transfer has been used for a wide range of purposes since the early 1980s (Wei *et al.*, 1981; Joyner and Bernstein, 1983; Miller *et al.*, 1983). A major stimulus for developing this technology was a desire for a method of highly efficient gene transfer for potential gene therapy applications in human diseases. An important catalyzing step was producing recombinant retroviruses without contaminating wild-type helper virus (Mann *et al.*, 1983; Mulligan, 1983). The major advantages of retroviral infection as a method of gene transfer include the fact that the recombinant proviral sequence transferred is integrated stably into the genome of the recipient cell as a single, randomly located integrant with a predictable molecular structure and no subsequent cytopathy. The efficiency of gene transfer is very high compared to other nonviral methods, at times approaching 100%. The purpose of this chapter is to review useful protocols and applications for constructing retroviral vectors, selecting packaging cell lines, estimating recombinant retroviral titers, and using recombinant retroviruses for infection and study of *in vivo* gene expression. The discussion provided here is based mostly on recombinant retroviral vectors and packaging cell lines derived from mouse leukemia virus (MuLV) genomic segments.

II. APPLICATIONS

Retrovirus-mediated gene transfer has been used in a wide range of applications. One of the first and widest applications has been gene transfer into cultured cells. This has been useful for developing and testing expression and titers of new vectors for eventual use *in vivo*. The major advantage is that a high number of cells can be transfected to usually contain a single copy of the transferred DNA construct. These cells can be cloned and studied as individual integration events with respect to expression of the gene. There has been much interest in using retroviral vectors for *in vivo* infection or for explant cell infection followed by transplantation and study of *in vivo* expression. This application has most widely involved bone marrow cells but has also involved hepatocytes, fibroblasts, keratinocytes, and various other cell lines. A related application has been the infection of mouse embryos as a method of gene transfer for transgenic mouse production. This can be done at 4- to 16-cell stages, so that mosaic transgenic mice are produced. This can be potentially useful when the desire is to produce germ line insertion mutations by retroviral integration.

III. GENERAL METHODS

A. Vector Design

Despite intensive study, the criteria for optimal vector design remain speculative. Many times one has to make the construct to see if it will work. Elements important to consider in the design of a retroviral expression vector include the promoter, enhancers, polyadenylation signals, selectable marker genes, the inserted gene(s), splicing/nonsplicing type vectors, wild-type retroviral genome components (e.g, gag^+ vectors), and size constraints of all elements included.

1. Promoters and Enhancers

Many expression vectors have the retroviral (e.g., Moloney MuLV) long terminal repeat (LTR) as the only promoter. This promoter is assisted by the enhancer sequences also in the LTRs, and it has been used successfully for constitutive expression in most cultured cells and in many cells *in vivo*. Expression in transplanted cells has been variable, as discussed later in this chapter (Section IV). The level of expression in the packaging cell line can be rate limiting for the eventual titer obtained with a given construct, since this expression is entirely dependent on the LTR for the transcript to be packaged as an RNA dimer genome. The level of subsequent expression from the LTR in the recipient cell line is more difficult to predict.

One can also put the inserted gene of interest under the control of another promoter, such as the natural promoter of the gene, arranged as an internal transcription unit. Appropriate orientation of the internal promoter to that of the LTR transcription direction should be considered. Generally, internal promoters can drive transcription appropriately, even in a tissue-specific manner, but expression is sometimes low. Some internal transcription units may have a higher incidence of deletions (Emerman and Temin, 1984; Hatzoglou *et al.*, 1990; Pathak and Temin, 1990), whereas with others there are no problems with deletions, transcription, or expression (Hantzopoulos *et al.*, 1989). Overall expression is affected at both the transcription/RNA level and the translation/functional protein level. There is a wide range of experience concerning these issues (Miller *et al.*, 1984; Williams *et al.*, 1986; Ledley *et al.*, 1987; Belmont *et al.*, 1988; Bowtell *et al.*, 1988; Moore *et al.*, 1989; Peng *et al.*, 1988; Wilson *et al.*, 1988a,b, 1990; Herman *et al.*, 1989). As a variation, so-called crippled LTR vectors may be designed in which the proviral retroviral sequence 5' end LTR contains a deletion negating any of its promoter/ enhancer function after genomic integration. Expression of the inserted gene is thus entirely dependent on the internal promoter (Yu *et al.*, 1986).

An extensive comparison of constructs to examine internal transcription units in both orientations with respect to the LTR promoter of the virus was reported by Belmont *et al.* (1988). They examined 20 different constructs containing a wide variety of promoters, neomycin resistance (*neo*) gene, and a cDNA for human adenosine deaminase (ADA) in various orientations. The one construct with the best titer and expression *in vivo* in bone marrow-derived stem cells was a simple modified N2 vector (Belmont *et al.*, 1988) consisting of LTR → ADA → LTR. These studies indicate that promoter considerations are critical, and the primary context for consideration must be expression in the recipient cell line.

2. Polyadenylation Signal Sequence

The 3' LTR will function as the necessary transcription termination and polyadenylation signals (PAS) (Cepko *et al.*, 1984) in many vectors. There has been concern about leaving in a PAS as part of the inserted cDNA or gene of interest transcribed from the LTR. In general, if the 5' LTR is the promoter for producing a retroviral full-length type transcript, it is advisable to remove or destroy any PAS signals within the insert. Not doing so has caused problems in some constructs that never produced a high titer and were not expressed in the recipient cell line. In constructing an inserted minigene to be transcribed independently of the LTR, an internal PAS may be needed. Possible risks are that transcription termination and poly(A) addition may reduce the overall recombinant retroviral titer owing to premature transcription termination with respect to the RNA genomic transcripts (Mulligan, 1983).

3. Inclusion of Selectable Markers

There may be advantages to including a built-in selectable marker such as guanine phosphoribosyltransferase (*gpt*), neomycin resistance (*neo*), or hygromycin resistance (*hygro*) genes in the vector. Dominant selectable markers obviously make the initial selection and cloning of the packaging cells convenient following gene transfer. The other major advantage of some selectable markers is that they allow determination of the titer of the recombinant retrovirus from the infectious dose used. Subsequent counting of colonies reflects the colony-forming units per milliliter (cfu/ml) of retrovirus-containing medium.

The major problem with these markers is that they can reduce titer and expression owing to interference from size and other nondefined characteristics. The size limitations are discussed in the following section. In the most optimal situation, expression of the gene of interest following gene transfer is simply used as the selectable marker. This, unfortunately, is not often possible. The initial selection problem can usually be overcome by cotransfecting the retroviral construct plasmid of interest with another plasmid containing a selectable marker such as pSV_2–Neo so that its molar ratio is low compared to that of the retroviral construct plasmid. We have used cotransfection ratios ranging from 1:4 to 1:10 (*neo* vector:retroviral vector). With this approach, any selection-resistant colonies are likely to have the retroviral vector plasmid as well.

4. Size Constraints

Retroviral constructs have limits on the size of the DNA fragments that are inserted between the LTRs. For the transcripts to be packaged properly, the total size must be neither too long nor too short. Natural retroviral genomes range from approximately 6 to 10 kb from LTR to LTR (Weiss *et al.*, 1985). The average size for the LTRs is less than 1 kb each. With some very small constructs a detectable titer of recombinant virus was never obtained, suggesting that they were too small. In the reverse situation, by removing a *neo* cassette from a pZIP NeoSV(B) type vector (Wood *et al.*, 1986a) to reduce its size from 6.3 to 4 kb, and keeping everything else constant, both the titer and expression increased by a few orders of magnitude. There are many anecdotal experiences in which the speculated problem with titer was that the construct was too big or complex because of insertions of several elements, including the gene of interest, selectable marker(s), promoters, enhancers, and known or cryptic splicing sequences. My experience has been that retroviral vector constructs in the 4- to 6-kb (LTR → insert → LTR) range work the best.

5. Splicing and Nonsplicing Vectors

Many of the early recombinant retroviral vectors, including pZIPNeoSV(X), pZIPNeoSV(B) (Cepko *et al.*, 1984), pLPL (Miller *et al.*, 1983), and pN2 (Eglitis *et al.*, 1986), were based on Moloney murine leukemia virus. These contain a splice site so that a gene placed in the natural position of the *gag* and *pol* genes is alternatively spliced out, and two different transcripts are produced from these vectors. Originally, it was thought that this type of vector had an advantage in that expression of the selectable marker gene would not depend on the expression of the *env* analog gene. With this design, it might be possible to modulate the level of expression of the cDNA insert, rather than obtaining only constitutive high-level expression, since this insert could be transcribed via additional controlling elements, whereas expression of the selectable marker would be mediated by the retroviral transcription signals (Mulligan, 1983). We speculated that better expression of the inserted gene of interest would be expected if it were in the position to be contained in both transcripts, such as is the *env* gene in the natural Moloney retrovirus, but we never found that. Our experience (Wood *et al.*, 1986b) and that of others (Belmont *et al.*, 1988) has been that, generally, the simpler the vector the better, and alternative splicing may not provide any clear advantage and may actually complicate expression of the vector.

6. Vectors Containing the *gag* Gene

One of the most successful recombinant retroviral vectors was N2 (Eglitis *et al.*, 1985). In many investigators' hands, it almost always produced a very high titer of recombinant retrovirus with a wide range of inserts in addition to and in the absence of its *neo* gene (Belmont *et al.*, 1988; Herman *et al.*, 1989). Its high-titer expression was likely due to a remaining 5' segment of the *gag* gene of the original retrovirus that seemingly provided for a highly stable and efficiently packaged recombinant RNA genome with the resulting high titer (Armentano *et al.*, 1987; Bender *et al.*, 1987; Adam *et al.*, 1988). The vector was also unfortunately notorious for generating wild-type helper virus along with the desired recombinant retrovirus (Miller and Buttimore, 1986). Recombination events with the packaging cell genome apparently allowed production of this fully infectious wild-type retrovirus. The increased likelihood of this vector undergoing recombination was thought to be due to the presence of the extended *gag* region, which provided a long region of homology conducive for recombination. This problem and its solutions are discussed more in the following section on packaging cells.

B. Packaging Cells

Initial selection of packaging cells is generally based on the recipient cell line species. Ecotropic packaging cell lines produce recombinant retroviruses that will infect only mouse and rat cells, whereas amphotropic packaging cell lines produce retroviruses that will infect a wide range of species including mouse, rat, human, monkey, and various domestic animal species (Weiss *et al.*, 1985). A wider infection range requires more careful biohazard containment owing to the possible infection of human cells. The actual biohazard present is unclear since the amphotropic virus used for these packaging cell lines is derived from a wild mouse retrovirus (Hartley and Rowe, 1976; Rasheed *et al.*, 1976), not a human pathogenic retrovirus, and murine amphotropic retroviruses are not an acute pathogen for primates (Sturm *et al.*, 1990).

The early, highly successful packaging cell lines such as the ecotropic packaging line known as ψ-2 (Mann *et al.*, 1983) have been widely used for several different vector constructs with high-titer production. The ψ-2 packaging cell line contains an integrated proviral genomic ecotropic sequence encoding for all of the proteins needed in trans to produce a retroviral particle; however, owing to a mutation in the ψ region needed for proper packaging of the dimer RNA transcript as the newly forming viral genome, it is unable to package its own genome into the viral particle. The only genome packaged would be one with the intact wild-type ψ sequence, like that found in the recombinant retroviral genome provided by gene transfer. This packaging cell line was made by transferring the appropriate proviral sequence containing all of the necessary genes as one single construct. This provided a target for recombination and complementation of the ψ mutation so that the viral gene-encoding sequences could be packaged, producing wild-type retroviruses as well as the desired recombinant retrovirus. Similar amphotropic lines were designed (Cone and Mulligan, 1984; Sorge *et al.*, 1984; Miller *et al.*, 1985).

The next prototype was a recombination-resistant packaging cell line, the PA317 amphotropic packaging cell line (Miller and Buttimore, 1986). It was made by making a series of deletions in the proviral genomic sequence such that multiple, highly unlikely, recombinations would have to occur to fully correct it to allow wild-type retrovirus production. Among several other variations, a highly used recombination-resistant packaging cell line for ecotropic virus was made by separating the proviral genomic genes onto two separate plasmids (Markowitz *et al.*, 1988a). For example, in a cell line known as GP + E 86, the *gag* and *pol* genes are contained and expressed from one plasmid, whereas the *env* gene is expressed from a second plasmid. This cell line has worked well to produce high titers of recombinant retrovirus without wild-type virus contamination for several different investigators. A similar amphotropic packaging cell line was made using the same principles (Markowitz *et al.*, 1988b). Also, both ecotropic and amphotropic packaging cell lines have been designed in which complementary frameshift mutations were introduced in the retroviral genes encoding the packaging functions in addition to the cis-acting alterations (ψ negative) (Danos and Mulligan, 1988).

In another approach, some investigators have used cross tropism infection to boost the recombinant retrovirus titer. In what has been referred to as "Ping-Pong" infection, they mix ecotropic and amphotropic packaging cell lines with the same recombinant vector so that they will cross infect to increase, at least by one, the functional recombinant proviral genome copy number (Lynch and Miller, 1991). Because retroviral infection by a virus with the same tropism is blocked at the recipient cell receptor, one must use the other cell line for infecting it. Then one can use selectable marker gene tricks to selectively kill off the undesired packaging cell line following some infection period. Likewise, one can use directly harvested virus of one tropism to infect the other packaging cell line.

C. Wild-Type Retrovirus Contamination Detection

There are several ways to evaluate for the presence of wild-type (helper virus) retrovirus contamination. Such contamination is often a major problem owing to potential biohazard concerns and confounding of experimental results, especially when trying to determine recombinant retroviral titer and evaluate *in vivo* expression. With wild-type retrovirus during *in vivo* experiments, one could have a continuing *in vivo* source of infection owing to a wild-type and recombinant retroviremia. This will not only complicate expression results because the recipient and expression cell pool is not stable, but it can also cause death of the animals from overt leukemia. This is usually seen clinically as death in mice with enlarged or engorged spleens and livers. The original tests for the presence of wild-type virus involved taking a test retrovirus sample and infecting a recipient cell line with it. The cell supernatant was then used to test either for the presence of reverse transcriptase (Goff *et al.*, 1981) or for infectivity by showing that a subsequent cell line could be infected with the medium. A "wild-type virus-free" virus harvest should not produce reverse transcriptase or a second infection event. The second infection event is usually detected by the use of indicator cell lines such as the rat fibroblast cell line (XC) for ecotropic wild-type virus (Rowe *et al.*, 1970) and a feline cell line (CCC-81) known as the S + L − assay (Fischinger *et al.*, 1975; Miller *et al.*, 1985). A more recent and more practical approach is to use a cell line with a *neo*-containing retroviral marker construct in which a wild-type virus would rescue the *neo*-tagged retroviral genome from the test recipient cell line (Belmont *et al.*, 1988; Moore *et al.*, 1989). A wild-type virus-free preparation would not produce neomycin-resistant colonies in subsequent test infections.

The general protocol for using this system is as follows. Supernatants from defective virus producer cell lines and serum samples from experimental animals are incubated with the indicator cell line SVB-3T3. The indicator cell line was produced by transfection of NIH 3T3 fibroblasts (ATCC, Rockville, MD) with the replication-defective vector pZIPNeoSV(B) (Cepko, *et al.*, 1984). When infected with replication-competent virus, these cells will package a *neo*-containing defective vector. After 2 weeks, the indicator cell line supernatant is harvested and used to infect Rat208F cells or 3T3 cells. These cells are then placed in G418 selection for examination for resistant colonies. The titer of the replication-competent virus is determined by serial dilutions of the initial supernatant or serum sample before infection onto the SVB-3T3 cells (Belmont *et al.*, 1988; Moore *et al.*, 1989). This is our routine procedure to test samples for wild-type virus. A recent variation on this method describes use of the polymerase chain reaction (PCR) to detect wild-type provirus in the test cell line (Scarpa *et al.*, 1991).

The availability of some of the latest, highly refined packaging cell lines has, it is hoped, taken care of these problems. It is probably wise, nevertheless, to continue monitoring the experimental viral producer cell lines.

D. Recombinant Retroviral Titer Determination

1. Gene Transfer of Plasmid Vector into Packaging Cell Line

We have most often used calcium phosphate-precipitated DNA-mediated gene transfer (Wood *et al.*, 1986a) based on the original procedure (Graham and Van der Eb, 1973). Most recently we have been using electroporation for transfecting the vector plasmid into the packaging cell lines. As noted above we usually either have a *Neo* gene within the vector or we cotransfect with SV$_2$-Neo. Other successful transfection methods include DEAE-dextran (Weintraub *et al.*, 1986) and lipofectin (Felgner *et al.*, 1987) procedures, but the main focus here is the previous two methods.

a. Method 1: Protocol for Calcium Phosphate-Precipitated DNA Gene Transfer The following solutions are needed:

2× HEPES-buffered phosphate, pH 7.15 (autoclaved)
   276 mM NaCl
   10 mM KCl
   1.7 mM Na$_2$HPO$_4$
   11.5 mM glucose
   42 mM HEPES
Calcium chloride, 500 mM (autoclaved)
Purified plasmid DNA (CsCl preparation quality)

1. Packaging cells, which are usually derived from mouse NIH 3T3 cells, should be grown to a density of approximately 1×10^6 per 100-mm cell culture dish. The cells should be seeded at this density the day prior to gene transfer. Most of the packaging cell lines grow well in either Dulbecco's modified essential medium (DMEM) or minimal essential medium (MEM) with 10% (v/v) newborn calf serum. One should check with the original source of the particular cell line for any specific instructions. It is also wise to periodically test all cell lines for mycoplasma contamination. We have always avoided use of antibiotics for standard growth of cells to promote less mycoplasma contamination. One should also prepare a negative control dish of cells receiving either no plasmid DNA or plasmid that has no selectable marker. It is often useful to also prepare a positive control plasmid that will reflect successful gene transfer and selection such as with SV_2-Neo. Control dishes should be treated exactly like the experimental dishes.

2. The plasmid DNA should be precipitated as follows. The plasmid DNA (10 μg/dish of cells) is diluted in autoclaved water to a total volume equal to one-fourth the final volume. The final total volume is usually 1 ml per dish, at 10 μg of precipitated DNA per milliliter. The diluted DNA is added to the calcium chloride solution, also in a total volume equal to one-fourth the final volume. One should now have the DNA in the calcium chloride in one-half total final volume. Then add dropwise an equal volume of the $2\times$ HEPES-buffered phosphate solution to produce an increasingly cloudy solution that should remain undisturbed for 30–45 min at room temperature.

3. The precipitated DNA (1 ml) now should be added to each dish of cells. We have found that simply adding this mixture directly to the medium (8–10 ml) works well. The cells at this density should be relatively spread out, and the medium will turn a cloudy, cherry red. One can easily see the precipitated DNA settle down onto the cells, and later some will be seen intracellularly. This medium with the DNA is generally left on the cells for at least 3 hr; then the medium is removed, and fresh growth medium without selection but containing gentamicin at 50–100 mg/ml, is added back.

b. Method 2: Electroporation   In many ways the basic principles for electroporation gene transfer are the same as for the calcium phosphate precipitation method except that an electrical pulse rather than phagocytosis drives the DNA into the cells. We electroporate cells in HBS (20 mM HEPES, pH 7.05, 137 mM NaCl, 5 mM KCl, 0.7 mM Na_2HPO_4, 6 mM glucose) using 40 μg DNA (linearized) and 2×10^7 cells/ml in a total volume of 400 μl. We have the Bio-Rad (Richmond, CA) gene pulser system and use the 0.4 cm electrode gap cuvettes with a pulse of 200 V and 960 μF. Other conditions can be used successfully, and parameters for the electroporator, the ingoing DNA vector, the cell line, and the medium may be optimized in each case. Following electroporation, cells are carefully removed from the cuvette and put into a 100-mm dish. Depending on the cell line, the cells should remain in the cuvette for 15 min before transfer to a dish.

c. Selection   Approximately 48 hr after gene transfer, using either method, the medium containing the selection agent should be added. We used G418 (Geneticin, GIBCO, Grand Island, NY) at a concentration ranging from 100 to 500 μg/ mg active ingredient. The G418 is usually sold as a powder that is approximately 50% active by weight. If this is made up in water as a concentrate to be added to the medium, it will be acidic; it can be made in phosphate-buffered saline (PBS), filter-sterilized, and stored at 4°C. There is no need for gentamicin after adding the G418 because it is a potent antimicrobial agent itself. At this point, the cells must have plenty of room for dividing, because various drug selections, including G-418, work effectively only on dividing cells. Therefore, one must either start with a relatively sparse cell density as described above or passage the cells prior to selection. The latter is less preferred because, if one wants to eventually derive single cell clones, the passaging process will produce daughter cell derived clones that cannot be distinguished from one another. If 3T3 cells become too concentrated during this selection process, they will form dense cell sheets and eventually roll up from the culture dish surface into cell balls that are undigestible with trypsin.

One will eventually want to have single cell derived clones to determine titer and to use for experiments. The easiest way to obtain such clones is to start with a relatively sparse number of cells for gene transfer so they do not require passage before selection. The goal is to end up with a few resistant colonies per dish after about 2 weeks of selection so that one can ring-clone the colonies to subsequently put them into individual wells of 24-well plates for individual characterization. Another way to obtain single cell derived clones is to make a series of limiting dilutions. For a given dish of resistant cells, serial dilutions are put into wells of 24-well plate so that no cells grow at one dilution, but at a lower dilution cells are found growing in the well. We have always found this more cumbersome than ring cloning.

2. Titer Determination

One often needs to know the titer of the recombinant retrovirus to determine its usefulness. Low-titer viruses (10^2-10^5 cfu/ml) are generally useful for cell culture gene transfer experiments, but one needs a much higher titer ($>10^6$ cfu/ml) to do *in vivo* experiments such as bone marrow infection for transplantation. Titer determination is valid only for the cell line used in determining it and is only an estimate for other cells. We have also found differences as large as orders of magnitude even between different 3T3 cell lines used for titering as measured by infectability and resultant titer produced by the same recombinant virus stock (see Table 1).

A major advantage of having a selectable marker within the retrovirus is that it

TABLE 1
Relative Titer of the Same Virus on Different Cell Lines

Sources of 3T3 Cells	G418-Resistant colonies (cfu/ml)
American Type Culture Collection	4
Laboratory A	123
Laboratory B	Confluent
Laboratory C (ψ-AM)	Confluent

can be used to readily determine a titer of the recombinant retrovirus by making dilutions before infecting, selecting, and counting colonies to reflect the colony-forming units per milliliter of retrovirus-containing medium. Without a selectable marker, one must guesstimate indirectly the titer based on methods such as infecting cells then doing quantitative PCR. The given PCR signal is then compared with some quantified infected cell standard derived from a virus previously titered by using colony counts. Another way to estimate titers is by antibody detection of infected cells by counting positively stained cell colonies grown in unselected conditions. Both of these methods are generally somewhat difficult to translate into standard units (cfu/ml).

a. General Titer Determination Protocol for Cells Grown in a Dish with Selectable Marker Genes

1. Harvesting recombinant retroviruses: The packaging cell line (now the producer line) to be tested should be grown to a standard density in 100-mm dishes. Retroviruses are harvested by removing entirely the previous growth medium, then adding fresh medium into which the retroviruses are extruded (Cepko *et al.*, 1984). Retroviral titer will vary depending on the cell number and stage of confluency. We have found that growing the producer cell line to the stage about 1–2 days before total confluency gives the best titer. There should be a small amount of space between cells, and the cells should still look plump. Near confluency, the cells will be very densely seeded to cover the entire surface of the dish and will appear more contracted than before. Allowing the retrovirus harvest medium to remain with the producer line for approximately 18 hr is usually optimum. At this time, the most virus will be extruded into the medium from the producer cell line, but there will not be a significant loss of viability. One should test and maximize the conditions for each new cell line, using these conditions only as a guide.

2. Infecting the target cells to determine titer: The best cell line to use for determining titer is the one you want to use for the experiment, but many

times this is not feasible nor practical. For example, pluripotential bone marrow stem cells must be titered by transplantation experiments. Following retroviral harvest, it is usually best to directly infect the target cells without a freeze/thaw step or substantial holding at room temperature. We generally filter our retroviral harvest medium with a syringe type, sterile, 0.45-μm pore filter to remove any producer cells or other debris.

Next, remove the medium from the target cell line (e.g., a 3T3 line), seeded the previous day at approximately 5×10^5 in a 100-mm dish. The filtered retrovirus stock (1 ml) or dilutions of it should now be applied to the cells. The final stock virus or its dilution should contain Polybrene at a working concentration of 8 μg/ml. Polybrene enhances the infection of cells by retroviruses. By using different dilutions, one will produce countable dishes of resistant cell colonies, each containing numbers of colonies in proportion to its respective dilution. For titering purposes, we can usually count effectively dishes having from 10 to 100 colonies. An appropriate back calculation allows us to determine the cfu/ml of the starting stock virus. We generally allow the 1-ml infection dose to remain on the cells for 3 hr, then add 9 ml more of growth medium without Polybrene. The cells are grown for 48 hr without selection, at which time they usually must be passaged. We trypsinize the cells as usual for passage, but we reseed each new dish with 5×10^5 cells, exactly like those we infected, so that cell divisions occurring since retroviral gene transfer are appropriately reflected in the resulting cfu/ml estimation (Wood *et al.*, 1986a). It usually takes from 10 to 14 days of selection to allow for formation of distinct colonies for counting. We generally stain the colonies for counting by using 0.1% methylene blue (w/v) in 50% water/50% methanol (by volume). After removal of the medium from the dish containing the colonies, one should put 2 ml of stain on the dish. This is allowed to stand at room temperature for around 5 min, then the staining solution can be removed by suction or pouring off. The dish can be destained with a normal saline rinse, which is also poured off. One should be careful when the colonies have grown heavily because they may come off the dish surface if staining and destaining are done too vigorously. This procedure gives a good estimate of retroviral titer from a given producer cell line.

3. Rapid screening for high-titer recombinant retrovirus: The above titering procedure, based on colony selection and counting, can be very labor intensive and time consuming, and it can require managing many dishes of cells. Methods devised to first screen producer cell lines can be more efficient. In one example from our work (Wood *et al.*, 1986a,b), we took advantage of a unique property of the gene of interest, argininosuccinate synthase. Argininosuccinate synthase (AS) is an enzyme of the mammalian urea cycle that converts citrulline, using aspartic acid and ATP, to argininosuccinic acid,

which is subsequently converted to arginine by argininosuccinate lyase. In cultured cells this arginine is readily available to be incorporated into protein. The system we devised to screen rapidly for high-titer recombinant AS retroviruses is based on infecting AS-deficient cells (rat XC cells). We then assay for [^{14}C] citrulline incorporation into precipitable protein, normalized to the incorporation of [^{3}H] leucine, as AS-independent amino acid (Wood *et al.*, 1986a,b). This method greatly accelerates the process of detecting only the very highest titer producer cell lines for subsequent selective titering. Details of this assay are described elsewhere, but it is presented here as an example that could have application to other recombinant retroviruses. For example, hypoxanthine phosphoribosyltransferase (HPRT) could be easily adapted for this type of assay. We have now used the AS rapid screening assay for several different recombinant retroviral constructs including those with β-globin, various enhancers, and other inserts using AS as a marker gene.

E. Evaluating Recombinant Retrovirus Gene Transfer and Expression

1. Evaluating Infected Cells for Provirus Molecular Structure: Genomic DNA Evaluation

The most straightforward approach to analyzing infected cell lines is done on single cell derived clones, much like what was described earlier for cloning of the retroviral producer cell lines. Demonstrating proviral integration and defining its molecular structure would likely be done by Southern blot analysis. The details of Southern blot techniques can be readily reviewed elsewhere (Southern, 1975); our purpose here is to describe specific analyses pertinent for retroviral structure. Generally, high molecular weight genomic DNA is prepared from the cultured cells and digested with restriction enzymes, and the DNA fragments are separated by agarose gel electrophoresis for Southern blotting. Information determined from Southern blot analysis includes the copy number of the retroviral proviral sequence, its characteristics as far as intact or deleted provirus structure, and its number of integration sites. Usually the inserted gene(s) is used as the probe. Using retroviral sequences as probes allows the detection of endogenous retroviral sequences (Jenkins *et al.*, 1982) that can confuse examination of the recombinant proviral sequence. One desired restriction digestion would have cleavage sites only in the LTRs such as demonstrated in the *Xba*I digestion of Fig. 1. This full-length type digestion allows one to evaluate the overall size of the provirus, permitting examination for possible deletions or rearrangements. One can also estimate the copy number per cell based on analysis of noninfected cell line genomic DNA spiked with various dilutions of the retroviral plasmid. The resulting signal is detected by autoradiography and

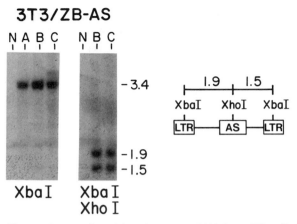

Figure 1. Southern blot of genomic DNA from 3T3 cell clones following infection with AS recombinant retrovirus. The AS cDNA was used for the probe. Letter designations refer to noninfected (N) and infected (A, B, C,) 3T3 clones.

evaluated by densitometry. In another useful digestion, we use a restriction enzyme that cleaves only once or not at all within the provirus. This allows the detection of the number of integration events represented. More than one integration is seen in nonclonally derived cells or where multiple infections occurred.

2. Evaluating Expression

a. Messenger RNA Evaluation It is often desirable to evaluate the transcripts produced from the recombinant proviral sequence. This is especially important when expression at either the packaging cell stage (i.e., titer) or the infected cell stage (i.e., protein concentration) is not as high as expected. For this evaluation, cells containing either the transfected plasmid or the infected recombinant provirus should be clonally derived. The cells to be evaluated should be grown in large cultures and the RNA isolated using standard techniques. We have had good results using the method of Chirgwin *et al.*, (1979) to produce high-quality RNA for Northern blots. Standard procedures for Northern blots (Wahl *et al.*, 1987) can be followed for evaluating the retroviral transcript size and abundance. One may find two different transcripts if splicing vectors are being used. In these cases the ratio between the full-length and the spliced message will be of interest. Also, the pattern seen in the producer cell lines may or may not be the same as that seen in the infected cell lines. In some cases an endogenous gene transcript may complicate a standard Northern analysis, and one may need to go to more specific analyses, such as RNA protection assays or primer extension analyses, in order to differentiate between endogenous mRNA and recombinant retroviral transcripts. These specific analyses can be devised using specific RNA probes or oligonucleotide primers such that one can readily compare the level of expression of the endogenous gene in contrast to that of the recombinant provirus.

b. Protein Evaluation There are usually no specific considerations for evaluating the actual gene product of interest produced from the recombinant retrovirus as compared to any other transgene. One can assay for the protein using Western or slot-blot analysis if an appropriate antibody is available. The final test is for the amount of functional activity, as measured by enzyme assay of the infected cells or by other functional assays, such as for activity of a produced hormone.

An example of recombinant retrovirus expression in clones of infected AS-deficient 3T3 cells is shown in Fig. 2. These results are from some of our own work (unpublished results of P. A. Wood, W. E. O'Brien, and A. L. Beaudet, 1987). Following cloning of 3T3 cells infected with an ecotropic ZNX-AS, ZB-AS recombinant retrovirus (Wood *et al.*, 1986b), and gene correction of human fibroblasts from patients with AS deficiency by amphotropic ZB AS retrovirus, AS activity was measured. In Fig. 2, the relative AS expression of different retroviral constructs and variation in their expression is compared to expression with AS activity of normal human fibroblasts and human liver. As shown in Fig. 2, the small ZB-AS retroviral vector provided for high expression in cultured 3T3 cells comparable to the activity found in human liver. We come back to this point later when we discuss infection and expression *in vivo* in bone marrow-derived cells.

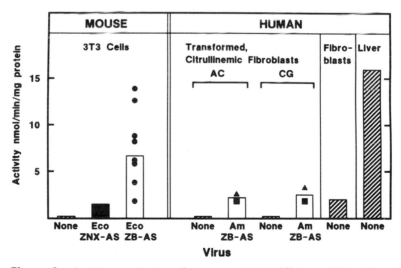

Figure 2. Argininosuccinate synthase expression in different cell lines infected with different recombinant AS retroviruses. The molecular structures are described in detail elsewhere (Wood *et al.*, 1986b). Using amphotropic ZB-AS, AS activity was restored to normal in human fibroblasts obtained from AS-deficient, citrullinemic patients.

IV. SPECIFIC APPLICATIONS FOR RETROVIRAL TRANSGENE EXPRESSION *IN VIVO*

A. *Retroviral Mediated Gene Transfer and Expression in Transplanted Bone Marrow*

One of the most commonly used and best developed systems for studying *in vivo* expression following retroviral mediated gene transfer is infection and subsequent transplantation of mouse bone marrow (Joyner and Bernstein, 1983; Williams *et al.*, 1984, 1986; Miller *et al.*, 1984; Dick *et al.*, 1985; Eglitis *et al.*, 1985; Keller *et al.*, 1985; Belmont *et al.*, 1986, 1988; Hock *et al.*, 1986; Hawley *et al.*, 1987; Magli *et al.*, 1987; McIvor *et al.*, 1987; Bowtell *et al.*, 1988; Dzierzak *et al.*, 1988; Karlsson *et al.*, 1988; Lim *et al.*, 1989; Wilson *et al.*, 1990). The basic goal of this system is successful infection of a pluripotential bone marrow stem cell that will repopulate the marrow of the transplanted animals and be expressed in the peripheral blood, in the bone marrow, and in any tissue with cells derived from bone marrow including liver Kupffer cells, microglia, and others. This strategy has met with various levels of success, and its true success has often been difficult to judge, especially in the context of gene therapy. An additional benefit of this approach is that retroviral integration also serves as a way to mark uniquely cell lineages based on the integration site. One can then follow cells throughout the differentiation pathways (Lemischka *et al.*, 1986).

As an example, I describe below our methods and experiences with developing retroviral mediated gene transfer of AS into bone marrow for transplantation and expression *in vivo* (Herman *et al.*, 1989). Important features of the system include successful cell culture, efficient retroviral infection, and successful bone marrow transplantation. I also describe effective ways to study the resulting gene expression and any resulting physiologic alterations.

1. Bone Marrow Culture, Infection, and Selection

a. Mice   The first important factor to consider when starting a series of experiments is the inbred strain, age, and infectious pathogen status of the mice. All of the experiments described here were done using either C57BL/6 mice from the colony of Dr. John Trentin (Baylor College of Medicine, Houston, TX), C57BL/6 NCr from the National Cancer Institute (NCI, Frederick, MD), or C57BL/6J from The Jackson Laboratory (Bar Harbor, ME). All three sublines have performed identically. We generally use mice that are between 8 and 16 weeks old for both bone marrow donors and recipients. We maintain the pathogen-free status of the mice using microisolator cages (Lab Products, Maywood, NJ), single mouse colony dedicated assistants, and routine health monitoring. The mice are essentially cared for like cultured cells; the cages are not opened except in a laminar flow hood, and the mice are not handled except with disinfected gloved hands or instruments. The mice must be free of common mouse infectious pathogens, such as mouse hepatitis virus, Sendai virus, mycoplasma, and other agents, in order for the mice to survive as planned from the transplantation. Experimental results are thus also not due to confounding, underlying, and many times inapparent infections.

Retroviral mediated gene transfer into bone marrow cells has been done in nonmurine species as well, including dogs (Kwok *et al.*, 1986; Eglitis *et al.*, 1988; Stead *et al.*, 1988) and nonhuman primates (Anderson *et al.*, 1986; Bodine *et al.*, 1990).

b. Bone Marrow Collection   Our general approach has been to collect aseptically the bone marrow cells from the femurs and tibias of humanely killed mice. We have not used 5-fluorouracil pretreatment of donor mice for any of our successful transplants. Harvest is done by dissecting the bones free of skin and muscle, cutting the ends of the bones off, and flushing out the marrow using a syringe and 27-gauge needle. We use the bone marrow culture medium as described in Table 2

TABLE 2
Complete Iscove's Modified Dulbecco's Medium (IMDM) for Bone Marrow Culture and Infection[a]

Component	Volume (ml) of component needed for a final volume of			
	13 ml	52 ml	65 ml	104 ml
Autoclaved, nanopure water	5.8	23.2	29	46.4
5 × IMDM (Gibco)	2.0	8.0	10	16.0
7.5% NaHCO$_3$ w/v	0.4	1.6	2.0	3.2
α-Thioglycerol (1/100)	0.0065	0.026	0.0325	0.052
Gentamicin	0.05	0.20	0.25	0.40
Bovine serum albumin solution 100 mg/ml	1.0	4.0	5.0	8.0
WEHI-conditioned medium[b]	1.0	4.0	5.0	8.0
5637-conditioned medium[c]	1.0	4.0	5.0	8.0
Fetal calf serum, heat inactivated	1.0	4.0	5.0	8.0
Soybean lipids 2 mg/ml	0.5	2.0	2.5	4.0
Human transferrin 30 mg/ml	0.1	0.4	0.5	0.8
Polybrene[d] (1 mg/ml) +/−	0.052	0.208	0.26	0.416

[a] Adapted from Belmont *et al.* (1988).
[b] WEHI-conditioned medium is prepared by coculture of WEHI3B(D−) cells (gift from J. Belmont) with plain 1× Iscove's medium as a source of interleukin-3 (IL-3).
[c] 5637-conditioned medium is prepared by coculture of cell line 5637 (ATCC) with plain Iscove's medium as a source of IL-1α and IL-6.
[d] You may want to remove a portion (e.g. ~20%) of the volume to remain Polybrene free for bone marrow harvest and resuspension/tail vein injection. Therefore, you will need to recalculate the amount to add based on 0.005 ml polybrene/ml of medium.

for the flushing, and we always maintain the bone marrow cells on ice between steps. We usually collect on average around $5-10 \times 10^7$ viable bone marrow cells per mouse. The cells are counted by standard trypan blue (0.4% w/v in PBS) exclusion staining on a standard hemocytometer. Counting is necessary to estimate the number of cells to put into the dishes for subsequent infection, selection, and culturing. The counting is done by mixing a sample of bone marrow cell suspension 1:1 with the trypan blue, then allowing the mixture to sit at room temperature for about 5 min. A drop of remixed suspension is then removed with a pipette and loaded onto the hemocytometer. We count the cells with an inverted cell culture microscope without any colorizing lenses by counting only the medium to large cells present having a pink appearance on focusing up and down. The major trypan-excluding cells to avoid counting are the red cells, which are noticeably smaller and much more numerous than the bone marrow cells. One should note also the relative percentage of blue-staining (dead) cells. The cells in the collection tube are now ready for distribution to the culture dishes.

   c. Bone Marrow Culture and Infection   We have found that coculture infection of the bone marrow cells for a few days before transplantation, with or without culture selection, always improves our efficiency of infection. Our general protocol is to seed approximately 5×10^5 recombinant retroviral producer cells per 100-mm dish on the day prior to bone marrow cell addition. The dishes will probably look very sparse to the uninitiated, but our goal is to have a rapidly growing feeder cell layer of producer cells turning out retroviruses for infection over a total of 4 days. The bone marrow cells are then added to the producer cultures containing 12 ml of the IMDM culture medium with Polybrene described in Table 2. The bone marrow cells will settle down onto the producer cells, and some will attach. We usually leave the cells to be infected like this until the following day. We then either let them remain in culture for a total of 4 days or add in selection medium such as G418 for an additional 2 days before transplantation. We grow all of our cells including the bone marrow cultures in 10% (v/v) CO_2.

2. Bone Marrow Cell Transplantation

   Following the culture period, the unattached bone marrow cells are harvested from the dishes and recounted. This is done by gently swirling the dish and removing the suspended cells with a pipette. One can gently back flush the producer cell surface and remove a few more cells, but it is not desirable to remove the producer cells. The bone marrow cells are collected and pooled as desired and recounted by trypan blue exclusion. If selection has been done, there will be a noticeable difference between the number of viable bone marrow cells and that of the nonselected ones. The cells will need to be centrifuged to concentrate them for transplantation. The cell density needed allows the total cell dose injected to be contained in 0.5 ml.

If the goal of the transplantation experiment is to study spleen colonies, one will desire a subsurvival dose in order to get about 5–10 colonies per spleen. We have found 5×10^5 viable cells per mouse to be optimal. For survival dosages we use over 5×10^6 per mouse.

The transplantation is done on lethally irradiated recipient mice so that the transplanted cells are given a favorable advantage to populate the bone marrow spaces. We inject all of the bone marrow cells by intravenous injection of the tail vein. This is best accomplished by carefully warming the mice under a heating lamp, putting them in a mouse restrainer, and injecting the cells using a tuberculin syringe and a 27-gauge/0.5-in. needle. The warming must be carefully monitored with a thermometer or timer such that the mice do not overheat. The warming allows for dilation of the tail vein for injection. After a little practice one can easily inject several mice in a few minutes with minimal restraint and stress. The counted cells must be resuspended each time when loading the syringe, and the syringe should be mixed just prior to injection.

Lethal irradiation doses vary with the strain of mouse. For example, we have used 1000 rads for C57BL/6 mice with good marrow engraftment without major radiation sickness such as intestinal damage. However, BALB/cByJ mice will tolerate only 750 rads under the same conditions; they still have occasional intestinal damage but will survive with the appropriate bone marrow transplant. Mice receiving a subsurvival bone marrow dosage usually die at 8–14 days post-transplant, depending on the actual dosage given. The usual cause of death in pathogen-free mice is hemorrhage into the gastrointestinal tract or brain. These mice appear to be depleted of functional platelets. Usually, if mice are infected during the transplant process, they will be found dead within 2–4 days, depending on the infectious agent and its dose.

We have found that examining and weighing the spleens can provide a useful early measure of transplantation success. A normal spleen in mice used as recipients but receiving no treatment will weigh 20–30 mg. Spleens also weigh about that with subsurvival bone marrow dosages; variation depends on the size and number of spleen colonies present. A few days to 14 days following lethal irradiation, mice will have a spleen that weighs only 10–15 mg. Spleens of mice given survival dosages of bone marrow cells will weigh 100–300 mg at 12–14 days post-transplant and will not have discernable spleen colonies. Spleen colony formation is an artifact of bone marrow transplantation and is useful in this context to examine single bone marrow cell infection events. The spleen colonies can be dissected free of other splenic tissue and examined for proviral DNA structure and expression.

3. Evaluating Expression

The basic principles outlined for evaluating expression in cultured cells above also apply to evaluating the transplanted mouse for recombinant retroviral integration and expression. Retroviral molecular structure, transcription, and protein expression and function can be examined in the context now of the intact animal. The first level of study is usually at the spleen colony, such that Southern blot analysis of integrated proviral sequences indicates the efficiency of infection during the bone marrow culturing procedures. In other words, if the recombinant retrovirus was of sufficient titer and infected the bone marrow cells well, then virtually every spleen colony examined will contain a provirus. With low-titer viruses, however, the percentage may be low (10%), or no sequences may be found. An example of the effects of G418 selection of bone marrow infected with a *neo*-containing virus as compared to those without G418 selection is demonstrated in Fig. 3. The use as a positive control of a well-established recombinant retrovirus that has worked well *in vivo* is invaluable to make sure the system is working properly, especially when employing a new, untried vector. We have used both MuLV NEO.1 and the MoTN vector (a gift from Alan Bernstein, University of Toronto; Bernstein *et al.*, 1986) for this purpose in ψ-2 cells. The *neo* vector control not only serves as a positive control for the infection and transplantation steps but also acts as a sham infection for a negative control for expression of other recombinant genes of interest. We

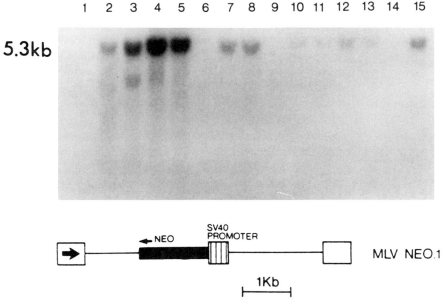

Figure 3. Effect on bone marrow cells of infection with a recombinant *neo*-containing retrovirus and either selection for 2 days with G418 (lanes 2–5) or no selection (lanes 6–15) prior to subsequent transplantation. Lane 1 is a spleen colony DNA from a noninfected transplanted mouse. There is a clear distinction by signal intensity in that the selection increases the overall copy number or promotes for retention of integrated single provirus copies in a higher percentage of the cells making up the spleen colonies. A *neo* probe was used following restriction enzyme digestion with *Xba*I.

TABLE 3
Summary of Mouse Blood Argininosuccinate Synthase Activity

Assay conditions (number of mice assayed)	Blood AS level (nmol/min/mg hemoglobin)
ZB-AS vector[a] (n = 14)	7.1 (10)
No ZB-AS vector[a] (n = 14)	3.2 (0.70)
Delta N2AS vector[a]	
No selection (n = 15)	26 (32)
Maximum selection[b] (n = 10)	18 (31)
No delta N2A5 vector[a] (n = 11)	2.1 (0.81)
Short term (nonreconstituting dose)[c]	
N2AS-B vector	21, 21, 12, 9.9
Nontransplanted controls	2.0, 2.6
Long term (reconstituting dose)[c]	
Nontransplanted controls	
9 weeks	3.8
25 weeks	1.4, 1.5, 1.2
55 weeks	1.2, 1.4
Sham-infected MoTN	
9 weeks	3.1
25 weeks	2.5, 2.6
N2AS-B vector	
9 weeks	32, 17, 22, 130
25 weeks	4.5, 2.8, 16, 100
55 weeks	—, —, —, 26

[a] Unpublished data [$\bar{x}$ (S.D.)] of P. A. Wood, G. E. Herman, C.-Y. Chao, J. J. Trentin, W. E. O'Brien, and A. L. Beaudet (1988).

[b] Bone marrow cells were grown in culture medium deficient in arginine, containing citrulline 400 μM and arginine analog-canavanine 100 μM. We hypothesized that this would select for AS positive cells, but there is no obvious effect.

[c] Unpublished data of P. A. Wood, N. Martin, S. McCune, and T. M. Townes (1990).

have found the most informative expression results by measuring recombinant enzyme (AS) activity in the circulating whole blood (see Table 3).

The premise of our work was to develop an *in vivo* system such that recombinant AS activity could be "artificially" expressed in circulating blood cells owing to expression from the LTR promoter. In human patients with the AS deficiency disease known as citrullinemia, we hypothesized that AS expression anywhere in the body, including circulating blood cells, would allow completion of the deficient step in urea synthesis outside the liver. This is because the substrates and products of AS are diffusible, and the product could readily return to the liver for completion of the urea cycle. AS is a cytosolic enzyme found normally in highest amount in liver and kidney. The data presented in Table 3 illustrate several examples of the problems and successes of infection and expression *in vivo* after initial cell culture evaluation.

The data in Table 3 demonstrate clearly that we have been able to obtain both expression *in vivo* of human AS in peripheral blood of mice with retroviral vectors and significantly improved expression with the delta N2-AS virus. One should note that the high-titer, highly expressed construct ZB-AS (Figs. 1 and 2) was not very effective for *in vivo* experiments. This demonstrates the requirement of evaluating the vector and packaging cell line in the ultimate recipient cell. Southern blot analysis has shown an infection efficiency of approximately 20–90% in spleen colonies using the various retroviral constructs. We attempted *in vitro* metabolic selection for AS expression prior to transplantation. This selection included culturing cells (postinfection) in medium deficient in arginine (Arg) but substituted with citrulline (Cit); there was no apparent selection on bone marrow cells whether they were infected with AS virus. A more stringent selection involves similarly using − Arg/ + Cit medium but adding 100 μM canavanine, an arginine analog. The recent experiments at the University of Alabama at Birmingham were done as before with the same type construct as the original delta N2AS but using a reconstructed form designated N2AS-B. Drawbacks to the AS retroviral experiments in mice include having to work around the endogenous mouse AS and not really knowing how much recombinant AS gene expression is therapeutically significant.

4. Perspectives of Expression

Because much of the work done has been ultimately directed at the possible gene therapy correction of human genetic diseases, one must evaluate the expression *in vivo* in this context. Let us look at AS deficiency and consider AS activity needed to be therapeutic for this inborn error of metabolism. As shown in Table 3, we have obtained *in vivo* the highest AS activity in blood at a level 60 times the low background activity in the mouse. Activity was readily detectable more than 1 year posttransplant at 10 times background activity.

How would this activity translate into any possible therapeutic benefit for a patient? If one calculates AS activity based on blood volume and enzyme activity at V_{max}, this mouse would have a total body AS activity potential of 39 nmol/min/ blood volume as compared to the AS activity contributed by a normal mouse liver of 3300 nmol/min/liver. Therefore, the recombinant AS activity contributed by the blood is 1.2% of that contributed by a normal mouse liver. Since we have speculated that one must provide at least 10% of normal body AS activity to correct an AS deficiency in a patient, is there any benefit to recombinant blood AS at this low level? The only true way to test this would be to evaluate it in an animal model or patient with AS deficiency; however, estimates can be made based on pharmacological doses of compounds such as phenylacetate (Brusilow and Horwich, 1989) that promote for urinary excretion of 2 mol of nitrogen that normally would be converted to a urea. The standard dosage provides the loss of 8 mmol/kg/day, and, at the level of AS activity provided in the mouse blood in the example, this would calculate out to be 2.25 mmol/kg/day. With this activity we are thus approaching the same benefit provided by the pharmacological dose of phenylacetate. There are several examples of expression of the recombinant gene product in blood after these gene therapy type experiments, but one must carefully evaluate and put into perspective what the expression really means.

B. Hepatocyte Retroviral Gene Transfer and Transplantation

There has been wide interest in developing strategies for retrovirus-mediated gene transfer into hepatocytes for subsequent transplantation as with bone marrow. This system has encountered a much less satisfactory success. The major problems are that hepatocytes *in vivo* are relatively nonmitotic, do not provide a good target cell to infect, and have stable retroviral gene integration. Hepatocytes have been infected with retroviruses (Wilson *et al.*, 1988a,b; Dichek *et al.*, 1991). Hepatocytes are, however, very resistant to culture and cannot be passaged. This makes hepatocyte harvest and culture with infection as done in the bone marrow system impractical.

Many variations have been tried to provide for *in vivo* growth following culture manipulations such as retroviral mediated gene transfer. Early attempts have included transplanting liver punch biopsies into liver parenchyma (Mukherjee and Krasner, 1973), intraportal transplantation of hepatocytes by injection into the portal vein (Matas *et al.*, 1975; Groth *et al.*, 1977), injection of hepatocytes into the peritoneal cavity or spleen (Minato *et al.*, 1984), injection of hepatocytes growing on a bead matrix into the peritoneal cavity (Demetriou *et al.*, 1986), and transplantation of other artificial matrices containing hepatocytes subjected to previous retroviral gene transfer (Anderson *et al.*, 1989). None of the systems has produced the desired goal of long-term growth of transferred hepatocytes. The two most successful approaches appear to be isolation of hepatocytes and transplantation into a vascularized organoid structure (Thompson *et al.*, 1988, 1989) and injection of hepatocytes into the spleen (Ponder *et al.*, 1991) such that the cells migrate to the liver through the portal circulation. The spleen injection of hepatocytes was demonstrated by β-galactosidase transgene (nonretroviral) expression. In addition, modified retroviruses have been used to target specific infection of hepatocytes in the intact animal (Hatzoglou *et al.*, 1990). Basically, hepatocyte transplantation or retrovirus-mediated gene transfer into hepatocytes with any reasonable efficiency has been very difficult to accomplish.

C. Infection and Transplantation of Other Cells

A wide variety of cells has been used for experiments to investigate retroviral gene transfer and *in vivo* expression following transplantation. Experiments have been done in which skin fibroblasts (Garver *et al.*, 1987; Selden *et al.*, 1987; St. Louis and Verma, 1988; Palmer *et al.*, 1989) have been infected, selected, and transplanted at various locations in animals. These transplants have been successful as long-term sources of cells expressing some desirable gene product, but a common problem has been developing an immune response to either the cells or the recombinant gene product. Various cell lines have been used as vehicles of retroviral mediated gene transfer for the possible correction of human disease. Endothelial cells have been infected in culture with recombinant retroviruses and subsequently transplanted onto denuded iliofemoral arteries (Nabel *et al.*, 1989; Wilson *et al.*, 1989); alternatively, direct gene transfer into endothelial cells in the arterial wall has been done (Nabel *et al.*, 1990). Other cells infected for transplantation include human epidermal cells (Morgan *et al.*, 1987).

D. Mouse Embryo Infection

There has been an interest in using recombinant retrovirus-mediated gene transfer into mouse embryos as a means of producing transgenic mice (Huszar *et al.*, 1985; Jahner *et al.*, 1985; van der Putten *et al.*, 1985). Advantages include transfer of a single randomly integrated gene copy with no need for microinjection procedures to develop potential gene line mosaics for insertion mutations of interest (Soriano and Jaenisch, 1986). Insertion mutations would be tagged by the retrovirus and could be potentially "fished" out and examined. The limitations are that constructs driven by LTRs are methylated and inactive as expressed transgenes, but retroviral vectors with internal promoters are expressed in a tissue-specific manner (Soriano *et al.*, 1986). Also, single insertion mutations of interest are few and far between; thus, it is not highly efficient without some selection scheme to get the desired mutant gene disruption.

Infection and production of mosaics allow one to follow cell lineages in the embryo during development (Soriano and Jaenisch, 1986) since the cell lineage is identified by a unique retroviral integration. These infections are generally done by coculturing the embryos on feeder layers of retroviral packaging cells as described for bone marrow infection by coculture. The embryos are usually cocultured at the 3 day postcoitus stage (Huszar *et al.*, 1985; Jahner *et al.*, 1985; van der Putten *et al.,* 1985) and are reimplanted into pseudopregnant foster mother mice. The transgenic mice are identified by the standard DNA detection methods described above, including slot blots, Southern blots, or PCR analyses of tail-tip genomic DNA.

V. SUMMARY

In this chapter we have described a wide range of applications for retroviral mediated gene transfer. This field has developed an enormous literature, and our goal here was to present some practical highlights of considerations and methods that we have found useful in our work. We have taken advantage of the laboratory manual format to present some of our protocols in a step-by-step guide.

ACKNOWLEDGMENTS

Much of our work described here was done at the Baylor College of Medicine in collaboration with Drs. Arthur L. Beaudet, William E. O'Brien, Gail E. Herman, John J. Trentin, and James Chao and was supported by National Institutes of Health Grant PO1-HD21452. The more recent work done at the University of Alabama at Birmingham was done in collaboration with Dr. Tim M. Townes, Ms. Nancy Martin, Mr. Steve McCune, and Mr. Doug Hamm and was supported in part by NIH Grant RR-02599. I thank Dr. Susan Farmer for expert editorial critique of the manuscript.

REFERENCES

Adam, M. A., and Miller, A. D. (1988). Identification of a signal in a murine retrovirus that is sufficient for packaging of nonretroviral RNA into virions. *J. Virol.* **62**, 3802–3806.

Anderson, K. D., Thompson, J. A., DiPietro, J. M., Montgomery, K. T., Reid, L. M., and Anderson W. F. (1989). Gene expression in implanted rat hepatocytes following retroviral-mediated gene transfer. *Somatic Cell Mol. Genet.* **15**, 215–227.

Anderson, W. F., Kantoff, P., Eglitis, M., McLachlin, J., Karson, E., Zwiebel, J., Nienhuis, A., Karlsson, S., Blaese, R. M., Kohn, D., Gilboa, E., Armentano, D., Zanjani, E. D., Flake, A., Harrison, M. R., Gillio, A., Bordignon, C., and O'Reilly, R. (1986). Gene transfer and expression in nonhuman primates using retroviral vectors. *Cold Spring Harbor Symp. Quant. Biol.* **51**, 1073–1081.

Armentano, D., Yu, S.-F., Kanoff, P. W., von Ruden, T., Anderson, W. F., and Gilboa, E. (1987). Effect of internal viral sequences on the utility of retroviral vectors. *J. Virol.* **61**, 1647–1650.

Belmont, J. W., Henkle-Tigges, J., Chang, S. M. W., Wagner-Smith, K., Kellems, R. E., Dick, J. E., Magli, M. C., Phillips, R. A., Bernstein, A., and Caskey, C. T. (1986). Expression of human adenosine deaminase in murine haematopoietic progenitor cells following retroviral transfer. *Nature (London)* **322**, 385–387.

Belmont, J. W., MacGregor, G. R., Wagner-Smith, K., Fletcher, F. A., Moore, K. A., Hawkins, D.,

Villalon, D., Chang, S. M. W., and Caskey, C. T. (1988). Expression of human adenosine deaminase in murine hematopoietic cells. *Mol. Cell. Biol.* **8**, 5116–5125.

Bender, M. A., Palmer, T. D., Gelinas, R. E., and Miller, A. D. (1987). Evidence that the packaging signal of Moloney murine leukemia virus extends into the *gag* region. *J. Virol.* **61**, 1639–1646.

Bernstein, A., Dick, J. E., Huszar, D., Robson, I., Rossant, J., Magli, C., Estrov, Z., Freedman, M., and Phillips, R. A. (1986). Genetic engineering of mouse and human stem cells. *Cold Spring Harbor Symp. Quant. Biol.* **51**, 1083–1091.

Bodine, D. M., McDonagh, K. T., Brandt, S. J., Ney, P. A., Agricola, B., Byrne, E., and Neinhuis, A. W. (1990). Development of a high-titer retrovirus producer cell line capable of gene transfer into rhesus monkey hematopoietic stem cells. *Proc. Natl. Acad. Sci. U.S.A.* **87**, 3738–3742.

Bowtell, D. D. L., Cory, S., Johnson, G. R., and Gonda, T. J. (1988). Comparison of expression in hematopoietic cells by retroviral vectors carrying two genes. *J. Virol.* **62**, 2464–2473.

Brusilow, S. W., and Horwich, A. L. (1989). Urea cycle enzymes, *In* "The Metabolic Basis of Inherited Disease" (C. R. Scriver, A. L. Beaudet, W. S. Sly, and D. Valle, eds.), pp. 629–663. McGraw-Hill, New York.

Cepko, C. L., Roberts, B. E., and Mulligan, R. C. (1984). Construction and applications of a highly transmissible murine retrovirus shuttle vector. *Cell (Cambridge, Mass.)* **37**, 1053–1062.

Chirgwin, J. M., Przybyla, A. E., MacDonald, R. J., and Rutter, W. J. (1979). Isolation of biologically active ribonucleic acid from sources enriched in ribonuclease. *Biochemistry* **18**, 5294–5299.

Cone, R. G., and Mulligan, R. C. (1984). High-efficiency gene transfer into mammalian cells: Generation of helper-free recombinant retrovirus with broad mammalian host range. *Proc. Natl. Acad. Sci. U.S.A.* **81**, 6349–6353.

Danos, O., and Mulligan, R. C. (1988). Safe and efficient generation of recombinant retroviruses with amphotropic and ecotropic host ranges. *Proc. Natl. Acad. Sci. U.S.A.* **85**, 6460–6464.

Demetriou, A. A., Levenson S. M., Novikoff, P. M., Novikoff, A. B., Roy Chowdhury, N., Whiting, J., Reisner, A., and Roy Chowdhury, J. (1986). Survival, organization, and function of microcarrier-attached hepatocytes transplanted in rats. *Proc. Natl. Acad. Sci. U.S.A.* **83**, 7475–7479.

Dichek, D. A., Bratthauer, G. L., Beg, Z. H., Anderson, K. D., Newman, K. D., Zwiebel, J. A., Hoeg, J. M., and Anderson, W. F. (1991). Retroviral vector-mediated *in vivo* expression of low-density-lipoprotein receptors in the Watanabe heritable hyperlipidemic rabbit. *Somatic Cell Mol. Genet.* **17**, 287–301.

Dick, J. E., Magli, M.-C., Huszar D., Phillips, R. A., and Bernstein, A. (1985). Introduction of a selectable gene into primitive stem cells capable of long-term reconstitution of the hematopoietic system of w/wᵛ mice. *Cell (Cambridge, Mass.)* **42**, 71–79.

Dzierzak, E. A., Papayannopoulou, T., and Mulligan, R. C. (1988). Lineage-specific expression of a human β-globin gene in murine bone marrow transplant recipients reconstituted with retrovirus-transduced stem cells. *Nature (London)* **331**, 35–41.

Eglitis, M. A., Kantoff, P. W., Gilboa, E., and Anderson, W. F. (1985). Gene expression in mice after high efficiency retroviral-mediated gene transfer. *Science* **230**, 1395–1398.

Eglitis, M. A., Kantoff, P. W., Jolly, J. D., Jones, J. B., Anderson, W. F., and Lothrop, C. D. (1988). Gene transfer into hematopoietic progenitor cells from normal and cyclic hematopoietic dogs using retroviral vectors. *Blood* **71**, 717–722.

Emerman, M., and Temin, H. M. (1984). High frequency deletion in recovered retrovirus vectors containing exogenous DNA with promoters. *J. Virol.* **50**, 42–49.

Felgner, P. L., Gadek, T. R., Holm, M., Roman, R., Chan, H. W., Wenz, M., Northrop, J. P., Ringold, G. M., and Danielson, M. (1987). Lipofection: A highly efficient, lipid-mediated DNA-transfection. *Proc. Natl. Acad. Sci. U.S.A.* **84**, 7413.

Fischinger, P. J., Nomura, S., and Bolognesi, D. P. (1975). A novel murine oncornavirus with dual eco- and xenotropic properties. *Proc. Natl. Acad. Sci. (U.S.A.)* **72**, 5150–5155.

Garver, R. I., Chytil, A., Courtney, M., and Crystal, R. G. (1987). Clonal gene therapy: Transplanted mouse fibroblast clones express human α1-antitrypsin gene *in vivo*. *Science* **237**, 762–764.

Goff, S., Traktman, P., and Baltimore, D. (1981). Isolation and properties of Moloney murine leukemia virus mutants: Use of a rapid assay for release of virion reverse transcriptase. *J. Virol.* **38**, 239–248.

Graham, F. L., and Van der Eb, A. J. (1973). A new technique for the assay of infectivity of human adenovirus 5 DNA. *Virology* **52**, 456–467.

Groth, C. G., Arborgh, B., Bjorken, C., Sundberg, B., and Lundgren, G. (1977). Correction of hyper-bilirubinemia in the glucuronyltransferase-deficient rat by intraportal hepatocyte transplantation. *Transplant. Proc.* **9**, 313–316.

Hantzopoulos, P. A., Sullenger, B. A., Ungers, G., and Gilboa, E. (1989). Improved gene expression upon transfer of the adenosine deaminase minigene outside the transcriptional unit of a retroviral vector. *Proc. Natl. Acad. Sci. U.S.A.* **86**, 3519–3523.

Hartley, J. W., and Rowe, W. P. (1976). Naturally occurring murine leukemia viruses in wild mice: Characterization of a new "amphotropic" class. *J. Virol.* **19**, 19–25.

Hartzoglou, M., Lamers, W., Bosch, F., Wynshaw-Boris, A., Clapp, D. W., and Hanson, R. W. (1990). Hepatic gene transfer in animals using retroviruses containing the promoter from the gene for phosphoenolpyruvate carboxykinase. *J. Biol. Chem.* **265**, 17285–17293.

Hawley, R. G., Covarrubias, L., Hawley, T., Mintz, B. (1987), Handicapped retroviral vectors efficiently transduce foreign genes into hematopoietic stem cells. *Proc. Natl. Acad. Sci. (U.S.A.)* **84**, 2406–2410.

Herman, G. E., Jaskoski, B., Wood, P. A., Trentin, J. J., O'Brien, W. E., and Beaudet, A. L. (1989). Expression of human argininosuccinate synthetase in murine hematopoietic cells *in vivo*. *Somatic Cell Mol. Genet.* **15**, 289–296.

Hock, R. A., and Miller, A. D. (1986). Retrovirus-mediated transfer and expression of drug resistance genes in human haematopoietic progenitor cells. *Nature (London)* **320**, 275–277.

Huszar, D., Balling, R., Kothary, R., Magli, M. C., Hozumi, N., Rossant, J., and Bernstein, A. (1985). Insertion of a bacterial gene into the mouse germ line using an infectious retrovirus vector. *Proc. Natl. Acad. Sci. U.S.A.* **82**, 8587–8591.

Jahner, D., Haase, K., Mulligan, R. C., Jaenisch, R. (1985). Insertion of the bacterial gpt gene into the germ line of mice by retroviral infection. *Proc. Natl. Acad Sci. (U.S.A.)* **82**, 6927–6931.

Jenkins, N. A., Copeland, N. G., Taylor, B. A., and Lee, B. K. (1982). Organization, distribution, and stability of endogenous ecotropic murine leukemia virus DNA sequences in chromosomes of *Mus musculus*. *J. Virol.* **43**, 26–36.

Joyner, A. L., and Bernstein, A. (1983). Retrovirus transduction: Generation of infectious retroviruses expressing dominant and selectable genes is associated with *in vivo* recombination and deletion events. *Mol. Cell. Biol.* **3**, 2180–2190.

Karlsson, S., Bodine, D. M., Perry, L., Papayannopoulou, T., and Nienhuis, A. W. (1988). Expression of the human β-globin gene following retroviral-mediated transfer into multipotential hematopoietic progenitors of mice. *Proc. Natl. Acad. Sci. U.S.A.* **85**, 6062–6066.

Keller, G., Paige, C., Gilboa, E., and Wagner, E. F. (1985). Expression of a foreign gene in myeloid and lymphoid cells derived from multipotent haematopoietic precursors. *Nature (London)* **318**, 149–154.

Kwok, W. W., Schuening, F., Stead, R. B., and Miller, A. D. (1986). Retroviral transfer of genes into canine hematopoietic progenitor cells in culture. A model for human gene therapy. *Proc. Natl. Acad. Sci. U.S.A.* **83**, 4552–4555.

Ledley, F. D., Darlington, G. J., Hahn, T., and Woo, S. L. C. (1987). Retroviral gene transfer into primary hepatocytes: Implications for genetic therapy of liver-specific functions. *Proc. Natl. Acad. Sci. U.S.A.* **84**, 5335–5339.

Lemischka, I. R., Raulet, D. H., and Mulligan, R. C. (1986). Developmental potential and dynamic behavior of hematopoietic stem cells. *Cell (Cambridge, Mass.)* **45**, 917–927.

Lim, B., Apperley, J. F., Orkin, S. H., and Williams, D. A. Long-term expression of human adenosine deaminase in mice transplanted with retrovirus-infected hematopoietic stem cells. *Proc. Natl. Acad. Sci. (U.S.A.)* **86**, 8892–8896.

Lynch, C. M., and Miller, A. D. (1991). Production of high-titer helper virus-free retroviral vectors by cocultivation of packaging cells with different host ranges. *J. Virol.* **65**, 3887–3890.

McIvor, R. S., Johnson, M. J., Miller, A. D., Pitts, S., Williams, S. R., Valerio, D., Martin, D. W., and Verma, I. M. (1987). Human purine nucleoside phosphorylase and adenosine deaminase: Gene transfer into cultured cells and murine hematopoietic stem cells by using recombinant amphotropic retroviruses. *Mol. Cell. Biol.* **7**, 838–846.

Magli, M.-C., Dick, J. E., Huszar, D., Bernstein, A., and Phillips, R. A. (1987). Modulation of gene expression in multiple hematopoietic cell lineages following retroviral vector gene transfer. *Proc. Natl. Acad. Sci. U.S.A.* **84**, 789–793.

Mann, R., Mulligan, R. C., and Baltimore, D. (1983). Construction of a retrovirus packaging mutant and it use to produce helper-free defective retrovirus. *Cell (Cambridge, Mass.)* **33**, 153–159.

Markowitz, D., Goff, S., and Bank, A. (1988a). A safe packaging line for gene transfer: Separating viral genes on two different plasmids. *J. Virol.* **62**, 1120–1124.

Markowitz, D., Goff, S., and Bank, A. (1988b). Construction and use of a safe and efficient amphotropic packaging cell line. *Virology* **167**, 400–406.

Matas, A. J., Sutherland, D. E. R., Mauer, S. M., Lowe, A., Simmons, R. L., and Najarian, J. S. (1976). Hepatocellular transplantation for metabolic deficiencies: Decrease of plasma bilirubin in Gunn rats. *Science* **192**, 892–894.

Miller, A. D., and Buttimore, C. (1986). Redesign of retrovirus packaging cell lines to avoid recombination leading to helper virus production. *Mol. Cell Biol.* **6**, 2895–2902.

Miller, A. D., Jolly, D. J., Friedmann, T., and Verma, I. M. (1983). A transmissible retrovirus expressing human hypoxanthine phosphoribosyltransferase (HPRT): Gene transfer into cells obtained from humans deficient in HPRT. *Proc. Natl. Acad. Sci. U.S.A.* **80**, 4709–4713.

Miller, A. D., Ong, E. S., Rosenfeld, M. G., Verma, I. M., and Evans, R. M. (1984). Infectious and selectable retrovirus containing an inducible rat growth hormone minigene. *Science* **225**, 993–998.

Miller, A. D., Law, M.-F., and Verma, I. M. (1985). Generation of helper-free amphotropic retroviruses that transduce a dominant-acting, methotrexate-resistant dihydrofolate reductase gene. *Mol. Cell Biol.* **5**, 431–437.

Minato, M., Houssin, D., Demma, I., Morin, J., Gigou, M., Szekely, A. M., and Bismuth, H. (1984). Transplantation of hepatocytes for treatment of surgically induced acute hepatic failure in the rat. *Eur. Surg. Res.* **16**, 162–169.

Moore, K. A., Fletcher, F. A., and Alford, R. L., Villalon, D. K., Hawkins, D. H., MacGregor, G. R., Caskey, C. T., and Belmont, J. W. (1989). Expression vectors for human adenosine deaminase gene therapy. *Genome* **31**, 832–839.

Morgan, J. R., Barrandon, Y., Green, H., and Mulligan, R. C. (1987). Expression of an exogenous growth hormone gene by transplantable epidermal cells. *Science* **237**, 1476–1479.

Mukherjee, A. B., and Krasner, J. (1973). Induction of an enzyme in genetically deficient rats after grafting normal liver. *Science* **183**, 68–69.

Mulligan, R. C. (1983). Construction of highly transmissible mammalian cloning vehicles derived from murine retroviruses. *In* "Experimental Manipulation of Gene Expression" (M. Inouye, ed.), pp. 155–173. Academic Press, Orlando, Florida.

Nabel, E. G., Plautz, G., Boyce, F. M., Stanley, J. C., and Nabel, G. J. (1989). Recombinant gene expression *in vivo* within endothelial cells of the arterial wall. *Science* **244**, 1342–1344.

Nabel, E. G., Plautz, G., and Nabel, G. J. (1990). Site-specific gene expression *in vivo* by direct gene transfer into the arterial wall. *Science* **249**, 1285–1286.

Palmer, T. D., Thompson, A. R., and Miller, A. D. (1989). Production of human factor IX in animals by genetically modified skin fibroblasts: Potential therapy for hemophilia B. *Blood* **73**, 438–445.

Pathak, V., and Temin, H. M. (1990). Broad spectrum of *in vivo* forward mutations, hypermutations, and mutational hotspots in a retroviral shuttle vector after a single replication cycle. Deletions and deletions with insertions. *Proc. Natl. Acad. Sci. (U.S.A.)* **87**, 6024–6028.

Peng, H., Armentano, D., MacKenzie-Graham, L., Shen, R.-F., Darlington, G., Ledley, F. D., and Woo, S. L. C. (1988). Retroviral-mediated gene transfer and expression of human phenylalanine hydroxylase in primary hepatocytes. *Proc. Natl. Acad. Sci. U.S.A.* **85**, 8146–8150.

Ponder, K. P., Gupta, S., Leland, F., Darlington, G., Finegold, M., Demayo, J., Ledley, F. D., Roy Chowdhury, J., and Woo, S. L. C. (1991). Mouse hepatocytes migrate to liver parenchyma and function indefinitely after intrasplenic transplantation. *Proc. Natl. Acad. Sci. U.S.A.* **88**, 1217–1221.

Rasheed, S., Gardner, M. B., and Chan, E. (1976). Amphotropic host range of naturally occurring wild mouse leukemia viruses. *J. Virol.* **19**, 13–18.

Rowe, W. P., Pugh, W. E., and Hartley, J. W. (1970). Plaque assay techniques for murine leukemia viruses. *Virology* **42**, 1136–1139.

Scarpa, M., Cournoyer, D., Munzy, D. M., Moore, K. A., Belmont, J. W., and Caskey, C. T. (1991). Characterization of recombinant helper retroviruses from Moloney-based vectors in ecotropic and amphotropic packaging cell lines. *Virology* **180**, 849–852.

Selden, R. F., Skoskiewicz, M. J., Howie, K. B., Russell, P. S., and Goodman, H. M. (1987). Implantation of genetically engineered fibroblasts into mice: Implications for gene therapy. *Science* **236**, 714–718.

Sorge, J., Wright, D., Erdman, V. D., and Cutting, A. E. (1984). Amphotropic retrovirus vector system for human cell gene transfer. *Mol. Cell. Biol.* **4**, 1730–1737.

Soriano, P., and Jaenisch, R. (1986). Retroviruses as probes for mammalian development: Allocation of cells to the somatic and germ cell lineages. *Cell (Cambridge, Mass.)* **46**, 19–29.

Soriano, P., Cone, R. D., Mulligan, R. C., and Jaenisch, R. (1986). Tissue-specific and ectopic expression of genes introduced into transgenic mice by retroviruses. *Science* **234**, 1409–1413.

Southern, E. M. (1975). Detection of specific sequences among DNA fragments separated by gel electrophoresis. *J. Mol. Biol.* **98**, 503–517.

Stead, R. B., Kwok, W. W., Storb, R., and Miller, A. D. (1988). Canine model for gene therapy: Inefficient gene expression in dogs reconstituted with autologous marrow infected with retroviral vectors. *Blood* **71**, 742–747, 1988.

St. Louis, D., and Verma, I. M. (1988). An alternative approach to somatic cell gene therapy. *Proc. Natl. Acad. Sci. U.S.A.* **85**, 3150–3154.

Sturm, S., Selegne, J., London, W., Blaese, R. M., and Anderson, W. F. (1990). Amphotropic murine leukemia retrovirus is not an acute pathogen for primates. *Hum. Gene Ther.* **1**, 15–30.

Thompson, J. A., Anderson, K. D., DiPietro, J. M., Zwiebel, J. A., Zametta, M., Anderson, W. F., and Maciag, T. (1988). Site-directed neovessel formation *in vivo*. *Science* **241**, 1349–1352.

Thompson, J. A., Haudenschild, C. C., Anderson, K. D., DiPietro, J. M., Anderson, W. F., and Maciag, T. (1989). Heparin-binding growth factor 1 induces the formation of organoid neovascular structures *in vivo*. *Proc. Natl. Acad. Sci. U.S.A.* **86**, 7928–7932.

van der Putten, H., Botteri, F. M., Miller, A. D., Rosenfeld, M. G., Fan, H., Evans, R. M., and Verma, I. M. (1985). Efficient insertion of genes into the mouse germ line via retroviral vectors. *Proc. Natl. Acad. Sci. U.S.A.* **82**, 6148–6152.

Wahl, G. M., Meinkoth, J. L., and Kimmel, A. R. (1987). Northern and Southern blots. *In* "Methods in Enzymology" (Kimmer, A. R., and S. L. Berger eds.), Vol. 152, pp. 572–580. Academic Press, Orlando, Florida.

Wei, C.-M., Gibson, M., Spear, P. G., and Scolnick, E. M. (1981). Construction and isolation of a transmissible retrovirus containing the *src* gene of Harvey murine sarcoma virus and the thymidine kinase gene of herpes simplex virus type 1. *J. Virol.* **39**, 935–944.

Weintraub, H., Cheng, P. F., and Conrad, K. (1986). Expression of transfected DNA depends on DNA topology. *Cell (Cambridge, Mass.)* **46**, 115–122.

Weiss, R., Teich, N., Varmus, H., and Coffin, J. (1985). "RNA Tumor Viruses," Vol. 2. Cold Spring Harbor, Cold Spring Harbor.

Williams, D. A., Lemischka, I. R., Nathan, D. G., and Mulligan, R. C. (1984). Introduction of new genetic material into pluripotent haematopoietic stem cells of the mouse. *Nature (London)* **310**, 476–480.

Williams, D. A., Orkin, S. H., and Mulligan, R. C. (1986). Retrovirus-mediated transfer of human adenosine deaminase gene sequences into cells in culture and into murine hematopoietic cells *in vivo. Proc. Natl. Acad. Sci. U.S.A.* **83**, 2566–2570.

Wilson, J. M., Jefferson, D. M., Roy Chowdhury, J., Novikoff, P. M., Johnston, D. E., and Mulligan, R. C. (1988a). Retrovirus-mediated transduction of adult hepatocytes. *Proc. Natl. Acad. Sci. U.S.A.* **85**, 3014–3018.

Wilson, J. M., Johnston, D. E., Jefferson, D. M., and Mulligan, R. C. (1988b). Correction of the genetic defect in hepatocytes from the Watanabe heritable hyperlipidemic rabbit. *Proc. Natl. Acad. Sci. U.S.A.* **85**, 4421–4425.

Wilson, J. M., Birinyl, L. K., Salomon, R. N., Libby, P., Callow, A. D., and Mulligan, R. C. (1989). Implantation of vascular grafts lined with genetically modified endothelial cells. *Science* **244**, 1344–1346.

Wilson, J. M., Danos, O., Grossman, M., Raulet, D. H., and Mulligan, R. C. (1990). Expression of human adenosine deaminase in mice reconstituted with retrovirus-transduced hematopoietic stem cells. *Proc. Natl. Acad. Sci. U.S.A.* **87**, 439–443.

Wood, P. A., Partridge, C. A., O'Brien, W. E., and Beaudet, A. L. (1986a). Expression of human argininosuccinate synthetase after retroviral-mediated gene transfer. *Somatic Cell Mol. Genet.* **12**, 493–500.

Wood, P. A., Herman, G. E., Chao, C.-Y., O'Brien, W. E., and Beaudet, A. L. (1986b). Retrovirus-mediated gene transfer of argininosuccinate synthetase into cultured rodent cells and human citrullinemic fibroblasts. *Cold Spring Harbor Symp. Quant. Biol.* **51**, 1027–1032.

Yu, S.-F., von Ruden, T., Kantoff, P. W., Garber, C., Seiberg, M., Ruther, U., Anderson, W. F., Wagner, E. F., and Gilboa, E. (1986). Self-inactivating retroviral vectors designed for transfer of whole genes into mammalian cells. *Proc. Natl. Acad. Sci. U.S.A.* **83**, 3194–3198.

Gene Transfer Technology: Alternative Techniques and Applications

Glenn M. Monastersky

Charles River Laboratories
Wilmington, Massachusetts 01887
    and
Transgenic Alliance
St. Germain sur l'Arbresle, France

I. INTRODUCTION OF TRANSGENES INTO GAMETES AND EMBRYOS

A. *Traditional Techniques*

The introduction of foreign DNA into cultured somatic cells has been accomplished using a wide variety of methods. Examples include calcium precipitation (Graham and van der Eb, 1973), infection with viral vectors (Mulligan *et al.*, 1979), electroporation (Neumann *et al.*, 1982; Toneguzzo *et al.*, 1986), and lipofection (Fraley *et al.*, 1980). Mouse embryos first were transformed by the introduction of SV40 DNA into blastocyst-stage embryos in 1974 (Jaenisch and Mintz, 1974), and germ line integration of Moloney murine leukemia retrovirus (M-MuLV) was achieved soon afterward (Jaenisch, 1976). The first transgenic mouse produced by pronuclear microinjection was reported in 1980 (Gordon *et al.*, 1980). Since then, much progress has been achieved in the development of innovative regulatory elements and vectors, but the technique of pronuclear microinjection remains the method of choice for creating transgenic animals in the vast majority of laboratories. Alternatives to the direct pronuclear injection of cloned DNA fragments have been proposed for transferring foreign genetic information into mammalian embryonic genomes and somatic cells. Several of these techniques will contribute to the application of transgenic technology in the development of human gene therapy protocols.

B. *Transfection of Gametes*

The spermatozoan is a cellular vehicle which is specifically designed to transfer foreign DNA into an oocyte. The capability to transform sperm cells and perform *in vitro* fertilization to produce transgenic embryos would serve as a rapid, cost-effective alternative to conventional methodologies (Fig. 1). It has not been determined if the nuclear DNA of the sperm head is present in a functional state that would permit the stable integration of foreign DNA, but many studies have indicated that sperm cells are capable of binding exogenous DNA and carrying it into the egg. In 1971, Brackett *et al.* reported that SV40 DNA was taken up by rabbit spermatozoa and subsequently transferred into oocytes. The washed, pelleted sperm cells were suspended in Krebs–Ringer phosphate buffer supplemented with penicillin and streptomycin. The sperm were diluted in IVF (*in vitro* fertilization) medium and exposed to intact SV40 particles (4×10^6 plaque-forming units per 10^6 spermatozoa; with or without a tritiated thymidine label) or to purified SV40 DNA ($1.0–10.0 \ \mu g$ per 10^7 spermatozoa). The treated sperm were washed and cultured in culture medium supplemented with 20% (v/v) heat-inactivated rabbit serum.

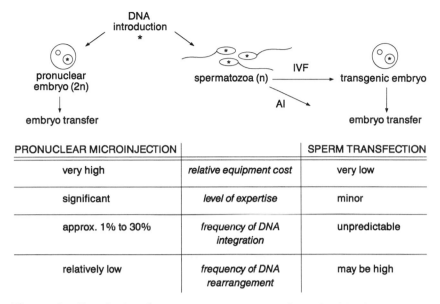

Figure 1. Transfection of spermatozoa versus pronuclear microinjection.

More than half of the treated spermatozoa retained their motility after 24 hr in culture. Motile treated sperm were incubated with CV-1 kidney cells in the presence of Sendai virus to induce fusion. Following washing and subsequent culture, unfused sperm were removed by trypsinization. In a separate experiment, SV40-treated sperm (3×10^6) were introduced into the uterine horns of superovulated docs. Early preimplantation embryos were recovered from the females, placed into cultures of CV-1 cells, and lysed.

It was observed that CV-1 cells fused with sperm cells which had been treated with intact SV40 exhibited cytopathic effects which were inhibited by SV40 antiserum. However, no labeled SV40 DNA was observed within the sperm. In contrast, approximately 30% of the sperm treated with naked labeled SV40 DNA were observed to contain internalized label in the postacrosomal region of the head. The sperm treated with naked DNA also produced cytopathic effects in CV-1 cells following fusion. Also, embryos derived from insemination with sperm exposed to SV40 DNA were shown to produce cytopathic effects in CV-1 cultures; T-antigen also was detected in these cultures. The conclusion of these studies was that intact SV40 particles were adsorbed to the surface of the sperm cells but were not internalized. However, infection with purified SV40 DNA, in conjunction with subsequent DNase treatment of the sperm surfaces, was reported to result in internalization of the DNA. Infection was observed to spread in culture from CV-1 cells containing sperm heads to neighboring cells. Finally, biologically active SV40 DNA was recovered from embryos derived from insemination of females with treated sperm.

In 1989, the publication of a paper by Lavitrano *et al.* created a great deal of excitement and controversy in the transgenic world. The authors reported that mouse sperm cells incubated with foreign DNA and used to inseminate cultured oocytes successfully produced a high frequency of transgenic animals. The transgenic DNA also was reported to have been inherited by progeny of the founders. In an initial experiment, capacitated mouse sperm that were incubated with radiolabeled plasmid DNA ($\sim$5 kb in length) at 37°C and at 0°C were shown to exhibit significant radioactivity within 30 min of culture. It was observed that the radioactivity was localized to a specific, unspecified portion of the sperm heads. Only live (i.e., non-formaldehyde-fixed) sperm were reported to take up the exogenous DNA.

In vitro fertilization (IVF) with DNA-treated sperm was performed using circular and linear plasmid DNA containing the chloramphenicol acetyltransferase (CAT) gene (Lavitrano *et al.*, 1989). IVF was performed in Whittingham's Tyrode solution lacking lactate and antibiotics and supplemented with 30 mg/ml bovine serum albumin (BSA; fraction V). Other alterations related to specific salt components of the medium were noted (Lavitrano *et al.*, 1989). Gametes were collected from B6D2 F1 mice, and the oocytes were harvested 14.5 hr after injection of human chorionic gonadotropin (HCG). Approximately 30% of 250 live-born animals were determined to be transgenic by Southern blot analysis. Evidence of DNA rearrangement was apparent, but subsequent sequence analysis of DNA obtained from transgenic genomic libraries revealed complete homology to SV40 sequences present within the original transgene vector. Study of transmission of the transgene to progeny indicated that inheritance of the circular plasmid form, relative to the linear form, tended more strongly to yield a mosaic pattern. Also, eukaryotic methylation patterns of the foreign sequence were cited as evidence of true transgene replication within the host rather than contamination with bacterially derived vector. Finally, F1 transgenic animals were shown to exhibit transgenic CAT activity, which was reported to be relatively higher in the circular DNA might be more prone to rearrangement. The CAT activity does indicate that the transgenic DNA retained functional integrity. Attempts to promote DNA uptake were unsuccessful with unfertilized oocytes and with embryos.

The results of Lavitrano *et al.* (1989) have not yet been reproduced, although a large number of laboratories have attempted similar experiments. The group subsequently indicated that their technology also was successful in the production of transgenic pig embryos (Gandolfi *et al.*, 1989). Many factors that might have accounted for the successful results have been discussed, including transgene size, the presence of phenol red in the medium, BSA concentration, and mouse strain. Brinster *et al.* (1989) repeatedly attempted to reproduce the results of Lavitrano *et al.* using several different combinations of variables. Even when using identical DNA and reagents acquired from Lavitrano, they were unable to produce a single transgenic mouse using the sperm transfection protocol. The review by Brinster *et al.* of these efforts also contains a compilation of several similarly unsuccessful efforts performed in other laboratories.

A subsequent report by members of the Spadafora group (Lavitrano *et al.*, 1992) investigated the reversible binding of plasmid DNA to epididymal sperm cells. DNA and other acidic molecules were reported to be taken up by sperm cells and localized in the area of the nucleus. The authors believe that ionic factors, possibly related to sperm head membrane proteins, are responsible for the binding of DNA molecules to the cells. Larger DNA molecules appeared to be preferentially bound rather than smaller DNA molecules. The size preference supports the involvement of the DNA charge in the uptake rather than the involvement of specific membrane pores. Seminal fluid was shown to inhibit the permeability of sperm cell membranes to DNA. This observation might be related to the relative impermeability of ejaculated sperm as compared to epididymal sperm.

Additional reports in the literature support the general idea of DNA uptake by spermatozoa (Hochi *et al.*, 1990; Castro *et al.*, 1991; Horan *et al.*, 1991). Spermatozoa from cattle and insects has been proved to bind exogenous DNA (Atkinson *et al.*, 1991). The sperm in these experiments were incubated for 1 hr at 30°C with radiolabeled linear DNA fragments. After DNase treatment, approximately 20% of the sperm retained label, which usually was found to be localized in the posterior head region. The frequency of DNA internalization was essentially equal in fresh motile sperm cells and in cold-shocked nonmotile cells. Sea urchin spermatozoa similarly were reported to carry foreign DNA containing the CAT gene into eggs (Arrezo, 1989). Positive CAT activity subsequently was observed in urchin blastula embryos that developed from the eggs.

C. Electroporation of Gametes

The process of using electrical current to penetrate the cell membrane and permit the entry of foreign DNA, that is, electroporation, was performed first on eukaryotic cells in 1982 (Neumann *et al.*, 1982). Electroporation techniques have not been successfully used to directly transform oocytes or embryos. However, electroporation may prove to be feasible for transforming spermatozoa. Gagne *et al.* (1991) reported the results of experiments in which frozen–thawed bovine sperm cells were subjected to electroporation in the presence of plasmic DNA. A relatively high proportion of electroporated sperm retained radiolabeled foreign DNA following DNase treatment. Optimal electrical currents were determined that permitted the survival of the sperm cells but did not induce an acrosome reaction. Motility and successful *in vitro* fertilization of oocytes was reported with the electroporated sperm. Autoradiography and polymerase chain reaction (PCR) amplification analysis of embryos produced from electroporated sperm found that approximately 20% of blastocyst cells contained the plasmid DNA. Approximately 10% of control embryos (i.e., produced by nonelectroporated sperm) possessed plasmid DNA. Fertilization rates were reported to be 85% for control sperm and 55% for electroporated sperm.

D. Embryonic Stem Cell-Mediated Transgenesis

Traditional incorporation of a functional transgene into the genome of a genetically deficient organism may yield dramatic phenotypic improvement. For example, the autosomal recessive defect of the mutant dwarf mouse (Eicher and Beamer, 1976) was overcome following the incorporation of a metallothionein–rat growth hormone transgene (Hammer *et al.*, 1984). The activity of human and mouse β-globin transgene similarly was reported to correct β-thalassemia in a murine model (Skow *et al.*, 1983; Costantini *et al.*, 1986). However, targeted gene incorporation as a consequence of homologous recombination promises to dramatically improve efforts to correct recessive and dominant genetic defects in somatic cells and in embryos. The potential of this system is illustrated by a study in which a human sickle β-globin gene was converted successfully to the normal α-globin allele by homologous recombination (Shesely *et al.*, 1991).

The technology for gene targeting has been developed using embryonic stem (ES) cell cultures. The biology of ES cells is described in Chapter 4. Targeted gene introduction results in transgene integration at a locus where all regulatory elements are intact. Also, insertional mutations and oncogene activations will not occur. Obviously, targeted insertion of a functional mutant gene or interruption of a normal gene can create powerful research models for development biology, cancer, and immunology. Many potential applications also exist for transgenic "pharming" strategies concerned with the production of functional proteins in blood and milk. Gene targeting in embryos and in adult somatic cells also may lead to molecular medicine strategies in which genes responsible for specific growth and disease processes may be modified. Human cells which closely resemble ES cells have been described (Pera *et al.*, 1987).

E. Transomic Technology

One method proposed to introduce large fragments of foreign DNA into mammalian embryos involves the direct microinjection of chromosome fragments into the embryonic nucleus (Richa and Lo, 1989). The success of this "transomic" technology would permit the transfer of intact gene clusters (e.g., immunological complexes). Also, entire minigene fragments, several kilobases in length, could be transplanted to minimize the effects of insertional loci on transgene expression. Chromosomal fragments may include more than 10 megabases of DNA. Chromosomal fragments are collected following dissection of stained metaphase spreads, and a single fragment is injected into the pronucleus using a micropipette. The transomic DNA may be identified within the recipient embryo by *in situ* hybridization.

Richa and Lo transferred centromeric chromosomal fragments from a normal human fibroblast (line MRC-5) into the pronuclei of (SJL/J × SWR/J) F1 embryos. Sixteen of 90 (18%) injected embryos survived microinjection and developed through the morula stage. Six of 12 embryos (50%) subjected to *in situ* hybridization with a probe recognizing human centromeric DNA exhibited a small number of cells which yielded positive hybridization signals. Four of 13 (31%) injected embryos which arrested at the 2- to 4-cell stage similarly yielded positive signals. It was not established if the transomic fragments were incorporated into the embryonic chromosomes or if the injected fragments were intact. An additional experiment conducted by Richa and Lo involved the transfer of blastocysts cultured from injected embryos into recipient female mice. The postimplantation embryos were collected on day 13 of gestation, and *in situ* hybridization was performed. Four of 8 (50%) embryos examined were found to yield strong hybridization signals representative of at least 100 cells per embryo. This result indicated that the injected fragments were maintained and replicated during development. The proportion of cells within the postimplantation embryos containing the transomic fragments was less than the proportion observed in preimplantation embryos. This suggests that mitosis may have been inhibited by the presence of the foreign fragments. The phenomenon of centromeric inactivation (Zinkowski *et al.*, 1986) may have been responsible for this observation. Further experiments involving noncentromeric chromosomal fragments were inconclusive.

II. SOMATIC CELL TRANSFORMATION

The successes of transgenic animal experimentation since the early 1980s have led to the design of animal models for human gene therapies. Human transgenic therapies are designed to transform somatic cells rather than germ cells. The capability to transform cells *in situ* is an attractive alternative to the *ex vivo* methods which involve the removal of cells from the body. *Ex vivo* gene therapy strategies involve intensive medical resources and therefore will not be available initially to large numbers of patients. The development of protocols in which genes may be introduced directly into intact tissues of the patient would obviate the need for hospital admissions and surgical procedures. Perhaps the most direct strategy would involve the introduction of transgene DNA or DNA expression vectors directly into the patient. In effect, localized regions of the patient would be transformed in a fashion similar to the transfection of cultured cells. More indirect protocols would utilize liposomes, transformed cells, or microcarrier vehicles to transfer therapeutic genes into intact tissues.

A. Viral Vectors

Many of the current protocols that have been proposed for human gene therapy involve the transformation of cellular vehicles with nonreplicating retroviral vectors. Helper-free retroviral vector systems with selectable markers of transgene expression appear to be very useful for somatic cell gene transfer. Derivatives of the Moloney murine leukemia virus are especially promising (Miller *et al.*, 1988). Retroviral vectors have been evaluated in many somatic cell transformation studies in which gene therapy strategies have been proposed. Transformation of two clinically important cell types, hepatocytes and hematopoietic stem cells, is discussed in Section II, D of this chapter.

The retroviral transfer of the human gene for granulocyte colony-stimulating factor (G-CSF) into murine adenocarcinoma cells (C-26) was reported to suppress the subsequent *in vivo* tumorigenicity of the cells (Colombo *et al.*, 1991). Therefore, tumor suppression and differentiation may be achieved *in situ* using specific retroviral vectors carrying antineoplastic transgenes. Retroviral vectors might be very useful to target brain tumors since the neoplastic cells may be selectively infected during division. Rat brain glioblastoma tumor cells have been targeted *in situ* by introducing a retrovirus packaging cell line to induce the infection of tumor cells (Short *et al.*, 1990). The introduction of retrovirally modified cells into the brain represents a potentially clinically useful therapeutic strategy which is discussed further in this chapter (see Section II, D, 3). Retrovirally transferred human tyrosine hydroxylase has been reported to produce dopamine in cultured murine anterior pituitary cells (Horellou *et al.*, 1989).

The main drawback with utilizing retroviruses as vehicles for gene therapy concerns the fact that their genes will integrate only within dividing cells (Miller *et al.*, 1990). Although the epithelium of the lung presents a highly accessible target tissue for *in situ* genetic transfer, the majority of airway luminal cells and alveolar cells are postmitotic. Therefore, adenoviruses, which naturally infect lung epithelium and are taken up by nonproliferating cells (Berkner, 1988), have been proposed as ideal vectors for pulmonary genetic therapy. An adenovirus vector carrying the ornithine carbamoyltransferase gene was reported to successfully overcome an ornithine carbamoyltransferase deficiency in spf-ash mutant mice for up to 1 year after intravenous introduction (Stratford-Perricaudet *et al.*, 1990). A replication-deficient adenovirus carrying the human α_1-antitrypsin gene under the control of the adenovirus major late promoter has been reported to produce human α_1-antitrypsin in cultured CHO cells and HeLa cells (Gilardi *et al.*, 1990). This expression vector was introduced *in vitro* and *in vivo* to the respiratory epithelium of the cotton rat (Rosenfeld *et al.*, 1991). The cotton rat (*Sigmodon hispidus*) has been described as a useful research animal in studies of respiratory viral infection. In these experiments, human α_1-antitrypsin was reported to be secreted from the respiratory epithelium both *in vitro* and *in vivo*, for up to 1 week.

The potential clinical relevance of these results is significant because the transgenic protein was efficiently secreted into the extracellular fluid of the lung, a site where therapeutic benefits would be desired. This type of direct *in situ* genetic therapy may be useful in treating inherited lung diseases including α_1-antitrypsin deficiency (Crystal, 1990) and cystic fibrosis. Cystic fibrosis is a lethal recessive disorder manifested by the expression of a mutant cystic fibrosis transmembrane conductance regulator (CFTR) protein (Riordan *et al.*, 1989). The defect has been shown to be correctable *in vitro* following the transfer of a retroviral vector carrying the cDNA encoding the normal CFTR product (Drumm *et al.*, 1990). An adenoviral vector carrying the cDNA for the normal human CFTR was introduced *in situ* into cotton rats (Rosenfeld *et al.*, 1992). Human CFTR gene expression was detected 2 days after introduction, and transgene transcripts were detected up to 6 weeks later. Human CFTR protein was identified in the recipient epithelial cells approximately 2 weeks after the onset of gene expression. This strategy appears to be an excellent candidate for *in situ* gene therapy, because CFTR mutations are thought to occur within the lung tissue (Boat *et al.*, 1989) and the normal gene product may be easily introduced directly to the epithelium.

It remains to be determined if clinically relevant levels of transgenic therapeutic compounds can be supplied by *in situ* adenoviral vectors or if adenoviral vectors are safe to administer to human patients. Safety issues include the potential for inducing cell transformation and the danger of recombination with other viruses. In addition, experiments must be performed to evaluate the short-term and long-term effects of the constitutive *in vivo* expression of proteins such as CFTR.

Herpesviruses also have been proposed as transgene delivery vectors. Herpes simplex virus type 1 (HSV-1) is a neurotropic virus that establishes a latent, nonlytic infection within neurons (Baringer and Sworeland, 1973; Cook *et al.*, 1974) and can transfer transgenes into the host cell genome (Shih *et al.*, 1984; Tackney *et al.*, 1984). Herpes simplex vectors have been evaluated for targeting the human hypoxanthine–guanine phosphoribosyltransferase (HPRT) gene to neurons *in vitro* (Palella *et al.*, 1988). HPRT-deficient B103 neuroma cells (Schubert *et al.*, 1974) were transformed with HSV vectors bearing the human HPRT gene and survived in selection medium (hypoxanthine/aminopterin/thymidine or HAT medium) for over 1 month. The recombinant herpesviruses did exhibit significant cytopathic activity.

Because the Lesch–Nyhan HPRT deficiency syndrome (Lesch and Nyhan, 1964; Seegmiller *et al.*, 1967) is characterized by a devastating neurological dysfunction, direct neuronal targeting represents a logical therapeutic approach for this disease. Direct gene therapy may be especially appropriate because the restoration of circulatory HPRT levels does not appear to alleviate the neurological component of the syndrome (Edwards *et al.*, 1984; Nyhan *et al.*, 1986).

A nonpathogenic deletion mutant (Desrosiers *et al.*, 1984) of the lymphotropic herpesvirus, Herpesvirus saimiri, was utilized to target a bovine growth hormone gene to cultured primate T cells (Desrosiers *et al.*, 1985). The virus-borne transgene was linked to the SV40 late-region promoter and introduced into a New World primate cell line (i.e., owl monkey kidney) which is permissive for the virus. Bovine growth hormone was produced by the cultured cells, and the characteristic latency of herpesviruses appeared to promote stable expression of the integrated transgene. The herpesvirus vectors remain replication-competent so that helper viruses are unnecessary.

B. Direct Introduction of Nucleotides

The expression of purified DNA and RNA sequences has been achieved following their direct injection into intact mouse skeletal muscle (Wolff *et al.*, 1990). Solutions containing RNA sequences or DNA plasmids carrying the CAT reporter gene were injected into the surgically exposed quadriceps muscles of BALB/c mice. CAT activity was detected in extracts of the treated tissues for up to 18 hr after RNA injection and for up to 48 hr after DNA injection. Expression levels, in certain instances, were reported to be equivalent to reporter gene expression in transfected cultured 3T3 fibroblasts. Subsequent experiments evaluated the expression of β-galactosidase in muscle tissue previously injected with *lacZ* DNA. Histological examination of treated tissue detected positive β-galactosidase staining in between 10 and 30% of the muscle cells within the injection site. It was calculated that approximately 1.5% (i.e., 60 of 4000 cells) of the entire quadriceps muscle exhibited transgene expression. The transfer of β-galactosidase along cellular processes or between adjacent cells was not discounted. In experiments in which firefly luciferase DNA or RNA was injected into muscle tissue, a dose–response relationship was reported; increased expression levels were observed with increasing amounts of transgene. Luciferase activity following DNA injection was apparent up to 60 days later. Following RNA injection, transgene activity was observed to peak within 1 day and was detectable for more than 2 days.

Luciferase protein and luciferase RNA transcripts were found to have a half-life of less than 24 hr in muscle tissue. Therefore, the persistent luciferase activity observed in the experiments (i.e., up to 60 days) suggests that integrated or extrachromosomal transgenes remained active within the treated tissues. Interpretation of Southern blot analyses of the expressing tissues led to the conclusion that an extrachromosomal circular form of the DNA plasmid was maintained up to 30 days following introduction. The possibility of limited chromosomal integration was not definitively excluded. Restriction enzyme analyses suggest that the foreign DNA plasmids did not replicate within the muscle cells. Although the authors claim that reporter gene activity was observed in liver, spleen, skin, lung, brain, and blood tissues following DNA introduction, expression in muscle cells was determined to be significantly greater than that in all other tissues. It was suggested that the unique sarcoplasmic reticulum and tubule communicating system of the muscle cell may enhance the uptake of exogenous nucleotides. Uptake of foreign nucleotides by damaged cells also may be important.

The introduction of transgenes into skeletal muscle represents a promising strategy for many types of human genetic therapies. The transient expression of genes encoding pharmaceutical proteins or immunological mediators within easily accessible muscle tissues is a feasible therapeutic protocol. Importantly, skeletal muscle presents a highly vascular site for the dissemination of foreign gene products which could act throughout the body. Also, direct injection of nucleotides appears to avoid the stimulation of an immune response. The direct genetic rescue of diseased muscle tissue (e.g., dystrophic) also appears to be a reasonable goal.

A somewhat novel type of genetic therapy, in which specific extracellular DNA is destroyed, has been proposed for the treatment of cystic fibrosis (Shak *et al.*, 1990; Hubbard *et al.*, 1992). Extracellular DNA from dead inflammatory cells is a significant component of the viscous secretions which obstruct the airways in cystic fibrosis (Lethem *et al.*, 1990). Recombinant human deoxyribonuclease I (rhDNase), administered by aerosol, was shown to digest efficiently the high molecular weight DNA within the purulent pulmonary secretions of cystic fibrosis patients. DNA digestion was correlated with significantly improved lung function.

C. *Microprojectile-Mediated DNA Delivery*

Perhaps the most radical method of *in situ* DNA introduction that has been proposed involves the use of a microprojectile bombardment device (Williams *et al.*, 1991). This technique was reported previously to have yielded the transformation of animal cells within a vacuum chamber (Zelenin *et al.*, 1989). The hand-held bombardment device utilizes a helium gas mechanism to propel microprojectiles (<5 μm in diameter) into the intact living tissue of anesthetized mice. The projectiles are coated with a calcium slurry of nonreplicative plasmid DNA and washed with ethanol.

Williams *et al.* (1991) report that mouse skin exhibited luciferase expression following bombardment with microprojectiles coated with a β-actin promoter/luciferase vector. Luciferase activity was reported to remain detectable for at least 4 days after transfection of leg skin and up to 10 days after transfection of the relatively thin skin of the ear. Approximately 10–20% of the epidermal cells within the leg skin target area were shown to express luciferase mRNA by *in situ* hybridization. However, less than 5% of dermal cells expressed transgene mRNA. Obviously, surgery would be required to provide access to deeper tissues. The ear bombardment protocol resulted in DNA deposition as deep as the cartilage tissue, and some superficial tissue damage was noted. DNA staining procedures showed that most of the microprojectiles retained their DNA coating following transit through the skin.

Transgene expression was detected in mouse liver for up to 23 days following bombardment with microprojectiles coated with a fatty acid-binding protein promoter/human growth hormone construct. The liver was surgically mobilized to facilitate bombardment. The microprojectile bombardment technique described did not yield integration or replication of the foreign gene. The procedure was used to transform somatic cells but has not been evaluated for the potential *in situ* transformation of the germ cells within the gonads. The unavoidable tissue damage associated with the technique might preclude the successful transfection of gonadal tissue. In addition, the possibility of infection or immune responses to the projectiles and/or DNA must be considered.

D. Transformed Cell-Mediated Gene Delivery

Somatic cells may be transformed *in situ* (Fig. 2a) or removed to receive transgenes *in vitro* prior to reimplantation into the host (Fig. 2b). The latter *ex vivo* strategies probably will require the utilization of autologous cells to prohibit rejection of the transplant. The use of allogeneic and syngeneic tissue implants might be feasible if the cell surface antigens could be modified by genetic engineering. The choice of the appropriate cell for modification depends on many factors including the specific gene expression desired, the transplantation site, and the ultimate somatic targets of the transgene expression. A chosen cell type should be easily accessible and must exhibit a satisfactory index of mitosis so that it may take up retroviruses or naked DNA and proliferate in culture. Also, *ex vivo* protocols require a cellular vehicle that will express the transgene after reimplantation and usually continue to proliferate *in vivo*. Finally, the normal physiological characteristics of a cell must be considered. In certain gene therapy strategies, it is desirable to use cells which have direct access to the bloodstream (e.g., endothelial cells and hepatocytes) or which are progenitors of more mature cell types (e.g., hematopoietic stem cells and myoblasts). Other issues include the regulation and duration of transgene expression *in vivo* and the prospect of immune responses to novel proteins to which the host may be immunologically naive.

1. Myoblasts

The *in situ* introduction of stably transformed cultured cells into tissues would appear to provide the potential for long-term expression of foreign genes within the patient. Myoblasts may be superior to other cell types because they are capable of crossing basal laminae (Hughes and Blau, 1990) and because they can be incorporated postnatally into multinucleated myofibers. Normal mouse myoblasts, expressing dystrophin, have been shown to fuse with the deficient mdx fibers of mdx dystrophic mice following introduction into dystrophic muscle (Partridge *et al.*, 1989).

The *in situ* introduction of genetically modified myoblasts has been reported to yield systemic delivery of a transgenic protein (Barr and Leiden, 1991; Dhawan *et al.*, 1991). The human growth hormone (hGH) gene was introduced into cultured murine myoblasts (C2C12 cell line) via either a retroviral vector or a plasmid vector. The same cells were transformed further by a second retroviral vector carrying the

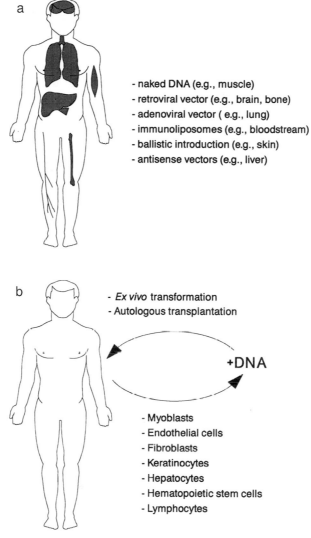

Figure 2. (a) Basic strategies for *in situ* gene therapy. (b) Transformation of somatic cells using *ex vivo* gene delivery protocols.

*lac*Z reporter gene. Synthesis and secretion of hGH from the transformed cells continued throughout the development of fused myotubes, and *in vitro* transgene expression was maintained equally in proliferating myoblasts and in differentiated muscle fibers. In the experiments reported by Dhawan *et al.* (1991), approximately 1×10^7 cells from a high-expression clone transformed with a retroviral vector were injected into the hind-limb musculature of individual C3H mice. Because the implanted C2C12 cells exhibited a different minor histocompatibility antigen profile from the host animals, continuous immunosuppression with cyclosporin (75 μg/g daily) was required. Significant expression of hGH was detected in the serum of animals for up to 35 days. In subsequent experiments, using a pool of cells from several clones, serum hGH was detected at high levels for up to 85 days following introduction of the transformed myoblasts. The ultimate serum hGH levels were correlated with the number of implanted transformed cells. Examination of β-galactosidase reporter gene activity suggested that implanted transformed myoblasts had been incorporated into existing myotubes. In contrast to observations following the implantation of transformed primary mouse or primary human myoblasts, implanted C2C12 clonal myoblasts were observed to proliferate for several months after introduction.

This technology would become much more feasible if methods evolve to permit the implanted cells to evade immune surveillance. Human myoblasts have been purified and expanded *in vitro* (Blau and Webster, 1981), and human clinical trials with transformed human myoblasts are being evaluated for use with Duchenne muscular dystrophy patients. Even though autologously derived primary cultures of myoblasts are preferable immunologically and should not form tumors, they may yield unpredictable transgene expression levels within the host patient. Also, critical threshold levels of incorporation of the implanted cells into host tissues must be attained to achieve a clinically significant improvement in certain disease states.

2. Endothelial Cells

Vascular endothelial cells would appear to serve as excellent cellular delivery vehicles for transgene products because they normally secrete a large variety of proteins directly into the bloodstream. Autologous endothelial cells could be transformed in an *ex vivo* procedure and reseeded into the vascular system of the patient. Blood-borne transgenic therapeutic agents could produce wide-ranging systemic effects such as insulin delivery and antineoplastic activity. More directly, transgenic proteins secreted by modified endothelial cells could elicit localized cardiovascular responses, such as thrombolysis, vasodilation, or angiogenesis.

In representative animal studies (Wilson *et al.*, 1989; Callow, 1990), canine endothelial cells lining the external jugular vein were removed, expanded *in vitro,* and transformed with M-MuLV retroviral vectors that carried the *lacZ* reporter gene encoding β-galactosidase. Transformed endothelial cells were seeded onto a synthetic graft which was implanted into one of the carotid arteries of the donor. A graft that carried nontransformed cultured endothelial cells was implanted into the contralateral carotid artery as a control. Grafts were removed after 5 weeks, and the transformed cells within the arterial intima were found to exhibit positive staining for β-galactosidase activity.

Seeded endothelial cells that have been recovered from grafts have been shown to retain several of the indicators of normal endothelial cell physiology, including production of prostacyclin, uptake of low-density lipoproteins (LDL), and the elaboration of characteristic endothelial cell antigens (Wilson *et al.*, 1989; Zwiebel *et al.*, 1989; Callow, 1990). Other animal models that have been studied include rabbits which received grafts of cells expressing rat growth hormone (Zwiebel *et al.*, 1989) and sheep which received grafts seeded with cells expressing human tissue plasminogen activator (Dichek *et al.*, 1989). Studies also have been performed in Yucatan miniswine, which exhibit artherogenetic properties similar to humans (Nabel *et al.*, 1989), with endothelial grafts expressing β-galactosidase. In the porcine models, transformed endothelial cells were instilled directly into arteries that had been locally denuded of host endothelial cells. The transformed cells were instilled into an arterial space created using a double-balloon catheter. Up to 11% of the transgenic cells introduced were found to have become attached to the host arterial wall. Reporter gene activity *in situ* was observed up to 4 weeks after introduction of the transformed cells. Localized injury of the arterial intima observed following catheterization could be reduced by appropriate transgenic products expressed by the donor cells themselves.

3. Fibroblasts and Keratinocytes

The fibroblasts and keratinocytes of the skin are easily obtained, cultured, and expanded *in vitro*. Connective tissue fibroblasts have been transformed successfully *in vitro* with retroviral vectors (Palmer *et al.*, 1987; Seldon *et al.*, 1987; Doehmer *et al.*, 1982) and may be transplanted into various sites of an autologous host, including the skin and brain. The growth of implanted fibroblasts may be enhanced by embedding the transplanted cells in a collagen matrix and including with each implant a Gelfoam implant treated with fibroblast growth factor (St. Louis and Verma, 1988). Autologous transplants of rat fibroblasts that expressed human growth hormone following calcium precipitation transfection were evaluated for several experimental factors (Chang *et al.*, 1990). Transformed cell expression was successful at intraperitoneal, intramuscular, and subcutaneous sites, and transgene expression was detectable for over 6 months.

Retrovirally transformed fibroblasts have been reported to express a human adenosine deaminase (ADA) transgene in ADA-deficient human cells (Palmer *et al.*, 1987) and to correct the heritable defect in fibroblasts from patients with type I gaucher disease (Sorge *et al.*, 1987). Rabbit fibroblasts deficient in LDL receptors were reported to express normal LDL receptors following retroviral transformation with the LDL receptor gene (Miyanohara *et al.*, 1988). Correction of the factor IX deficiency in hemophiliac cells has been achieved using transformed fibroblasts (Palmer *et al.*, 1989; Axelrod *et al.*, 1990). *In vitro* expression of the human LDL receptor gene has been reported in experiments in which retroviral vectors were utilized to infect rabbit fibroblasts (Miyanohara *et al.*, 1988). Functional LDL receptor activity was induced in fibroblasts collected from the genetically hyperlipidemic Watanabe rabbit, which exhibit a dramatic deficiency of LDL receptors (Watanabe, 1980).

The generation of an immune response to the transgene product secreted from autologous transplants has been discussed (Palmer *et al.*, 1987; St. Louis and Verma, 1988; Chang *et al.*, 1990). Long-term *in vivo* expression of transplanted fibroblast transgenes may be improved by selecting efficient "housekeeping" promoters such as the dihydrofolate reductase (DHFR) promoter (Scharfmann *et al.*, 1991). Implants that correct a genetic defect by supplying a novel protein to the host system probably will initiate immunological responses unless the loss of gene function has occurred after birth. Experimental methods have been proposed to deal with the immune responses to transgene expression from autologous transplants of rat fibroblasts (Bennett and Chang, 1990). The immunosuppression of host rat antibody production to human growth hormone secretion from autologous fibroblast transplants was achieved with cyclosporin A (20 mg/kg/day).

The clinical utility of introducing transformed cells into the immunologically privileged brain has been proposed (Gage *et al.*, 1987; Breakefield, 1989; Shimohama *et al.*, 1989; Culver *et al.*, 1992). The brain is characterized as an immunologically privileged site with a virtual absence of normal cell division. Transformed rat fibroblast cells which produce L-dopa have been described for the potential gene therapy of Parkinson's disease (Wolff *et al.*, 1989; Horellou *et al.*, 1990).

The implantation of transformed fibroblast cells has been reported to prevent neuronal degeneration in host animal brains following the creation of specific lesions (M. B. Rosenberg *et al.*, 1988). Experimental rats were prepared by completely transecting the fimbria-fornix unilaterally by aspiration. This region transmits cholinergic fibers from neurons within the basal forebrain to target cells within the hippocampus. Once deprived of their targets, many of the septal cholinergic neurons exhibit retrograde degeneration owing to the cessation of the retrograde transport of nerve growth factor (NGF) from the hippocampus. The degeneration of NGF-deprived neurons can be prevented by chronic infusion of exogenous NGF. Rat clonal fibroblasts (line 208F) were infected with a retroviral vector containing the neomycin-resistance gene and the mouse NGF gene. A clone exhibiting a high level of NGF secretion *in vitro* was selected, and a suspension containing approximately 4×10^5 cells was injected into the lesion cavity and into the lateral ventricle ipsilateral to the lesion site. The lesion was covered with Gelfoam, and the wound was closed. Controls were represented by the contralateral unlesioned brain tissue within each experimental animal and by lesioned animals in which nontransformed fibroblasts were implanted into the lesion site.

Two weeks after surgery, the brains were removed from the rats for histochemical examination. The presence of fibronectin, a fibroblast-specific marker, was detected in all experimental animals, indicating survival of the implanted cells. Staining for choline acetyltransferase (ChAT) was performed to evaluate the survival of cholinergic neuron cell bodies, which was expressed as the percentage of remaining cholinergic cells in the septum ipsilateral to the lesion relative to the intact contra-lateral septum. Neuron survival was reported to be 92% in lesioned animals bearing NGF-secreting implanted cells and 49% in lesioned animals bearing nontrans-formed implanted cells. Animals with NGF-secreting implants also exhibited in-creased numbers of acetylcholinesterase (AChE) positive cells and fibers.

The epidermal keratinocyte is highly proliferative cell type that has been uti-lized to grow large amounts of cultured epithelium for use in skin grafting proce-dures (Gallico et al., 1984). Although keratinocytes are poorly vascularized in situ, the accessibility and transplant characteristics of these skin cells favor their evalua-tion as gene therapy vehicles. In one study, a retroviral vector was used to transfer the human growth hormone gene into cultured human foreskin keratinocytes (Mor-gan et al., 1987). Transgene-expressing epidermis was transplanted to athymic nude mice, but subsequent secretion of hGH into the serum was undetectable. Ob-viously, gene therapy protocols are envisioned in which transformed epithelia may be expanded in culture and grafted to autologous sites from which the grafts would deliver transgenic therapeutic proteins to the patient.

4. Hepatocytes

Hepatocytes have been transformed in vitro using adenoviral vectors (Friedman et al., 1986), electroporation (Tur-Kaspa et al., 1986), and SV40 (Isom and Georg-off, 1984). However, retroviral vectors appear to be among the most promising vehicles for inducing predictable transgene expression in hepatocytes (Wolff et al., 1987). Retroviral vectors also have been used to transfer genes into hepatocytes, both in vitro (Wood et al., 1986; Ledley et al., 1987; Peng et al., 1988; Friedman et al., 1989; Huber et al., 1991) and in vivo (Hatzoglou et al., 1990; Ferry et al., 1991). Retroviral vectors carrying the LDL receptor gene also were shown to pro-vide a genetic correction in hepatocytes from Watanabe rabbits (Wilson et al., 1988).

Hepatocytes may be transformed in ex vivo procedures and returned to the host. The in vivo survival of transplanted hepatocytes and prolonged transgene expression from the transformed cells have been reported in many studies (Jirtle and Michalo-poulos, 1982; Demetriou et al., 1986; Jaffe et al., 1988). Ex vivo methods increase the efficiency of retroviral vector uptake because the cultured hepatocytes undergo dedifferentiation and proliferation that permit susceptibility to infection (Wolff et al., 1987; Friedmann et al., 1989). One method of reimplantation of transformed hepatocytes utilizes collagen-coated microcarrier beads introduced into the perito-neum (Demetriou et al., 1986). The spleen appears to be a satisfactory alternative site for introduction of transformed hepatocytes (Kusano and Mito, 1982, Gupta et al., 1987). The survival and location of intrasplenically transplanted hepatocytes were evaluated in rats by following the fate of transgenic cells expressing the sur-face antigen (HBsAg) of the hepatitis B virus (Gupta et al., 1991). More than half of the transplanted hepatocytes translocated to the liver, where they assimilated and functioned normally for over 1 year. Serum HBsAg was detectable throughout the entire year of the study, and the number of transplanted cells found in the host liver 48 hr after intrasplenic introduction was equivalent to the quantity of cells present after 1 year. Approximately 15% of the transplanted cells remained within the spleen, and smaller numbers of hepatocytes were found in the lungs and pancreas.

In vivo transgene expression in the liver has been achieved following the injection of polyoma virus (Dubensky *et al.*, 1984) and liposomes containing DNA into the portal vein (Nicolau *et al.*, 1983; Soriano *et al.*, 1983). The use of retroviral vectors to target hepatocytes *in situ* also has been evaluated. In one study, retroviral vectors carrying the amino-3'-glycosylphosphotransferase [neomycin phosphotransferase (*neo*)] gene or the bovine growth hormone (bGH) gene linked to the phosphoenolpyruvate carboxykinase (PEPCK) promoter were introduced into fetal and adult rats (Hatzoglou *et al.*, 1990). The PEPCK promoter directs transgene expression primarily to the liver and kidneys (McGrane *et al.*, 1988) and is responsive to many factors including cAMP and glucocorticoids (Wynshaw-Boris *et al.*, 1984) and insulin (Hatzoglou *et al.*, 1988). The vector was administered to fetal animals by *in utero* intraperitoneal injection and to adults by direct injection into the portal vein after the performance of a partial hepatectomy. Hepatectomy induces regenerative cell division that enhances the uptake of retroviral vectors by the adult liver (Wolff *et al.*, 1987; Wu *et al.*, 1989). Developing fetal hepatocytes may be infected readily with retroviruses. In these experiments, proviral DNA was integrated in the genome of the hepatic cells and was expressed for up to 8 months.

About 50% of the experimental fetuses and approximately 10% of the partially hepatectomized experimental adults exhibited liver transgene expression. Because approximately 40% of the cells of the adult liver are nonhepatic cells (Panduro *et al.*, 1986), the specific cell types that express the transgenes need to be determined to fully evaluate these results. Infection of fetal liver, which contains hematopoietic cells, also produced the colonization of the bone marrow with transformed cells. However, these cells did not express the transgene, probably because the PEPCK promoter may be inactive in differentiated bone marrow. An alternative method for introducing retroviral vectors to partially hepatectomized rats has been described in which the remaining lobes of the liver are surgically excluded from the circulation and perfused directly with the vectors (Ferry *et al.*, 1991). In these types of experiments, reporter gene expression was detected in the regenerated livers for up to 3 months.

A hepatocyte receptor-mediated method for the *in situ* targeting of transgene-bearing plasmids to the liver has been described (Wu and Wu, 1987). Hepatocytes elaborate receptors that specifically bind galactose-terminal (asialo-) glycoproteins and internalize them via endocytosis. It was proposed that noncovalently binding exogenous DNA to asialoglycoproteins would induce internalization of the DNA by hepatocytes *in vivo*. The DNA was bound electrostatically to an asialoglycoprotein, specifically asialoorosomucoid, by the polycation, poly (L-lysine). The soluble asialoorosomucoid/poly (L-lysine) conjugate was complexed with a plasmid carrying the CAT reporter gene. Positive transgene activity was detected in HepG2 hepatoma clonal cells following exposure of the cultured cells to the DNA conjugate. In subsequent *in vivo* experiments, the asialoglycoprotein/CAT plasmid complex was injected intravenously into Sprague-Dawley rats, and liver-specific reporter gene expression was detected 24 hr later (Wu and Wu, 1988). In more advanced experiments, the conjugate was joined to a CAT gene linked to the mouse albumin promoter, and the entire complex was injected intravenously into rats (Wu *et al.,* 1989). Liver-specific CAT expression was detected within 24 hr but not more than 96 hr after injection. If liver regeneration was induced by partial hepatectomy, reporter gene activity could be detected in liver homogenates for up to 11 weeks after injection. Also, the transgene was shown to be integrated into the host genome in the regenerating livers. This system was tested further by delivering the human albumin gene *in vivo* to partially hepatectomized Nagase analbuminemic rats (Wu *et al.,* 1991). Subsequent molecular and immunological analyses found that human albumin mRNA and circulating serum human albumin were present within 48 hr of injection. Serum albumin expression peaked within about 2 weeks and was consistent for more than 4 weeks after injection.

5. Hematopoietic Stem Cells

Pluripotent hematopoietic stem cells generate the complete array of lymphoid and myeloid lineage progenitor cells (Fig. 3a). The accessibility of autologous hematopoietic cells and methods for destroying the remaining marrow tissue present an obvious target for *ex vivo* gene therapy strategies in which the entire hematopoietic system may be replaced (Fig. 3b). Hematopoietic stem cells from murine bone marrow have been transformed *ex vivo* and used to repopulate the bone marrow compartments of recipient animals (Joyner *et al.,* 1983; Miller *et al.,* 1984; Keller *et al.,* 1985). The *in vitro* expression of the *neo* gene and a mutant dihydrofolate reductase gene in human hematopoietic progenitor cells has been reported following retrovirus-mediated gene transfer (Hock and Miller, 1986). The expression of a retrovirally delivered *neo* transgene also has been achieved in the irradiated bone marrow of the cat (Lothrop *et al.,* 1991). Several canine models have been developed to evaluate the transplantation of transformed bone marrow as well (Kwok *et al.,* 1986; Eglitis *et al.,* 1988).

In vitro stem cell cultures may be enriched to provide colonies of the following specific cell lineages (Fig. 3a): erythroid cells (BFU-E; Stephenson *et al.,* 1971), megakaryocytes (CFU-Meg; Metcalf *et al.,* 1975), eosinophils (CFU-Eo; Metcalf *et al.,* 1974), granulocytes/monocytes (CFU-GM; Pike and Robinson, 1970), and mixed colonies of erythroid and myeloid cells (CFU-MIX; Johnson and Metcalf, 1977). Studies in mice have yielded transformed bone marrow that expresses human α- and β-globins (Dzierzak *et al.,* 1988; Bender *et al.,* 1989; Li *et al.,* 1990)

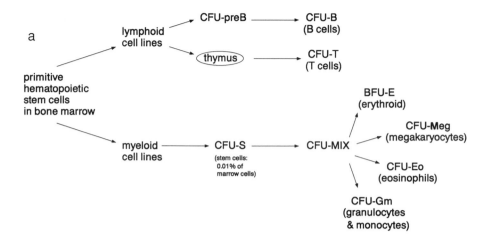

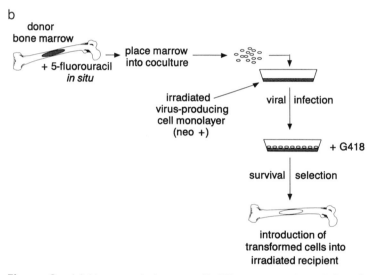

Figure 3. (a) Hematopoietic stem cell differentiation. CFU, Colony-forming units; BFU, blood-forming units. (b) General strategy for transformation of hematopoietic stem cells.

and human ADA (Belmont *et al.*, 1986; Lim *et al.*, 1989). The lymphokines interleukin-3 and interleukin-6 have been reported to synergistically enhance the retroviral infection frequency of mouse hematopoietic stem cells (Bodine *et al.*, 1989). The transfer of retrovirus-transformed stem cells into stem cell-deficient mutant mice (*W/W*) (Russell, 1979) that exhibit severe hereditary macrocytic anemia has produced much information concerning the hierarchy of hematopoietic stem cell development (Dick *et al.*, 1985).

The transfer of the human HPRT gene into murine (Miller *et al.*, 1984) and human (Miller *et al.*, 1983; Gruber *et al.*, 1985) hematopoietic stem cells has been reported and may lead to the development of a therapy for the treatment of the Lesch–Nyhan HPRT deficiency syndrome (Lesch and Nyhan, 1964; Seegmiller *et al.*, 1967). The calcium phosphate-mediated transfection of bone marrow cells from C57BL/6 (*H-2*^{b}) mice with the *E-α^d* gene resulted in the novel cell surface expression of I-E antigens on bone marrow-derived macrophages (BMDM) (Pullen and Schook, 1986). These mice normally do not express the I-E molecule, a class II major histocompatibility complex (MHC) antigen, which contributes to glycoproteins found on macrophages, B cells, and activated T cells (Mengle-Gaw and McDevitt, 1985). The expression of the transgene was determined to be dependent on the presence of γ-interferon. These studies showed that the responses of the cells of the immune recognition system can be modified accurately following the transformation of stem cells. Human hematopoietic stem cells also have been transformed successfully using electroporation-mediated DNA introduction (Toneguzzo and Keating, 1986).

An example of potentially useful animal models produced by transgenic bone marrow transplantation was described in which mice expressed the human multidrug-resistance (MDR) gene (Galski *et al.*, 1989; Mickisch *et al.*, 1991). The transgenic mice promise to be valuable in the elucidation of factors involved in the MDR phenomenon and in the design of novel systemic chemotherapeutic protocols. Large animal models have been very useful to evaluate hematopoietic cell gene therapy protocols. Monkey models have been developed to test the transfer of transformed autologous bone marrow cells that express human ADA (Kantoff *et al.*, 1987). Fetal sheep circulating hematopoietic progenitor cells have been collected *in utero* (Kantoff *et al.*, 1989). The cells were transformed *ex vivo* and were observed to express the *neo* gene following *in utero* autologous reinfusion. These results suggest that prenatal gene therapy may be feasible.

The basic *ex vivo* transformation of stem cells usually begins with the coculture of aspirated bone marrow with recombinant retrovirus-producing cells. The donors may be pretreated with 5-fluorouracil to enrich the available population of stem cells (Hodgson and Bradley, 1979). The aspirate is placed into culture with a 50–80% confluent monolayer of irradiated virus-producing cells. The culture medium may contain Polybrene, transferrin, horse serum, hydrocortisone, BSA, and conditioned medium supplementation (Keller *et al.*, 1985). Transformed cells bearing the *neo* gene may be selected based on survival in medium supplemented with G418 and injected into irradiated hosts.

6. Lymphocytes

Circulating B lymphocytes have the potential to differentiate into mature plasma cells if they are stimulated properly. Lymphocytes obviously may be collected with ease, transformed *in vitro*, and reinjected into the bloodstream. Pre-B-cell proliferative disorders usually are due to the inhibition of this differentiative response. The expression of transgenes in malignant B cells has been discussed as a novel gene therapy strategy (Casali and Notkins, 1989), especially to halt the progress of lymphocytic leukemia (Tremisi and Bich-Thuy, 1991). The utilization of immunoglobulin enhancers and promoters, interleukin-2 (IL-2), and γ-interferon could serve to regulate transgene expression in the malignant lymphocytes.

The retrovirus-mediated transformation of peripheral blood lymphocytes from patients with a variant of severe combined immunodeficiency (SCID) caused by ADA deficiency (O'Reilly *et al.*, 1989) has been reported (Kantoff *et al.*, 1986; Ferrari *et al.*, 1991). The function of ADA-deficient lymphocytes transformed with the human ADA gene was evaluated in immunodeficient hymozygous *bg/nu/xi*[d] mice (BNX mice) (Mosier *et al.*, 1988). The long-term survival of the transplanted transgenic lymphocytes was accompanied by a dramatic restoration of immune function in the immunodeficient animals. Circulating human immunoglobulin and antigen-specific T cells were observed.

Tumor-infiltrating lymphocytes (TILs) are lymphocytes that recognize tumor-specific antigens and infiltrate solid tumors (Rosenberg *et al.*, 1986). These cells are involved in natural host defenses against neoplastic growth, and the infusion of TILs accompanied by IL-2 has been shown to induce significant tumor regression in patients with malignant melanoma (S. A. Rosenberg *et al.*, 1988). Enriched cultures of TILs may be obtained by culturing dissociated cell suspensions of solid tumors in medium supplemented with IL-2 (S. A. Rosenberg *et al.*, 1988). The TILs may be transformed *in vitro* with retroviral vectors to express functional anti-neoplastic proteins or marker genes that enable the tracking of TIL migration *in vivo*. Transgenic TILs carrying the *neo* marker transgene were infused into human patients diagnosed with advanced metastatic malignant melanoma (Rosenberg *et al.*, 1990). Transplanted cells were detected within tumors and in the circulation of the patients for up to 2 months, during which the *neo* transgene was functional.

E. Lipofection

For quite some time, liposomes (i.e., phospholipid vesicles) have been utilized to carry materials across mammalian cell membranes following phagocytosis or fusion (Papahadjopoulos *et al.*, 1974). The introduction of transgenes using liposome vehicles is designed to accomplish the targeted *in situ* transformation of cells in a highly tissue-specific and temporally accurate fashion. One of the more promising cationic lipid compounds that has been evaluated for lipofection purposes is Lipofectin (Chang and Brenner, 1989). Lipofectin is composed of dioleoxylphosphatidylethanolamine (DOPE) and *N*-[1-(2,3-dioleoxyl)propyl]-*N*,*N*,*N*-trimethylammonium chloride (DOTMA). These reagents have been observed to form positively charged liposomes which spontaneously complex with polynucleotides. The liposomes fuse with the negatively charged cell membrane and transfer their DNA component into the cytoplasm. Lipofectin is relatively nontoxic compared to other transfection reagents such as calcium phosphate and DEAE-dextran and therefore may be used to accomplish *in situ* transformation (Felgner and Ringold, 1989). In addition, lipofection techniques may yield transfection frequencies up to 100-fold greater than those achieved with other techniques.

Successful *in situ* liposome-mediated transfection has been reported in the developing nervous system of the *Xenopus* embryo (Holt *et al.*, 1990). Four progressive experimental procedures were performed to evaluate the lipofection techniques, which yielded transfection frequencies between approximately 0.1 and 5% of total cells. Transgene DNA activity was detected for several weeks in whole embryos, suggesting that integration of the transgene might have occurred. The possible integration of plasmic DNA was not investigated, and the possibilities of long-term extrachromosomal expression or prolonged liposome activity within the embryo have not been discounted. Initially, surgically collected embryonic head parts (embryonic stage 20–24) were incubated at room temperature for several hours in a solution of Lipofectin and plasmids carrying the luciferase gene (3:1 Lipofectin:DNA was the optimal ratio). The optimal *in vitro* incubation time was found to be approximately 20 hr. Within 48 hr, luciferase activity was detected within the cultured brain primordia in the head parts. Control tissue, incubated in a plasmid solution lacking Lipofectin, was negative. Lipofection efficiency was enhanced greatly if the extracellular matrix of the tissue was removed enzymatically prior to incubation. Trypsin was reported to be more effective than collagenase, ficin, or elastase.

In a second series of experiments, either the entire neuroepithelium or only the optic primordia were exposed by skin removal. Whole embryos were incubated in DNA–Lipofectin solution following these manipulations. Control embryos with intact skin did not exhibit reporter gene activity within the brain following incubation. Denuded embryos exhibited relatively high levels of luciferase activity, and the localized skin removal from the optic regions successfully restricted the targeting of the transgene to the eye primordia.

In more advanced experiments, the DNA–Lipofectin solution was injected directly into the lumen of the eye vesicle of the embryo. The solution was supplemented with fast green dye and was observed to spread along the lumen of the entire neural tube. Transfection efficiency for these embryonic brains was severalfold lower than the frequencies reported for isolated brain tissue. Finally, excised optic primordia were incubated in DNA–Lipofectin solution and returned to the donor embryos. Luciferase expression within the regrafted manipulated eyes was restricted to the retinas and to retinal ganglion cell processes projecting to the optic tectum within the midbrain.

The future utility of the *in situ* lipofection techniques might include developmental studies of embryonic morphogenesis and gene expression, and the techniques may facilitate the generation of useful models for neurodegenerative disease syndromes. Transient or long-term gene therapy strategies also may utilize liposome-mediated targeted transfection technology. In addition, lipofection might be used for *in situ* transfer of functional mRNA across cell membranes.

The major drawback of liposome-mediated gene delivery is the nonspecific cellular uptake of the liposome vehicles. This problem should be avoided by the synthesis of immunoliposomes created by covalently attaching specific antibody molecules to the surfaces of the liposomes (Huang *et al.*, 1980, 1983; Martin *et al.*, 1981). The antibody is designed to recognize a surface antigen that is unique to the target cell type, and the binding of the immunoliposome will trigger uptake of the liposomal contents via endocytosis. The endocytosed material probably is delivered into the lysosomal compartment, and acidic compounds, such as nucleic acids, should be partitioned into the cytoplasm owing to the acidic environment of the lysosomes. In studies of liposomes conjugated to polyclonal antierythrocyte antibody, a higher affinity for erythrocytes was reported compared to unconjugated antibody because the immunoliposome represents a multivalent complex of antibodies (Heath *et al.*, 1981). Liposomes bearing several monoclonal antibodies may be the most efficient targeting vectors because they are less likely to be inhibited by soluble antigens and can be custom-designed to bind to cells that exhibit a unique combination of nonunique antigens. Immunoliposome technology has been evaluated for targeting viral particles. In one study, immunoliposomes containing antiviral compounds were targeted to the gp80 envelope glycoprotein of the avian myeloblastosis virus (AMV) (Kalvakolanu and Abraham, 1991).

Alternative methods of directing liposomes to specific targets have been proposed. Liposomes covalently linked to low-density lipoproteins (LDL) were reported to exhibit high affinity for leukemic lymphocytes *in vitro* (Vidal *et al.*, 1985). Specific lymphocyte cell death was achieved when the LDL–liposomes were loaded with the protein synthesis inhibitor hygromycin B. Liposomes bearing bovine brain gangliosides will efficiently bind Sendai virus (HVJ), and these complex vehicles have been reported to promote the successful transfer of liposome contents into cultured cells (Kaneda *et al.*, 1987). Reconstituted envelopes of Sendai virus have been utilized to transfer macromolecules into cells (Uchida *et al.*, 1977). Influenza virions have been reported to be as efficient as Sendai virus in the promotion of liposome-enclosed DNA transfer into cultured cells (Lapidot and Loyter, 1990). Influenza virus binding and fusion are mediated by the viral hemagglutinin (HA) glycoprotein, which is resistant to proteolysis and activated by acidic pH. Influenza-mediated DNA transfer should result in the liberation of DNA from the endosomal compartment following cell membrane sialic acid-mediated endocytosis.

F. Microcarrier-Assisted Delivery of Transformed Cells

One approach for delivering transformed cells to specific organs utilizes collagen-coated dextran microbeads. Donor hepatocytes attached to microbeads have been shown to colonize the liver following intraperitoneal injection (Demetriou *et al.*, 1986). Primary rat hepatocytes were incubated with microbeads in Dulbecco's minimal essential medium (MEM) supplemented with 20% fetal calf serum. In one series of experiments, two strains of mutant rats were used as recipients of donor hepatocytes: (1) hyperbilirubinemic Wistar Gunn rats, which have an inherited deficiency of bilirubin–uridine diphosphate glucoronosyltransferase (UDPGT) and therefore lack conjugated bilirubin in their bile, and (2) analbuminemic Nagase rats (NAR), which synthesize only trace amounts of albumin. Introduction of microcarrier-attached normal rat hepatocytes was shown to yield increased plasma albumin in the NAR recipients and conjugated bilirubin in the bile of the Gunn rat recipients for up to 3 weeks. Although concentrations of transplanted hepatocytes were observed attached to the peritoneum up to 8 weeks after introduction, minimal inflammatory responses and adhesions were discovered. Hepatocytes that were introduced without microcarriers were not observed to survive or to affect the deficient physiology of the mutant recipients.

III. STRATEGIES IN MOLECULAR MEDICINE

A. *Antisense Suppression of Endogenous Genes*

Many human diseases are caused by the expression of a mutant gene or genes that are not appropriately expressed. Many other diseases are caused by viral or bacterial genomic expression. Antisense technology appears to offer a promising therapeutic strategy in which synthetic oligonucleotides are utilized to inhibit specific protein synthesis, pre-mRNA processing, or DNA replication (Fig. 4). Basic antisense RNA strategy introduces small DNA constructs designed to produce mRNA transcripts that are complementary to an mRNA produced by the target gene. The antisense mRNA will hybridize to the pathogenic RNA molecule and effectively suppress translation of the specific protein at the ribosome. Antisense suppression of gene expression would completely prohibit the onset or progression of a disease. This technology would selectively target the genetic basis of the pathology, whereas most treatments are directed, usually in a nonspecific fashion, toward the symptoms of a disease. Also, antisense protocols might enable the eradication of diseases for which vaccines or therapeutics do not exist.

The initial proof that antisense gene suppression might be realistic was reported in 1978 following experiments in which the expression of Rous sarcoma virus (RSV) genes was successfully blocked (Stephenson and Zamecnik, 1978). Clinical exploitation of this technology is promising. For example, the translation of mRNA has been suppressed for the human immunodeficiency virus (HIV) *tat* gene (Stevenson and Iversen, 1989), the influenza virus (Leiter *et al.*, 1990), and the human

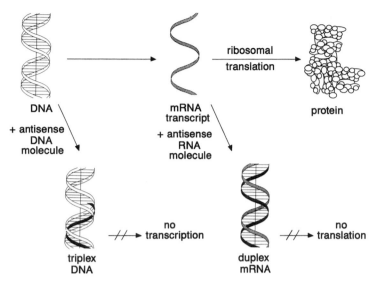

Figure 4.   Strategies of antisense inhibition of gene expression.

T-lymphotropic virus type 1 (HTLV-1) (von Ruden and Gilboa, 1989). Antisense suppression also appears to be promising for the regulation of oncogene activity. Several oncogenes have been targeted, including *ras* (Brown *et al.,* 1989) and c-*myc* (Prochownik *et al.,* 1988). The accessibility of unprotected mRNA molecules in the cytoplasm of the cell is a positive aspect of antisense therapy. Also, antisense technology suggests powerful research strategies that may be used to identify the proteins which cause specific pathologies. For example, transgenic mice exhibiting a "shiverer" phenotype have been created following the microinjection of vectors expressing an antisense transcript to myelin basic protein (MBP) (Katsuki *et al.,* 1988). The phenotype was correlated with a decrease in normal endogenous MBP mRNA and reduced myelination within the central nervous system.

Even more striking was the report that virus-induced disease was inhibited in transgenic mice which expressed an antisense viral gene (Han *et al.,* 1991). The transgenic mice possessed an antisense sequence of the proviral packaging gene for M-MuLV, directed by lymphocyte-specific regulatory elements. All of the mice that expressed the antisense transgene failed to develop leukemia following infection at birth with M-MuLV. Approximately 30% of control nontransgenic mice developed leukemia following viral infection. Additional experiments conducted *in vitro* with NIH 3T3 cells transformed with the antisense sequence showed that virus-infected cells produced unpackaged viral RNA.

Strategies for selectively suppressing gene expression also are being developed in which double-stranded DNA molecules are targeted. The formation of triplex DNA is accomplished by binding of the complementary oligonucleotide to the major groove region of the DNA double helix (Dervan, 1989). Initial human therapeutic applications of antisense technology probably will involve the *ex vivo* treatment of cells followed by reintroduction into the patient. The ultimate goal would be to develop topical or systemic treatment modalities. One significant impediment to antisense treatment is the potential for eliciting an immune response to the foreign DNA. Within the patient, immunological reactions to the antisense constructs potentially could negate the benefits of the therapy by producing systemic lupus erythematosus. The initial human clinical test of antisense therapy will evaluate the efficacy of an anti-human papilloma virus (HPV) compound. In 1992, the U.S. Food and Drug Administration approved clinical trials of Isis 2105, an antisense oligonucleotide which is designed to inhibit HPV particle replication following direct injection into affected skin regions. Other promising antisense antiviral compounds have been designed to inhibit herpes simplex virus, HIV, influenza virus, cytomegalovirus, and Epstein-Barr virus. Antisense strategies also are being proposed to combat cancer by blocking oncogene expression and to treat inflammatory disease by inhibiting the elaboration of specific cellular adhesion factors.

B. Intracellular Immunization

The concept of intracellular immunization is based on the interference of viral activity within infected cells by the expression of a dominant negative mutation (Herskowitz, 1987; Baltimore, 1988). It is expected that the production of mutant viral proteins will interfere significantly with the production of functional wild-type virus particles (Fig. 5). Cultured mouse cells have been rendered resistant to herpes

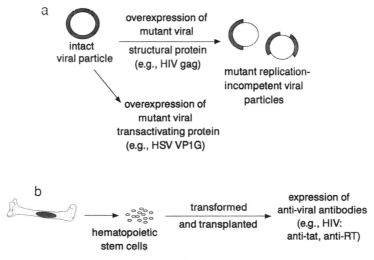

Figure 5. Examples of intracellular immunization strategies. (a) Dominant interference with viral replication. (b) Generation of targeted antiviral immune response.

simplex virus infection following the expression of a mutated form of the viral transactivating protein VP16 (Friedman *et al.*, 1988). The success of this strategy is especially promising if the protein concerned is multimeric (e.g., the HIV-1 *gag* proteins) because mutant monomers frequently will interact with wild-type monomers. Therefore, even moderate levels of mutant protein synthesis might be clinically effective.

A different type of intracellular immunization would utilize the expression of specific antiviral proteins to resist the spread of a viral infection. Transgenic chickens have been produced that express a viral envelope glycoprotein which confers resistance to avian leukosis virus (Salter and Crittenden, 1989). Another transgenic chicken model was developed in which the hemagglutinin–neuraminidase (HN) gene was transferred into chick embryo cells with a retroviral vector (Morrison *et al.*, 1990). Chickens that expressed the transgene were susceptible to infection with Newcastle disease virus and exhibited viral replication but were free of disease. Transgenic mice which constitutively express human β-interferon have been reported to exhibit limited resistance to pseudorabies virus (Chen *et al.*, 1988). Another transgenic model has been described in which dramatic resistance to influenza virus was achieved in animals expressing the anti-influenza protein Mx1 (Arnheiter *et al.*, 1990). Mice that possess the autosomal dominant Mx1$^+$ alleles of the Mx1 gene will resist infection with influenza A and B viruses, whereas mice that have the Mx1$^-$ alleles are susceptible to infection (Lindemann, 1964; Lindemann *et al.*, 1963).

Experiments in which the Mx1 cDNA was constitutively expressed in transgenic mice failed to produce viral resistance. Therefore, a strategy in which the transgene was driven by the presence of the influenza virus was developed. An Mx1 cDNA construct was created which incorporated the Mx1 promoter that responds both to interferon and to the influenza virus (Hug *et al.*, 1988). The construct was microinjected into the pronuclei of Mx1$^-$ mouse embryos, namely, (C57BL/6 $\times$ C3H) F1 $\times$ (C57BL/6 $\times$ C3H) F1 embryos. Homozygous Mx1$^+$ mice (i.e., A2G) were utilized as controls. Mx1 expression in the transgenic mice was evaluated by culturing peritoneal macrophages in the presence of interferon and was found to be inducible. The *in vivo* administration of interferon also induced Mx1 production in similar to endogenous Mx expression in A2G mice and was constitutive in skeletal muscle tissues. Transgenic Mx1 mice were shown to survive following intracerebral inoculation with the neurotrophic A/NWS influenza virus. Immunofluorescence analysis of the brain tissue discovered that Mx1-expressing cells are present in large quantities in the immediate vicinity of infected ependymal cells. This observation indicates that interferon production and Mx1 expression are localized to loci of viral infection and implies that the area of infection may be effectively surrounded by protective antiviral cells. Mx1 protein was observed to be localized to the nuclei of different types of glia and neurons lacking virus and to some, but not all, cells exhibiting viral antigens.

The spread of the influenza infection was restricted in the Mx1 transgenic animals and in the A2G mice, in contrast to nontransgenic Mx1⁻ animals which exhibited extensive spread of the infection. High dose infection of transgenic mice produced much more effective Mx1 expression and viral resistance than lower doses. This phenomenon might be due to a rapid initial transgenic response to infection, yielding high levels of Mx1 protein expression which would be sufficient to inhibit viral spread. Similar observations support this conclusion (Sabin, 1952; Lindemann *et al.*, 1963).

A complex strategy for achieving intracellular immunity to HIV or retrovirus-induced leukemia has been proposed in which hematopoietic stem cells would be engineered to produce antibodies to specific viral proteins (Faraji-Shadan *et al.*, 1990; Fig. 5b). To immunize against HIV, for example, naive B cells would be collected from an infected patient and exposed *in vitro* to the virus-specific reverse transcriptase (RT) or *tat* proteins. Specific mRNA from these cells would be used to generate cDNA molecules encoding anti-RT or anti-*tat* immunoglobulins. The cDNAs would be modified to target their proteins intracellularly and used to transfect the hematopoietic stem cells of the patient in an *ex vivo* protocol. The transformed stem cells would be transplanted autologously and, ideally, would give rise to lymphocytes that would express resistance to HIV.

Transgenic mice have been developed as useful models of genetically engineered viral immunity. Transgenic mice that express the mouse mammary tumor virus (MMTV) open reading frame protein (ORF) gene were reported to resist infection with milk-borne MMTV (Golovkina *et al.*, 1992). It has been determined that the ORF protein is associated with the minor lymphocyte-stimulating loci (i.e., self-superantigens; Janeway, 1990) of the MMTV. A specific subset of T cells (V-β-14⁺) is deleted in the ORF transgenic mice, and this event precludes MMTV infection. These observations indicate that MMTV infection involves specific cells of the immune system and that animals with endogenous MMTV should be resistant to infection with exogenous virus. The HIV virus might encode a self-superantigen (Imberti *et al.*, 1991), and the deletion of specific T-cell populations or the transgenic expression of superantigens may lead to the development of successful HIV gene therapies. Another transgenic mouse model for viral resistance was described in which the animals expressed a metallothionein/human β_1-interferon construct (MT/HuIFN-β_1; Chen *et al.*, 1988). Fibroblast-produced β-interferons are involved in resistance to viral infections. Transgenic mice that expressed the human interferon yielded sera which protected cultured human cells from infection with vesicular stomatitis virus. The animals exhibited significantly enhanced resistance to infection with pseudorabies virus.

C. Virus-Mediated Molecular Surgery

Virus-directed enzyme/prodrug therapy (VDEPT; Huber *et al.*, 1991) represents a very promising strategy designed to selectively kill neoplastic cells *in situ*. This

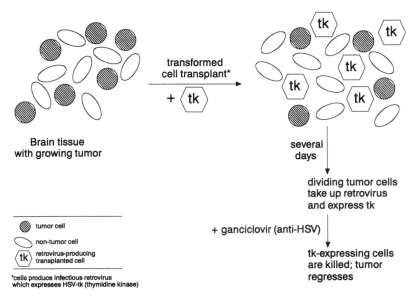

Figure 6. Virus-directed enzyme/prodrug therapy.

strategy utilizes retrovirus-mediated transgenes to selectively target neoplastic cells based upon the transcriptional differences between cancer cells and normal cells (Fig. 6). The retroviral vector includes a transgene encoding a nonmammalian enzyme that is capable of metabolically transforming a nontoxic prodrug into a cytotoxic anabolite. If the transgene is linked to a tumor-specific regulatory sequence, then transgene expression and the activation of the cytotoxic metabolite will be restricted to neoplastic cells. Neighboring normal cells should be unaffected.

One VDEPT strategy (Huber *et al.*, 1991) exploited the selective toxicity of 6-methoxypurine arabinonucleoside [9-(β-D-arabinofuranosyl)-6-methoxy-9H-purine (araM)] for cells infected with the varicella-zoster virus (VZV; Averett *et al.*, 1991). The selective distinction that determines toxicity in VZV-infected cells relies on the efficient interaction between araM and VZV thymidine kinase (TK) compared to mammalian nucleoside kinases. The araM substrate is metabolized in cells expressing the VZV-TK enzyme. Monophosphorylated araM eventually yields a cytotoxic metabolite, adenine arabinonucleoside triphosphate (araATP). Therefore, the selective expression of the viral genome within tumor cells treated with the prodrug araM should provide a highly accurate chemotherapeutic gene therapy. In experiments designed to test gene therapy for hepatocellular carcinoma, clonal hepatoma cells were transformed with retroviral vectors carrying the VZV-TK gene under the transcriptional control of the hepatoma-associated α-fetoprotein promoter or the human albumin promoter (Huber *et al.*, 1991). The subsequent application of araM to transformed hepatoma cells expressing VZV-TK validated the potential utility of this strategy. The araM was metabolized predictably and produced a selective cytotoxicity.

The therapeutic success of using transplanted transgenic somatic cells is promising to treat disease in the immunologically privileged brain (Widner and Brundin, 1988). Also, the inoperability of many brain tumors and the impermeability of the blood–brain barrier underscore the urgent need for effective molecular therapies. A different type of virus-mediated strategy has been described in which transgenic murine fibroblasts expressed the herpes simplex virus thymidine kinase (HSV-TK) gene (Culver *et al.*, 1992). The transformed cells were implanted directly into rat cerebral glioma tumors. During 5 days of *in situ* production of TK-bearing retroviral vectors, glioma cells were transformed selectively, and nondividing neural cells were not infected. The treated rats received injections of the anti-herpes drug gancyclovir, which induced dramatic regression of the brain tumors. An interesting "bystander" phenomenon was described for the complete regression of experimental tumors containing less than 50% HSV-TK-expressing cells. Normal neighboring cells were minimally affected even when underlying nontransformed tumor cells were destroyed. The bystander effect was observed both for experimental and established tumors. The detrimental effects of the uptake and expression of escaped amphotropic retroviruses at body sites other than the brain must be considered when administering gancyclovir. In these studies, no cytopathic events were detected in distal rapidly proliferating tissues such as intestinal epithelium, thymus, or bone marrow. Treated rats were observed to remain tumor-free for greater than 100 days after the administration of gancyclovir.

D. Potential Gene Therapy Targets

The creation of the first somatic cell transgenic humans has been reported. The initial projects chosen for human gene therapy trials concerned the correction of adenosine deaminase (ADA) deficiency (Anderson, 1990) and the manipulation of tumor-infiltrating lymphocytes (Rosenberg *et al.*, 1990). In June 1992, the National Institutes of Health recombinant DNA advisory committee (RAC) approved a protocol for the clinical trial of inserting viral genes into human brain tumors (discussed in Section III, C above). ADA deficiency is considered to present an excellent candidate for gene therapy. It represents a single gene defect which might be fully corrected by replacement therapy and, importantly, does not respond to any current treatments. The lethal immune dysfunction resulting from ADA deficiency may be overcome by the transplantation of bone marrow or lymphocytes which express a normal ADA transgene (Kantoff *et al.*, 1987). Other purine metabolic disorders that may yield to gene therapy include purine nucleoside phosphorylase (PNP) deficiency, which causes a severe immune dysfunction, and hypoxanthine–guanine phosphoribosyltransferase (HPRT) deficiency, the cause of Lesch–Nyhan syndrome.

Other likely candidates for human gene therapy include Gaucher disease type I (glucocerebrosidase deficiency) and phenylketonuria (phenylalanine hydroxylase deficiency) (Ledley *et al.*, 1986). Functional correction of cell-specific enzyme deficiencies, in which accurate cellular targeting of a replacement protein is required, represents a more difficult challenge. Examples of this type of disease would be Tay-Sachs disease (hexosaminidase A deficiency). Ideal candidate genes for gene therapy would be involved in recessive diseases rather than dominant disorders. Loss of function genetic problems will be corrected more easily than dominant mutations until homologous recombination methodology is perfected. The ideal correctable genetic defect also should be a single-gene defect that yields a monomeric protein. The ideal gene therapy should not depend on precise regulation of transgene expression: overexpression must not be detrimental. Successful gene therapy will depend on reliable transgene expression *in situ*. Even though constitutive expression of therapeutic genes may be satisfactory, the development of reliably inducible transgene regulatory elements will elicit a steep increase in human gene therapy trials.

Experiments directed toward potential gene therapy for classic genetic diseases such as cystic fibrosis and muscular dystrophy have been discussed in this chapter. Hypercholesterolemia is another disorder that eventually may be treated with gene therapy. One proposed protocol would involve partial hepatectomy and the introduction of retrovirus-transformed hepatocytes expressing the gene for the low-density lipoprotein receptor (Wilson and Chowdhury, 1990). Other cardiovascular gene therapies utilizing vascular endothelial cells were discussed previously in this chapter. Cardiovascular diseases may be treated by the localized transgenic expression of thrombolytic agents (i.e., tissue plasminogen activator) or by promoting or inhibiting molecules implicated directly in hypertension, such as atrial natriuretic factor.

The correction of metabolic disorders, especially diabetes and obesity, also should be possible using straightforward replacement therapies. Transformed fibroblasts, hepatocytes, and endothelial cells should be able to deliver therapeutic systemic levels of various transgenic proteins including insulin, vasopressin, and lipolytic factors.

Hemoglobinopathies, such as the thalassemia disorders which involve defective β-globin chain synthesis, are another group of classic genetic disorders which may be corrected by transgene therapy. However, immune responses must be expected if the production of a novel protein corrects a loss of function mutation in an individual described as "cross-reacting-material negative" (CRM$^-$). Hemophilia is another candidate for gene therapy for which current treatment is considered suboptimal. Exact regulation of transgenic clotting factors, especially factors VIII and IX, should not be necessary to yield increased hemostatic performance. Skeletal muscle, endothelial cells, hepatocytes, fibroblasts, and bone marrow cells each may provide adequate replacement levels of the clotting factors. The large size of genomic transgenes like factor VIII may be a problem for conventional retroviral vector-mediated therapies.

Additional genetic disease candidates for therapy should include α_1-antitrypsin-mediated emphysema and cystic fibrosis. Genetic factors in the development of cancer include protooncogenes, metabolic enzymes, and tumor suppressor genes. Antineoplastic gene therapies may involve the transplantation into patients of *ex vivo* transformed tumor cells expressing IL-2 or tumor necrosis factor. Other strategies would sensitize the immune system to tumor cells by elaborating viral proteins on tumor cell surfaces and killing the cells with gancyclovir. The expression of specific histocompatability antigens on tumor cells also has been proposed. Similar strategies may be successful for sensitizing the immune system to HIV-1 virus proteins.

ACKNOWLEDGMENTS

The author thanks graphic artist Robert Zaccardi for preparing the figures that accompany this chapter.

REFERENCES

Anderson, W. F. (1990) September 14, 1990: The beginning. *Hum. Gene Therapy* **1**, 371–372.

Arnheiter, H., Skuntz, S., Noteborn, M., Chang, S., and Meier, E. (1990). Transgenic mice with intracellular immunity to influenza virus. *Cell (Cambridge, Mass.)* **62**, 51–61.

Arrezo, F. (1989). Sea urchin sperm as a vector of foreign genetic information. *Cell Biol. Int. Rep.* **13**, 391–404.

Atkinson, P. W., Hines, E. R., Beaton, S., Matthaei, K. I., Reed, K. C., and Bradley, M. P. (1991). Association of exogenous DNA with cattle and insect spermatozoa *in vitro*. *Mol. Reprod. Dev.* **29**, 1–5.

Averett, D. R., Koszalka, G. W., Fyfe, J. A., Roberts, G. B., Purifoy, D. J. M. and Krenitsky, T. A. (1991). 6-Methoxypurine arabinoside as a selective and potent inhibitor of varicella zoster virus. *Antimicrob. Agents Chemother.* **35**, 851–857.

Axelrod, J. H., Read, M., Brinkhous, K. M., and Verma, I. M. (1990). Phenotypic correction of factor IX deficiency in skin fibroblasts of hemophiliac dogs. *Proc. Natl. Acad. Sci. U.S.A.* **87**, 5173–5177.

Baltimore, D. (1988). Intracellular immunization. *Nature (London)* **335**, 395–396.

Baringer, J. R., and Sworeland, P. (1973). Recovery of herpes-simplex virus from human trigeminal ganglions. *N. Engl. J. Med.* **228**, 648–650.

Barr, E., and Leiden, J. M. (1991). Systemic delivery of recombinant proteins by genetically modified myoblasts. *Science* **254**, 1507–1509.

Belmont, J. W., Henkel-Tigges, J., Chang, S. M., Wager-Smith, K., Kellems, R. E., Dick, J. E., Magli, M. C., Phillips, R. A., Bernstein, A., and Caskey, C. T. (1986). Expression of human adenosine deaminase in murine hematopoietic progenitor cells following retroviral transfer. *Nature (London)* **322**, 385–387.

Bender, M. A., Gelinas, R. E., and Miller, A. D. (1989). A majority of mice show long-term expression of a human β-globin gene after retrovirus transfer into hematopoietic stem cells. *Mol. Cell. Biol.* **9**, 1426–1434.

Bennett, V. J., and Chang, P. L. (1990). Suppression of immunological response against a novel gene product delivered by implants of genetically modified fibroblasts. *Mol. Biol. Med.* **7**, 471–477.

Berkner, K. L. (1988). Development of adenovirus vectors for the expression of heterologous genes. *BioTechniques* **6**, 616–629.

Blau, H. M., and Webster, C. (1981). Isolation and characterization of human muscle cells. *Proc. Natl. Acad. Sci. U.S.A.* **78**, 5623–5627.

Boat, T. F., Welsh, M. J., and Beaudet, A. L. (1989). Cystic fibrosis. *In* "The Metabolic Basis of Inherited Disease" (C. R. Scriver, A. L. Beaudet, W. S. Sly, and D. Valle, eds.), pp. 2649–2680. McGraw Hill, New York.

Bodine, D. M., Karlsson, S., and Nienhuis, A. W. (1989). Combination of interleukins 3 and 6 preserves stem cell function in culture and enhances retrovirus mediated gene transfer into hematopoietic stem cells. *Proc. Natl. Acad. Sci. U.S.A.* **86**, 8897–8901.

Brackett, B. G., Baranska, W., Sawicki, W., and Koprowski, H. (1971). Uptake of heterologous genome by mammalian spermatozoa and its transfer to ova through fertilization. *Proc. Natl. Acad. Sci. U.S.A.* **68**, 353–357.

Breakefield, X. O. (1989). Combining CNS transplantation and gene transfer. *Neurobiology (Copenhagen)* **1**, 339–371.

Brinster, R. L., Sandgren, E. P., Behringer, R. R., and Palmiter, R. D. (1989). No simple solution for making transgenic mice. *Cell (Cambridge, Mass.)* **59**, 239–241.

Brown, D., Yu, Z. P., Miller, P., Blake, K., Wei, C., Kung, H. F., Black, R. J., Ts'o, P. O., and Chang, E. H. (1989). Modulation of *ras* expression by anti-sense, nonionic deoxyoligonucleotide analogs. *Oncogene Res.* **4**, 243–249.

Callow, A. D. (1990). The vascular endothelial cell as a vehicle for gene therapy. *J. Vasc. Surg.* **11**, 793–798.

Casali, P., and Notkins, A. L. (1989). Probing the human B-cell repertoire with EBV: polyreactive antibodies and CD5+ B lymphocytes. *Annu. Rev. Immunol.* **7**, 513–535.

Castro, F. O., Hernandez, O., Uliver, C., Solano, R., Milanes, C., Aguilar, A., Perez, A., De Armas, R., Herrera, L., and De La Fuente, J. (1991). Introduction of foreign DNA into the spermatozoa of farm animals. *Theriogenology* **34**, 1099–1110.

Chang, A. C. Y., and Brenner, D. G. (1989). Cationic liposome-mediated transfection: a new method for the introduction of DNA into mammalian cells. *Focus* **10**, 66–69.

Chang, P. L., Capone, J. P., and Brown, G. M. (1990). Autologous fibroblast implantation. Feasibility and potential problems in gene replacement therapy. *Mol. Biol. Med.* **7**, 461–470.

Chen, X.-Z., Yun, J. S., and Wagner, T. E. (1988). Enhanced viral resistance in transgenic mice expressing the human β_1-interferon. *J. Virol* **62**, 3883–3887.

Colombo, M. P., Ferrari, G., Stoppacciaro, A., Parenza, M., Rodolfo, M., Mavilio, F., and Parmiani, G. (1991). Granulocyte colony-stimulating factor gene transfer suppresses tumorigenicity of a murine adenocarcinoma *in vivo*. *J. Exp. Med.* **173**, 889–897.

Cook, M. L., Bastone, V. B., and Stevens, J. G. (1974). Evidence that neurons harbor latent herpes simplex virus. *Infect. Immun.* **9**, 946–951.

Costantini, F., Chada, K., and Magram, J. (1986). Correction of murine β-thalassemia by gene transfer into the germ line. *Science* **233**, 1192–1194.

Crystal, R. G. (1990). α-Antitrypsin deficiency, emphysema and liver disease. Genetic basis and strategies for therapy. *J. Clin. Invest.* **85**, 1343–1352.

Culver, K. W., Ram, Z., Wallbridge, S., Ishii, H., Oldfield, E. H., and Blaese, R. M. (1992). *In vivo* gene transfer with retroviral vector-producer cells for treatment of experimental brain tumors. *Science* **256**, 1550–1552.

Demetriou, A. A., Whiting, J. F., Feldman, D., Levenson, S. M., Chowdhury, N. R., Moscioni, A. D., Kram, M., and Chowdhury, J. R. (1986). Replacement of liver function in rats by transplantation of microcarrier-attached hepatocytes. *Science* **233**, 1190–1192.

Dervan, P. B. (1989). Oligonucleotide recognition of double-helical DNA by triple-helix formation. *In* "Oligodeoxynucleotides: Antisense Inhibitors of Gene Expression" (J. S. Cohen, ed.), pp. 197–210. CRC Press, Boca Raton, Florida.

Desrosiers, R. C., Burghoff, R. L., Bakker, A., and Kamine, J. (1984). Construction of replication-competent *Herpesvirus saimiri* deletion mutants. *J. Virol.* **49**, 343–348.

Desrosiers, R. C., Kamine, J., Bakker, A., Silva, D., Woychik, R. P., Sakai, D. D., and Rottman, F. M. (1985). Synthesis of bovine growth hormone in primates by using a herpesvirus vector. *Mol. Cell. Biol.* **5**, 2796–2803.

Dhawan, J., Pan, L. C., Pavlath, G. K., Travis, M. A., Lanctot, A. M., and Blau, H. M. (1991). Systemic delivery of human growth hormone by injection of genetically engineered myoblasts. *Science* **254**, 1509–1512.

Dichek, D. A., Neville, R. F., Zwiebel, J. A., Freeman, S. M., Leon, M. B., and Anderson, W. F. (1989). Seeding of intravascular stents with genetically engineered endothelial cells. *Circulation* **80**, 1347–1353.

Dick, J. E., Magli, M. C., Huszar, D., Phillips, R. A., and Bernstein, A. (1985). Introduction of a selectable gene into primitive stem cells capable of long-term reconstitution of the hematopoietic system of *W/W*v mice. *Cell (Cambridge, Mass.)* **42**, 71–79.

Doehmer, J., Barinaga, M., Vale, W., Rosenfeld, M. G., Verma, I. M., and Evans, R. M. (1982). Introduction of rat growth hormone gene into mouse fibroblasts via a retroviral DNA vector: Expression and regulation. *Proc. Natl. Acad. Sci. U.S.A.* **79**, 2268–2272.

Drumm, M. L., Pope, H. A., Cliff, W. H., Rommens, J. M., Marvin, S. A., Tsui, L.-C., Collins, F. S., Frizzell, R. A., and Wilson, J. M. (1990). Correction of the cystic fibrosis defect *in vitro* by retrovirus-mediated gene transfer. *Cell (Cambridge, Mass.)* **62**, 1227–1233.

Dubensky, T. W., Campbell, B. A., and Villareal, L. P. (1984). Direct transfection of viral and plasmid DNA into the liver and spleen of mice. *Proc. Natl. Acad. Sci. U.S.A.* **81**, 7529–7533.

Dzierzak, E. A., Papayannopoulou, T., and Mulligan, R. C. (1988). Lineage-specific expression of a human β-globin gene in murine bone marrow transplant recipients reconstituted with retrovirus-transduced stem cells. *Nature (London)* **331**, 35–41.

Edwards, N. L., Jeryc, W., and Fox, I. H. (1984). Enzyme replacement in the Lesch–Nyhan syndrome with long-term erythrocyte transfusions. *Adv. Exp. Med. Biol.* **165A**, 23–26.

Eglitis, M. A., Kantoff, P. W., Jolly, J. D., Jones, J. B., Anderson, W. F., and Lothrop, C. D., Jr. (1988). Gene transfer into hematopoietic progenitor cells from normal and cyclic hematopoietic dogs using retroviral vectors. *Blood* **71**, 717–722.

Eicher, E. M., and Beamer, W. G. (1976). Inherited ateliotic dwarfism in mice. Characteristics of the mutation, little, on chromosome 6. *J. Hered.* **67**, 87–91.

Faraji-Shadan, F., Stubbs, J. D., and Bowman, P. D. (1990). A putative approach for gene therapy against human immunodeficiency virus (HIV). *Med. Hypotheses* **32**, 81–84.

Felgner, P. L., and Ringold, G. M. (1989). Cationic liposome-mediated transfection. *Nature (London)* **337**, 387–388.

Felgner, P. L., Gadek, T. R., Holm, M., Roman, R., Chan, H. W., Wenz, M., Northrop, J. P., Ringold, G. M., and Danielsen, M. (1987). Lipofection: A highly efficient, lipid-mediated DNA-transfection procedure. *Proc. Natl. Acad. Sci. U.S.A.* **84**, 7413–7417.

Ferrari, G., Rossini, S., Giavazzi, R., Maggioni, D., Nobili, N., Soldati, M., Ungers, G., Mavilio, F., Gilboa, E., and Bordignon, C. (1991). An *in vivo* model of somatic cell gene therapy for human severe combined immunodeficiency. *Science* **251**, 1363–1366.

Ferry, N., Duplessis, O., Houssin, D., Danos, O., and Heard, J.-M. (1991). Retroviral-mediated gene transfer into hepatocytes *in vivo*. *Proc. Natl. Acad. Sci. U.S.A.* **88**, 8377–8381.

Fraley, R., Subrami, S., Berg, P., and Papahadjopoulos, D. (1980). Introduction of liposome-encapsulated SV40 DNA into cells. *J. Biol. Chem* **255**, 10431–10435.

Friedman, A. D., Triezenberg, S. J., and McKnight, S. L. (1988). Expression of a truncated viral trans-activator selectively impedes lytic infection by its cognate virus. *Nature (London)* **335**, 452–454.

Friedman, J. M., Babiss, L. E., Clayton, D. F., and Darnell, J. E. (1986). Cellular promoters incorporated into the adenovirus genome: Cell-specificity of albumin and immunoglobulin expression. *Mol. Cell. Biol.* **6**, 3791–3797.

Friedmann, T., Xu, L. Wolff, J., Yee, J.-K., and Miyanohara, A. (1989). Retrovirus vector-mediated gene transfer into hepatocytes. *Mol. Biol. Med.* **6**, 117–125.

Gage, F. H., Wolff, J. A., Rosenberg, M. B., Xu, L., Yee, J.-K., Shults, C., and Friedman, T. (1987). Grafting genetically modified cells to the brain: Possibilities for the future. *Neuroscience (Oxford)* **23**, 795–807.

Gagne, M. B., Pothier, F., and Sirard, M.-A. (1991). Electroporation of bovine spermatozoa to carry foreign DNA in oocytes. *Mol. Reprod. Dev.* **29**, 6–15.

Gallico III, G. G., O'Connor, N. E., Compton, C. C., Kehinde, O., and Green, H. (1984). Permanent coverage of large burn wounds with autologous cultured human epithelium. *N. Engl. J. Med.* **311**, 448–451.

Galski, H., Sullivan, M., Willingham, M. C., Chin, K.-V., Gottesman, M. M., Pastan, I., and Merlino, G. T. (1989). Expression of a human multidrug resistance cDNA (MDR1) in the bone marrow of transgenic mice: Resistance to daunomycin-induced leukopenia. *Mol. Cell. Biol.* **9**, 4357–4363.

Gandolfi, F., Lavitrano, M., Camaioni, A., Spadafora, C., Siracusa, G., and Lauria, A. (1989). The use of sperm-mediated gene transfer for the generation of transgenic pigs. *J. Reprod. Fertil. Abstr. Ser.* **4**, 10.

Gilardi, P., Courtney, M., Pavirani, A., and Perricaudet, M. (1990). Expression of human α_1-antitrypsin using a recombinant adenovirus vector. *FEBS Lett.* **267**, 60–62.

Golovkina, T. V., Chervonsky, A., Dudley, J. P., and Ross, S. R. (1992). Transgenic mouse mammary tumor virus superantigen expression prevents viral infection. *Cell (Cambridge, Mass.)* **69**, 637–645.

Gordon, J. W., Scangos, G. A., Plotkin, D. J., Barbosa, J. A., and Ruddle, F. H. (1980). Genetic transformation of mouse embryos by microinjection of purified DNA. *Proc. Natl. Acad. Sci. U.S.A.* **77**, 7380–7384.

Graham, F. L., and van der Eb, A. J. (1973). A new technique for the assay of infectivity of human adenovirus 5 DNA. *Virology* **52**, 456–467.

Gruber, H. E., Finley, K. D., Hershberg, R. M., Katzman, S. S., Laikind, P. K., Seegmiller, J. E., Friedmann, T., Yee, J.-K., and Jolly, D. J. (1985). Retroviral vector-mediated gene transfer into human hematopoietic progenitor cells. *Science* **230**, 1057–1061.

Gupta, S., Johnstone, R., Darby, H., Selden, C., Price, Y., and Hodgson, H. J. F. (1987). Transplanted isolated hepatocytes: Effect of partial hepatectomy on proliferation of long-term syngeneic implants in rat spleen. *Pathology* **19**, 28–30.

Gupta, S., Aragona, E., Vemuru, R. P., Bhargava, K. K., Burk, R. D., and Chowdhury, J. R. (1991). Permanent engraftment and function of hepatocytes delivered to the liver: Implications for gene therapy and liver repopulation. *Hepatology (N.Y.)* **14**, 144–149.

Hammer, R. E., Palmiter, R. D., and Brinster, R. L., (1984). Partial correction of murine hereditary growth disorder by germ-line incorporation of a new gene. *Nature (London)* **311**, 65–67.

Han, L., Yun, J. S., and Wagner, T. E. (1991). Inhibition of Moloney murine leukemia virus-induced leukemia in transgenic mice expressing antisense RNA complementary to the retroviral packaging sequences. *Proc. Natl. Acad. Sci. U.S.A.* **88**, 4313–4317.

Hatzoglou, M., Park, E., Wynshaw-Boris, A., Kaung, H. C., and Hanson, R. W. (1988). Hormonal regulation of chimeric genes containing the phosphoenolpyruvate carboxykinase promoter regulatory region in hepatoma cells infected by murine retroviruses. *J. Biol. Chem.* **263**, 17798–17808.

Hatzoglou, M., Lamers, W., Bosch, F., Wynshaw-Boris, A., Clapp, D. W., and Hanson, R. W. (1990). Hepatic gene transfer in animals using retroviruses containing the promoter from the gene for phosphoenolpyruvate carboxykinase. *J. Biol. Chem.* **265**, 17285–17293.

Heath, T. D., Fraley, R. T., and Papahadjopoulos, D. (1981). Antibody targeting of liposomes: Cell specificity obtained by conjugation of F(ab')$_2$ to vesicle surface. *Science* **210**, 599–601.

Herskowitz, I. (1987). Functional inactivation of genes by dominant negative mutations. *Nature (London)* **329**, 219–222.

Hochi, S., Ninomiya, T., Mizuno, A., Honma, M., and Yuki, A. (1990). Fate of exogenous DNA carried into mouse eggs by spermatozoa. *Anim. Biotechnol.* **1**, 25–30.

Hock, R. A., and Miller, A. D. (1986). Retrovirus-mediated transfer and expression of drug resistance genes in human haematopoietic progenitor cells. *Nature (London)* **320**, 275–277.

Hodgson, G. S., and Bradley, T. R. (1979). Properties of haematopoietic stem cells surviving 5-fluorouracil treatment: Evidence for a pre-CFU-S cell? *Nature (London)* **281**, 381–382.

Holt, C. E., Garlick, N., and Cornel, E. (1990). Lipofection of cDNAs in the embryonic vertebrate central nervous system. *Neuron* **4**, 203–214.

Horan, R., Powell, R., McQuaid, S., Gannon, F., and Houghton, J. A. (1991). The association of foreign DNA with porcine spermatozoa. *Arch. Androl.* **26**, 83–92.

Horellou, P., Guibert, B., Leviel, V., and Mallet, J. (1989). Retroviral transfer of a human tyrosine hydroxylase cDNA in various cell lines: Regulated release of dopamine in mouse anterior pituitary AtT-20 cells. *Proc. Natl. Acad. Sci. U.S.A.* **86**, 7233–7237.

Horellou, P., Marlier, L., Privat, A., and Mallet, J. (1990). Behavioural effect of engineered cells that synthesize L-DOPA or dopamine after grafting into the rat neostriatum. *Eur. J. Neurosci.* **2**, 116–119.

Huang, A., Huang, L., and Kennel, S. J. (1980). Monoclonal antibody covalently coupled with fatty acid—A reagent for *in vitro* liposome targeting. *J. Biol. Chem.* **255**, 8015–8018.

Huang, A., Kennel, S. J., and Huang, L. (1983). Interactions of immunoliposomes with target antigens. *J. Biol. Chem.* **248**, 14034–14040.

Hubbard, R. C., McElvaney, N. G., Birrer, P., Shak, S., Robinson, W. W., Jolley, C., Wu, M., Chernick, M. S., and Crystal, R. G. (1992). A preliminary study of aerosolized recombinant human deoxyribonuclease I in the treatment of cystic fibrosis. *N. Engl. J. Med.* **326**, 812–815.

Huber, B. E., Richards, C. A., and Krenitsky, T. A. (1991). Retroviral-mediated gene therapy for the treatment of hepatocellular carcinoma: An innovative approach for cancer therapy. *Proc. Natl. Acad. Sci. U.S.A.* **88**, 8039–8043.

Hug, H., Costas, M., Staeheli, P., Aebi, M., and Weissmann, C. (1988). Organization of the murine *Mx* gene and characterization of its interferon- and virus-inducible promoter. *Mol. Cell. Biol.* **8**, 3065–3079.

Hughes, S. M., and Blau, H. M. (1990). Migration of myoblasts across basal lamina during skeletal muscle development. *Nature (London)* **345**, 350–353.

Imberti, L., Sottini, A., Bettinardi, A., Puoti, M., and Primi, D. (1991). Selective depletion in HIV infection of T cells that bear specific T cell receptor V-beta sequences. *Science* **254**, 860–862.

Isom, H. C., and Georgoff, I. (1984). Quantitative assay for albumin-producing liver cells after simian virus 40 transformation of rat hepatocytes maintained in chemically defined medium. *Proc. Natl. Acad. Sci. U.S.A.* **81**, 6378–6382.

Jaenisch, R. (1976). Germ line integration and Mendelian transmission of the exogenous Moloney leukemia virus. *Proc. Natl. Acad. Sci. U.S.A.* **73**, 1260–1264.

Jaenisch, R., and Mintz, B. (1974). Simian virus 40 DNA sequences in DNA of healthy adult mice derived from preimplantation blastocysts injected with viral DNA. *Proc. Natl. Acad. Sci. U.S.A.* **71**, 1250–1254.

Jaffe, V., Darby, H., Selden, C., and Hodgson, H. J. F. (1988). Growth and proliferation of transplanted hepatocytes in rat pancreas. *Transplantation* **45.2**, 497.

Janeway, C. A., Jr. (1990). Self superantigens? *Cell (Cambridge, Mass.)* **63**, 659–661.

Jirtle, R. L., and Michalopoulos, G. (1982). Effects of partial hepatectomy on transplanted hepatocytes. *Cancer Res.* **42**, 3000–3004.

Johnson, G. R., and Metcalf, D. (1977). Pure and mixed erythroid colony formation *in vitro* stimulated by spleen conditioned medium with no detectable erythropoietin. *Proc. Natl. Acad. Sci. U.S.A.* **74**, 3878–3882.

Joyner, A., Keller, G., Phillips, R. A., and Bernstein, A. (1983). Retrovirus transfer of a bacterial gene into mouse hematopoietic progenitor cells. *Nature (London)* **305**, 556–558.

Kalvakolanu, D. V. R., and Abraham, A. (1991). Preparation and characterization of immunoliposomes for targeting of antiviral agents. *BioTechniques* **11**, 218–225.

Kaneda, Y., Uchida, T., Kim, J., Ishiura, M, and Okada, Y. (1987). The improved efficient method for introducing macromolecules into cells using HVJ (Sendai virus) liposomes with gangliosides. *Exp. Cell Res.* **173**, 56–69.

Kantoff, P. W., Flake, A. W., Eglitis, M. A., Scharf, S., Bond, S., Gilboa, E., Erlich, H., Harrison, M. R., Zanjani, E., and Anderson W. F. (1989). *In utero* gene transfer and expression: A sheep transplantation model. *Blood* **73**, 1066–1074.

Kantoff, P. W., Gillio, A., McLachlin, J., Bordignon, C., Eglitis, M. A., Kernan, N. A., Moen, R. C., Kohn, D. B., Yu, S.-F., Karson, E., Karlsson, S., Zwiebel, J. A., Gilboa, E., Blaese R. M., Nienhuis, A. W., O'Reilly, R. J., and Anderson, W. F. (1987). Expression of human adenosine deaminase in non-human primates after retroviral mediated gene transfer. *J. Exp. Med.* **166**, 219–234.

Kantoff, P. W., Flake, A. W., Eglitis, M. A., Scharf, S., Bond, S., Gilboa, E., Erlich, H., Harrison, M. R., Zanjani, E., and Anderson W. F. (1989). *In utero* gene transfer and expression: A sheep transplantation model. *Blood* **73**, 1066–1074.

Katsuki, M., Sato, M., Kimura, M., Yokoyama, M., Kobayashi, K., and Nomura, T. (1988). Conversion of normal behavior to shiverer by myelin basic protein antisense cDNA in transgenic mice. *Science* **241**, 593–595.

Keller, G., Paige, C., Gilboa, E., and Wagner, E. F. (1985). Expression of a foreign gene in myeloid and lymphoid cells derived from multipotent haematopoietic precursors. *Nature (London)* **318**, 149–154.

Kusano, M., and Mito, M. (1982). Observations on the fine structure of long survived isolated hepatocytes inoculated into rat spleen. *Gastroenterology* **82**, 616–628.

Kwok, W. W., Scheming, F., Stead, R. B., and Miller, A. D. (1986). Retroviral transfer of genes into canine hematopoietic progenitor cells in culture: A model for human gene therapy. *Proc. Natl. Acad. Sci. U.S.A.* **83**, 4552–4555.

Lapidot, M., and Loyter, A. (1990). Fusion-mediated microinjection of liposome-enclosed DNA into cultured cells with the aid of influenza virus glycoproteins. *Exp. Cell Res.* **189**, 241–246.

Lavitrano, M., Camaioni, A., Fazio, V. M., Dolci, S., Farace, M. G., and Spadafora, C. (1989). Sperm cells as vectors for introducing foreign DNA into eggs: Genetic transformation of mice. *Cell (Cambridge, Mass.)* **57**, 717–723.

Lavitrano, M., French, D., Zani, M., Frati, L., and Spadafora, C. (1992). The interaction between exogenous DNA and sperm cells. *Mol. Reprod. Dev.* **31**, 161–169.

Ledley, F. D., Grenett, H. E., McGinnis-Shelmutt, M., and Woo, S. L. C. (1986). Retroviral mediated gene transfer of human phenylalanine hydroxylase into NIH 3T3 and hepatoma cells. *Proc. Natl. Acad. Sci. U.S.A.* **83**, 409–413.

Ledley, F. D., Grenett, H. E., Bartos, D. P., and Woo, S. L. (1987). Retroviral-mediated transfer and expression of α-anti-trypsin in cultured cells. *Gene* **61**, 113–118.

Leiter, J. M. E., Agrawal, S., Palese, P., and Zamecnik, P. C. (1990). Inhibition of influenza virus replication of phosphorothioate oligodeoxyribonucleotides. *Proc. Natl. Acad. Sci. U.S.A.* **87**, 3430–3434.

Lesch, M., and Nyhan, W. L. (1964). A familial disorder of uric acid metabolism and central nervous system function. *Am. J. Med.* **36**, 561–570.

Lethem, M. I., James, S. L., Marriott, C., and Burke, J. F. (1990). The origin of DNA associated with mucus glycoproteins in cystic fibrosis sputum. *Eur. Respir, J.* **3**, 19–23.

Li, C. L., Dwarki, V. J., and Verma, I. M. (1990). Expression of human α- and mouse/human hybrid β-globin genes in murine hematopoietic stem cells transduced by recombinant retroviruses. *Proc. Natl. Acad. Sci. U.S.A.* **87**, 4349–4353.

Lim, B., Apperley, J. F., Orkin, S. H., and Williams D. A. (1989). Long-term expression of human adenosine deaminase in mice transplanted with retrovirus-infected hematopoietic stem cells. *Proc. Natl. Acad. Sci. U.S.A.* **86**, 8892–8896.

Lindemann, J. (1964). Inheritance of resistance to influenza virus in mice. *Proc. Soc. Exp. Biol. Med.* **116**, 506–509.

Lindemann, J., Lane, C. A., and Hobson, D. (1963). The resistance of A2G mice to myxoviruses. *J. Immunol.* **90**, 942–951.

Lothrop, C. D., Al-Lebban, Z. S., Niemeyer, G. P., Jones, J. B., Peterson, M. G., Smith, J. R., Baker, H. J., Morgan, R. A., Eglitis, M. A., and Anderson, W. F. (1991). Expression of a foreign gene in cats reconstituted with retroviral vector infected autologous bone marrow. *Blood* **78**, 237–245.

McGrane, M. M., deVente, J., Yun, J., Bloom, J., Park, E., Wynshaw-Boris, A., Wagner, T., Rottman, F. M., and Hanson, R. W. (1988). Tissue-specific expression and dietary regulation of a chimeric phosphoenolpyruvate carboxykinase/bovine growth hormone gene in transgenic mice. *J. Biol. Chem,* **263**, 11443–11451.

Martin, F. J., Hubbell, W. J., and Papahadjopoulos, D. (1981). Immunospecific targeting of liposomes to cells: A novel and efficient method for covalent attachment of Fab fragments via disulfide bonds. *Biochemistry* **20**, 4229–4238.

Mengle-Gaw, L., and McDevitt, H. O. (1985). Genetics and expression of mouse Ia antigens. *Annu. Rev. Immunol.* **3**, 367–396.

Metcalf, D., Parker, J., Chester, H. M., and Kincade, P. W. (1974). Formation of eosinophil-like granulocytic colonies by mouse bone marrow cells *in vitro. J. Cell. Physiol.* **84**, 275–289.

Metcalf, D., MacDonald, H. R., Odartchenko, N., and Sordat, B. (1975). Growth of mouse megakaryocyte colonies *in vitro. Proc. Natl. Acad. Sci. U.S.A.* **72**, 1744–1748.

Mickisch, G. H., Merlino, G. T., Galski, H., Gottesman, M. M., and Pastan, I. (1991). Transgenic mice that express the human multidrug-resistance gene in bone marrow enable a rapid identification of agents that reverse drug resistance. *Proc. Natl. Acad. Sci. U.S.A.* **88**, 547–551.

Miller, A. D., Jolly, D. J., Friedmann, T., and Verma, I. M. (1983). A transmissible retrovirus expressing human hypoxanthine phosphoribosyltransferase (HPRT) gene: Gene transfer into cells obtained from humans deficient in HPRT. *Proc. Natl. Acad. Sci. U.S.A.* **80**, 4709–4713.

Miller, A. D., Eckner, R. J., Jolly, D. J., Friedmann, T., and Verma, I. M. (1984). Expression of a retrovirus encoding human HPRT in mice. *Science* **225**, 630–632.

Miller, A. D., Bender, M. A., Harris, E. A. S., Kaleko, M., and Gelinas, R. E. (1988). Design of retrovirus vectors for transfer and expression of the human β-globin gene. *J. Virol.* **62**, 4337–4345. (Erratum: *J. Virol.* **63**, 1493.)

Miller, D. G., Adam, M. A., and Miller, A. D. (1990). Gene transfer by retrovirus vectors occurs only in cells that are actively replicating at the time of infection. *Mol. Cell. Biol.* **10**, 4239–4242. (Erratum: *Mol. Cell. Biol.* **12**, 433.)

Miyanohara, A., Sharkey, M. F., Witztum, J. L., Steinberg, D., and Friedmann, T. (1988). Efficient expression of retroviral vector-transduced human low density lipoprotein (LDL) receptor in LDL receptor-deficient rabbit fibroblasts *in vitro. Proc. Natl. Acad. Sci. U.S.A.* **85**, 6538–6542.

Morgan, J. R., Barrandon, Y., Green, H., and Mulligan, R. C. (1987). Expression of an exogenous growth hormone gene by transplantable human epidermal cells. *Science* **237**, 1476–1479.

Morrison, T., Hinshaw, V. S., Sheerar, M., Cooley, A. J., Brown, D., McQuain, C., and McGinnes, L. (1990). Retroviral expressed hemagglutinin–neuraminidase protein protects chickens from Newcastle disease virus induced disease. *Microb. Pathogen.* **9**, 387–396.

Mosier, D. E., Gulizia, R. J., Baird, S. M., and Wilson, D. B. (1988). Transfer of a functional human immune system to mice with severe combined immunodeficiency. *Nature (London)* **335**, 256–259.

Mulligan, R. C., Howard, B. H., and Berg, P. (1979). Synthesis of rabbit β-globin in cultured monkey kidney cells following infection with a SV40 β-globin recombinant genome. *Nature (London)* **277**, 108–111.

Nabel, E. G., Plautz, G., Boyce, F. M., Stanley, J. C., and Nabel, G. J. (1989). Recombinant gene expression *in vivo* within endothelial cells of the arterial wall. *Science* **244**, 1342–1344.

Neumann, E., Schaefer-Ridder, M., Wang, Y., and Hofschneider, P. H. (1982). Gene transfer into mouse lyoma cells by electroporation in high electric fields. *EMBO J.* **1**, 841–845.

Nicolau, C., LePape, A., Soriano, P., Fargette, F., and Juhe, M.-F. (1983). *In vivo* expression of rat insulin after intravenous administration of the liposome-entrapped gene for rat insulin I. *Proc. Natl. Acad. Sci. U.S.A.* **80**, 1068–1072.

Nyhan, W. L., Parkman, R., Page, T., Gruber, H. E., Pyati, J., Jolly, D., and Friedmann, T. (1986). Bone marrow transplantation in Lesch–Nyhan disease. *Adv. Exp. Med. Biol.* **195A**, 167–170.

O'Reilly, R. J., Keever, C. A., Small, T. N., and Brochstein, J. (1989). The use of HLA-non-identical T-cell-depleted marrow transplants for correction of severe combined immunodeficiency disease. *Immunodefic. Rev.* **1**, 273–309.

Palella, T. D., Silverman, L. J., Schroll, C. T., Homa, F. L., Levine, M., and Kelley, W. N. (1988). Herpes simplex virus-mediated human hypoxanthine–guanine phosphoribosyltransferase gene transfer into neuronal cells. *Mol. Cell. Biol.* **8**, 457–460.

Palmer, T. D., Hock, R. A., Osborne, W. R., and Miller, A. D. (1987). Efficient retrovirus-mediated transfer and expression of a human adenosine deaminase gene in diploid skin fibroblasts from an adenosine deaminase-deficient human. *Proc. Natl. Acad. Sci. U.S.A.* **81**, 1055–1059.

Palmer, T. D., Thompson, A. R., and Miller, A. D. (1989). Production of human factor IX in animals by genetically modified skin fibroblasts: Potential therapy for haemophilia B. *Blood* **73**, 438–445.

Panduro, A., Shalaby, F., Weiner, F. R., Biempica, L., Zern, M. A., and Shafritz, D. A. (1986). Transcriptional switch from albumin to α-fetoprotein and changes in transcription of other genes during carbon tetrachloride induced liver regeneration. *Biochemistry* **25**, 1414–1420.

Papahadjopoulos, D., Mayhew, E., Poste, G., and Smith, S. (1974). Incorporation of lipid vesicles by mammalian cells provides a potential method for modifying cell behaviour. *Nature (London)* **252**, 163–165.

Partridge, T. A., Morgan, J. E., Coulton, G. R., Hoffman, E. P., and Kunkel, L. M. (1989). Conversion of mdx myofibres from dystrophin-negative to positive by injection of normal myoblasts. *Nature (London)* **337**, 176–179.

Peng, H., Armentano, D., MacKenzie-Graham, L., Shen, R.-F., Darlington, G., Ledley, F. D., and Woo, S. L. C. (1988). Retroviral-mediated gene transfer and expression of human phenylalanine hydroxylase in primary mouse hepatocytes. *Proc. Natl. Acad. Sci. U.S.A.* **85**, 8146–8150.

Pera, M. F., Blasco-Lafita, M. J., and Mills, J. (1987). Cultured stem cells from human testicular teratomas: the nature of human embryonal carcinoma and its comparison with two types of yolk-sac carcinoma. *Int. J. Cancer* **40**, 334–343.

Pike, B. L., and Robinson, W. A. (1970). Human bone marrow colony growth in agar-gel. *J. Cell. Physiol.* **76**, 77–84.

Prochownik, E. V., Kukowska, J., and Rodgers, C. (1988). c-*myc* antisense transcripts accelerate differentiation and inhibit G1 progression in murine erythroleukemia cells. *Mol. Cell. Biol.* **8**, 3683–3695.

Pullen, J. K., and Schook, L. B. (1986). Bone marrow-derived macrophage expression of endogenous and transfected class II MHC genes during differentiation *in vitro*. *J. Immunol.* **137**, 1359–1365.

Richa, J., and Lo, C. W. (1989). Introduction of human DNA into mouse eggs by injection of dissected chromosome fragments. *Science* **245**, 175–177.

Riordan, J. R., Rommens, J. M., Kerem, B.-S., Alon, N., Rozmahel, R., Grzelczak, Z., Zielenski, J., Lok, S., Plavsic, N., Chou, J.-L., Drumm, M. L., Iannuzzi, M. C., Collins, F. S., and Tsui, L.-C. (1989). Identification of the cystic fibrosis gene: Cloning and characterization of complementary DNA. *Science* **245**, 1066–1073.

Rosenberg, M. B., Friedmann, T., Robertson, R. B., Tuszynski, M., Wolff, J. A., Breakefield, X. O., and Gage, F. H. (1988). Grafting genetically modified cells to the damaged brain: Restorative effects of NGF expression. *Science* **242**, 1575–1577.

Rosenberg, S. A., Spiess, P., and Lafreniere, R. (1986). A new approach to the adoptive immunotherapy of cancer with tumor-infiltrating lymphocytes. *Science* **233**, 1318–1321.

Rosenberg, S. A., Packard, B. S., Aebersold, P. M., and Solomon, D. (1988). Use of tumor-infiltrating lymphocytes and interleukin-2 in the immunotherapy of patients with metastatic melanoma: A preliminary report. *N. Engl. J. Med.* **319**, 1676–1680.

Rosenberg, S. A., Aebersold, P., Cornetta, K., Kasid, A., Morgan, R. A., Moen, R., Karson, E. M., Lotze, M. T., Yang, J. C., Topalian, S. L., Merino, M. J., Culver, K., Miller, A. D., Blaese, R. M., and Anderson, W. F. (1990). Gene transfer into humans—Immunotherapy of patients with advanced melanoma, using tumor-infiltrating lymphocytes modified by retroviral gene transduction. *N. Engl. J. Med.* **323**, 570–578.

Rosenfeld, M. A., Siegfried, W., Yoshimura, K., Yoneyama, K., Fukayama, M., Stier, L. E., Paakko, P. K., Gilardi, P., Stratford-Perricaudet, L. D., Perricaudet, M., Jallat, S., Pavirani, A., Lecocq, J.-P., and Crystal, R. G. (1991). Adenovirus-mediated transfer of a recombinant α_1-antitrypsin gene to the lung epithelium *in vivo*. *Science* **252**, 431–434.

Rosenfeld, M. A., Yoshimura, K., Trapnell, B. C., Yoneyama, K., Rosenthal, E. R., Dalemans, W., Fukayama, M., Bargon, J., Stier, L. E., Stratford-Perricaudet, L., Perricaudet, M., Giggino, W. B., Pavirani, A., Lecocq, J.-P., and Crystal, R. G. (1992). *In vivo* transfer of the human cystic fibrosis transmembrane conductance regulator gene to the airway epithelium. *Cell (Cambridge, Mass.)* **68**, 143–155.

Russell, E. S. (1979). Hereditary anemia of the mouse: A review for geneticists. *Adv. Genet.* **20**, 357–459.

Sabin, A. B. (1952). Genetic, hormonal and age factors in natural resistance to certain viruses. *Ann. N.Y. Acad. Sci.* **54**, 936–944.

St. Louis, D., and Verma, I. M. (1988). An alternative approach to somatic cell gene therapy. *Proc. Natl. Acad. Sci. U.S.A.* **85**, 3150–3154.

Salter, D. W., and Crittenden, L. B., (1989). Artificial insertion of a dominant gene for resistance to avian leukosis virus into the germ line of the chicken. *Theor. Appl. Genet.* **77**, 457–461.

Scharfmann, R., Axelrod, J. H., and Verma, I. M. (1991). Long-term *in vivo* expression of retrovirus-mediated gene transfer in mouse fibroblast implants. *Proc. Natl. Acad. Sci. U.S.A.* **88**, 4626–4630.

Schubert, D., Heinemann, S., Carlisle, W., Tarikas, H., Kimes, B., Patrick, J., Steinbach, J. H., Culp, W., and Brandt, B. L. (1974). Clonal cell lines from the rat central nervous system. *Nature (London)* **249**, 224–227.

Seegmiller, J. E., Rosenbloom, F. M., and Kelley, W. N. (1967). Enzyme defect associated with a sex-linked human neurological disorder and excessive purine synthesis. *Science* **155**, 1682–1684.

Seldon, R. F., Skoskiewicz, M. J., Howie, K. B., Russell, P. S., and Goodman, H. M. (1987). Implantation of genetically engineered fibroblasts into mice: Implications for gene therapy. *Science* **236**, 714–718.

Shak, S., Capon, D. J., Hellmiss, R., Marsters, S. A., and Baker, C. L. (1990). Recombinant human DNase I reduces the viscosity of cystic fibrosis sputum. *Proc. Natl. Acad. Sci. U.S.A.* **87**, 9188–9192.

Shesely, E. G., Kim, H.-S., Shehee, W. R., Papayannopoulou, T., Smithies, O., and Popovich, B. W. (1991). Correction of a human β^s-globin gene by gene targeting. *Proc. Natl. Acad. Sci. U.S.A.* **88**, 4294–4298.

Shih, M.-F., Arsenakis, M., Tiollais, P., and Roizman, B. (1984). Expression of hepatitis B virus S gene by herpes simplex virus type 1 vectors carrying alpha- and beta-regulated gene chimeras. *Proc. Natl. Acad. Sci. U.S.A.* **81**, 5867–5870.

Shimohama, S., Rosenberg, M. B., Fagan, A. M., Wolff, J. A., Short, M. P., Breakefield, X. O., Friedman, T., and Gage, F. H. (1989). Grafting genetically modified cells into the rat brain: Char acteristics of *E. coli* galactosidase as a reporter gene. *Mol. Brain Res.* **5**, 271–278.

Short, M. P., Choi, B. C., Lee, J. K., Malick, A., Breakefield, X. O., and Martuza, R. L. (1990). Gene delivery to glioma cells in rat brain by grafting of a retrovirus packaging cell line. *J. Neurosci. Res.* **27**, 427–433.

Skow, L. C., Burkhart, B. A., Johnson, F. M., Popp, R. A., Popp, D. M., Goldberg, S. Z., Anderson,

W. F., Barnett, L. B., and Lewis, S. E. (1983). A mouse model for β-thalassemia. *Cell (Cambridge, Mass.)* **34**, 1043–1052.

Sorge, J., Kuhl, W., West, C., and Beutler, A. (1987). Complete correction of the enzymatic defect of type I Gaucher disease fibroblasts by retroviral-mediated gene transfer. *Proc. Natl. Acad. Sci. U.S.A.* **84**, 906–909.

Soriano, P., Dijkstra, J., Legrand, A., Spanjer, A., Londos-Gagliardi, D., Roerdink, F., Scherphof, G., and Nicolau, C. (1983). Targeted and non-targeted liposomes for *in vivo* transfer to rat liver cells of a plasmid containing the preproinsulin I gene. *Proc. Natl. Acad. Sci. U.S.A.* **80**, 7128–7131.

Stephenson, J. R., Axelrod, A. A., McLeod, D. L., and Shreeve, M. M. (1971). Induction of colonies of hemoglobin-synthesizing cells by erythropoietin *in vitro*. *Proc. Natl. Acad. Sci. U.S.A.* **68**, 1542–1546.

Stephenson, M. L., and Zamecnik, P. C. (1978). Inhibition of Rous sarcoma viral RNA translation by a specific oligodeoxyribonucleotide. *Proc. Natl. Acad. Sci. U.S.A.* **75**, 285–288.

Stevenson, M., and Iversen, P. L. (1989). Inhibition of human immunodeficiency virus type 1-mediated cytopathic effects by poly(L-lysine)-conjugated synthetic antisense oligodeoxyribonucleotides. *J. Gen. Virol.* **70**, 2673–2682.

Stratford-Perricaudet, L. D., Levrero, M., Chasse, J.-F., Perricaudet, M., and Briand, P. (1990). Evaluation of the transfer and expression in mice of an enzyme-encoding gene using a human adenovirus vector. *Hum. Gene Therapy* **1**, 241–256.

Tackney, C., Cachianes, G., and Silverstein, S. (1984). Transduction of the Chinese hamster ovary *aprt* gene by herpes simplex virus. *J. Virol.* **52**, 606–614.

Toneguzzo, F., and Keating, A. (1986). Stable expression of selectable genes introduced into human hematopoietic stem cells by electric field mediated DNA transfer. *Proc. Natl. Acad. Sci. U.S.A.* **83**, 3496–3499.

Toneguzzo, F., Hayday, A., and Keating, A. (1986). Electric field-mediated DNA transfer: Transient and stable gene expression in human and mouse lymphoid cells. *Mol. Cell. Biol.* **6**, 703–706.

Tremisi, P. J., and Bich-Thuy, L. T. (1991). Restoration of high immunoglobulin gene expression in chronic lymphoid leukemia: A possible application for gene therapy. *Cell. Immunol.* **135**, 326–334.

Tur-Kaspa, R., Teicher, L., Levine, J., Skoultchi, A., and Shafritz, D. A. (1986). Use of electroporation to introduce biological active foreign genes into rat primary hepatocytes. *Mol. Cell. Biol.* **6**, 716–718.

Uchida, T., Yamaizumi, M., and Okada, Y. (1977). Reassembled HVJ (Sendai virus) envelopes containing non-toxic mutant proteins of diphtheria toxin show toxicity to mouse L cells. *Nature (London)* **266**, 839–840.

Vidal, M., Sainte-Marie, J., Philippot, J. R., and Bienvenue, A. (1985). LDL-mediated targeting of liposomes to leukemic lymphocytes *in vitro*. *EMBO J.* **4**, 2461–2467.

von Rüden, T., and Gilboa, E. (1989). Inhibition of human T-cell leukemia virus type 1 replication in primary human T cells that express antisense RNA. *J. Viol.* **63**, 677–682.

Watanabe, Y. (1980). Serial breeding of rabbits with hereditary hyperlipidemia (WHHL rabbit). *Atherosclerosis* **36**, 261–268.

Widner, H., and Brundin, P. (1988). Immunological aspects of grafting in the mammalian central nervous system. A review and speculative synthesis. *Brain Res. Rev.* **13**, 287–324.

Williams, R. S., Johnston, S. A., Riedy, M., DeVit, M. J., McElligott, S. G., and Sanford, J. C. (1991). Introduction of foreign genes into tissues of living mice by DNA-coated microprojectiles. *Proc. Natl. Acad. Sci. U.S.A.* **88**, 2726–2730.

Wilson, J. M., and Chowdhury, J. R. (1990). Prospects for gene therapy of familial hypercholesterolemia. *Mol. Biol. Med.* **7**, 223–232.

Wilson, J. M., Johnstone, D. E., Jefferson, D. M., and Mulligan, R. C. (1988). Correction of the genetic defect in hepatocytes from the Watanabe heritable hyperlipidemic rabbit. *Proc. Natl. Acad. Sci. U.S.A.* **85**, 4421–4425.

Wilson, J. M., Birinyi, L. K., Salomon, R. N., Libby, P., Callow, A. D., and Mulligan, R. C. (1989). Implantation of vascular grafts lined with genetically modified endothelial cells. *Science* **244**, 1344–1346.

Wolff, J. A., Yee, J.-K., Skelley, H. F., Moores, J. C., Respess, J. G., Friedman, T., and Leffert, H. (1987). Expression of retrovirally transduced genes in primary cultures of adult rat hepatocytes. *Proc. Natl. Acad. Sci. U.S.A.* **84**, 3344–3348.

Wolff, J. A., Fisher, L. J., Xu, L., Jinnah, H. A., Langlais, P. J., Iuvone, P. M., O'Malley, K. L., Rosenberg, M. B., Shimohama, S., Friedman, T., and Gage, F. H. (1989). Grafting fibroblasts genetically modified to produce L-dopa in a rat model of Parkinson disease. *Proc. Natl. Acad. Sci. U.S.A.* **86**, 9011–9014.

Wolff, J. A., Malone, R. W., Williams, P., Chong, W., Acsadi, G., Jani, A., and Felgner, P. L. (1990). Direct gene transfer into mouse muscle *in vivo*. *Science* **247**, 1465–1468.

Wood, P. A., Herman, G. E., Chao, C. Y., O'Brien, W. E., and Beaudet, A. L. (1986). Retrovirus-mediated gene transfer of argininosuccinate synthetase into cultured rodent cells and human citrullinemic fibroblasts. *Cold Spring Harbor Symp. Quant. Biol.* **51**, 1027–1032.

Wu, G. Y., and Wu, C. H. (1987). Receptor-mediated *in vitro* gene transformation by a soluble DNA carrier system. *J. Biol. Chem.* **262**, 4429–4432.

Wu, G. Y., and Wu, C. H. (1988). Receptor-mediated gene delivery and expression *in vivo*. *J. Biol. Chem.* **263**, 14621–14624.

Wu, C. H., Wilson, J. M., and Wu, G. Y. (1989). Targeting genes: Delivery and persistent expression of a foreign gene driven by mammalian regulatory elements *in vivo*. *J. Biol. Chem.* **264**, 16985–16987.

Wu, G. Y., Wilson, J. M., Shalaby, F., Grossman, M., Shafritz, D. A., and Wu, C. H. (1991). Receptor-mediated gene delivery *in vivo*. Partial correction of genetic analbuminemia in Nagase rats. *J. Biol. Chem.* **266**, 14338–14342.

Wynshaw-Boris, A., Lugo, T. G., Short, J. M., Fournier, R. E. K., and Hanson, R. W. (1984). Identification of a cAMP regulatory region in the gene for rat cytosolic phosphoenolpyruvate carboxykinase (GTP). Use of chimeric genes transfected into hepatoma cells. *J. Biol. Chem.* **259**, 12161–12169.

Zelenin, A. V., Titomirov, A. V., and Kolesnikov, V. A. (1989). Genetic transformation of mouse cultured cells with the help of high-velocity mechanical DNA injection. *FEBS Lett.* **244**, 65–67.

Zinkowski, R. P., Vig, B. K., and Broccoli, D. (1986). Characterization of kinetochores in multicentric chromosomes. *Chromosoma* **94**, 243–248.

Zwiebel, J. A., Freeman, S. M., Kantoff, P. W., Cornetta, K., Ryan, U.S., and Anderson, W. F. (1989). High-level recombinant gene expression in rabbit endothelial cells transduced by retroviral vectors. *Science* **243**, 220–222.

7

Molecular Approaches Involved in Mammalian Gene Transfer: Analysis of Transgene Integration

Brad T. Tinkle, Charles J. Bieberich, and Gilbert Jay
Department of Virology
Jerome H. Holland Laboratory
American Red Cross
Rockville, Maryland 20855

I. INTRODUCTION

The purpose of this chapter is to outline the approaches and techniques used to evaluate the integration of a transgene, as to its presence, intactness, copy number, number of integration sites, as well as the orientation of tandem copies. Several approaches may be available to evaluate each of the points, and the merit of one approach over another may be dictated by circumstances unique to each transgenic project. A clear understanding of the logic behind each strategy may help the investigator to choose the most appropriate techniques.

221

II. DETECTION OF THE TRANSGENE

The identification of a transgenic animal is based on the detection of some aspect of the transgene. In many cases, the engineered transgene contains unique nucleic acid sequences. For example, it may contain a viral transcriptional promoter used for conferring tissue specificity (Stewart et al., 1984; Overbeek et al., 1986; Feigenbaum et al., 1992), a bacterial coding sequence that will produce an enzyme used for facilitating the detection of promoter activity (Overbeek et al., 1986; Goring et al., 1987), or yeast heterochromatic sequences, including telomeric and centromeric regions, that will allow replication of yeast artificial chromosomes containing large segments of mammalian DNA (Schedl et al., 1992). These unique DNA segments can obviously facilitate the identification of a transgenic animal by nucleic acid hybridization. In situations where the transgene is derived from the same host species and hence does not contain any unique sequences, one has to exploit alternative strategies.

One strategy for detecting a transgene with a similar or identical endogenous counterpart is to exploit restriction fragment length polymorphisms (RFLPs) that distinguish between the two. For example, in the construction of the transgene, one may have deleted an intragenic restriction enzyme site that is present in the endogenous gene. As a consequence, one can distinguish between the transgene and the endogenous counterpart by an RFLP analysis using that particular enzyme. A cDNA sequence, when placed in a transgene, should be easily distinguishable from the endogenous gene, which may contain one or more introns. If the transgene is a long segment of genomic DNA that has not been altered, its detection can most easily be achieved by identifying a flanking restriction enzyme site within the chromosomal DNA near the site of transgene integration. Ideally, such a restriction enzyme should either cleave once or have just a few cleavage sites within the transgene. Detection of the RFLP is facilitated by choosing a DNA probe that is located at one end of the transgene fragment. Hence, by using a restriction enzyme site common to the endogenous and transgenic sequences, and the nearest flanking site outside of the common region, one should be able to generate distinctive RFLPs.

Because detection of the transgene is almost invariably the basis for identification of a transgenic animal, choice of methodologies must be considered at the time the transgene is being constructed. If an approach is not obvious, a molecular "tag" must be introduced into the transgene to allow its detection on integration into the host cell. The molecular tag could be a foreign segment of DNA that will facilitate detection by hybridization. In the past, injected transgenes often contained vector sequences that were exploited for identifying transgenic animals. However, it was discovered that vector sequences could greatly down-regulate transgene expression (Townes et al., 1985; Hammer et al., 1987), an observation which has prompted the suggestion to remove as much of the plasmid sequences prior to the injection of the transgene into single-cell embryos. Instead of introducing a segment of DNA, one may consider deleting a region from within the transgene, such as a part of an intronic sequence. However, this approach carries with it the risk that an important cis-acting element which is responsible for regulating either the site or level of expression will be removed. Given these concerns, perhaps the least risky tag is an altered convenient restriction enzyme site. It is often possible to change a restriction site or generate a new one by changing one or two bases within a gene. Whatever the approach, consideration for the successful screening of transgenic animals must play a significant role in the logic behind the design of a transgene.

Genotypic screening for the physical presence of transgene DNA sequences may not be necessary if the transgenic animal has a distinctive phenotypic change that is detectable early on; for example, changes in the texture of the skin of a newborn such as scaliness, blistering, or epidermal thickening (Fuchs *et al.*, 1992). In situations where one does not detect a phenotype induced by the transgene but there is a need to identify transgenic animals as quickly as possible, one can consider deliberately inducing an inconsequential phenotypic change that is readily detectable. This is possible because of the observation that when two linear pieces of DNA are coinjected into single-cell embryos, the two transgenes frequently cointegrate at the same chromosomal site (Einat *et al.*, 1987; Lo *et al.*, 1988). Given this observation, if one were to introduce a transgene of interest together with, for example, the tyrosinase gene into albino mice, animals born with pigmented skin and eyes will likely have integrated not only the tyrosinase gene but also the transgene of interest (Schedl *et al.*, 1992; Yokoyama *et al.*, 1990; see also Chapter 3). This strategy can facilitate identification of newborn transgenic mice, and it may be extremely useful when a transgene induces perinatal death. However, one has to be cautious that the transcriptional promoter of the marker construct does not interfere with the tissue specificity of the other transgene. Additionally, because cointegration does not occur in all cases, one may still have to resort to genotypic screening for confirmation.

III. PARAMETERS FOR EVALUATING A TRANSGENE

The questions that must be asked of a potential transgenic animal, apart from the presence of a transgene, are as follows: Is the transgene complete and functional? How many integration sites exist per cell? How many copies are present at a chromosomal location? How are multiple copies of the transgene arranged at a particular chromosomal site? Can the transgene be inherited as a stable genetic element?

Although the mechanism of transgene integration into the host chromosome is not well understood, only rarely is it by homologous recombination (Brinster *et al.*, 1989). Frequently, the transgene is found as multimers at the site of integration (Palmiter *et al.*, 1982). It is not known whether the transgene multimerizes prior to integration or is amplified at the site of integration. Whichever the case, sequences on both ends of the injected fragment are frequently lost. Incomplete copies of the transgene may arise by different mechanisms and have sequences with different extents of representation. Therefore, it is imperative to analyze carefully the intactness of the injected transgene at a chromosomal site.

The determination of the number of copies of the transgene is not just a laboratory exercise. Theoretically, a single intact copy may be sufficient. However, the expression level may correlate with gene dosage; a higher copy number may result in increased expression (Grosveld et al., 1987; Greaves et al., 1989). In general, this observation does not hold true for all transgenes; an unusually high copy number may, in fact, result in low expression (Bieberich et al., 1986). A high copy number may not only affect the expression level but the genetic stability of the transgenic locus. A large number of tandem repeating elements may cause intrachromosomal recombination, resulting in rearrangement, deletion, breakage, or translocation (Covarrubias et al., 1986; Mahon et al., 1988). It is our experience that two to five copies are ideal for most purposes.

Once the transgene of a founder animal has been characterized, selective breeding can be used to establish a stable transgenic line. Any founder animal with more than one integration site should be bred to allow segregation of the two loci and the establishment of independent lines. An intact transgene does not assure expression. Its expression may be "silenced" by the surrounding chromatin structure. Therefore, determining the expression level may also be helpful in selecting which transgenic line to maintain. More than one line should be used to confirm any phenotype that may develop and to control for possible integration-site artifacts. Foreknowledge of the parameters concerning the analysis of a transgene is critical in the design of a transgenic project.

IV. ANALYSIS OF THE TRANSGENE

With the knowledge of the unique characteristics of the transgene, the investigator can utilize different methods for the identification and confirmation of transgene integration. Some of the techniques are described below. Specific examples are used to illustrate and compare the various methods.

A. Southern Blot Hybridization

Mice are usually weaned at about 3 weeks of age, at which time they can be marked for identification by ear punching, toe clipping, or tattooing. DNA for transgene analysis can be obtained by cutting off the terminal end of the tail. Tail tissue is then minced and digested with proteinase K (Hogan et al., 1986). DNA can be recovered by ethanol precipitation after phenol–chloroform extraction.

Typically, 10–20 μg of DNA is digested with a restriction enzyme, electrophoresed through an agarose gel, denatured, neutralized, blotted to a membrane such as nitrocellulose, and hybridized to a transgene-specific probe. In Fig. 1A, we

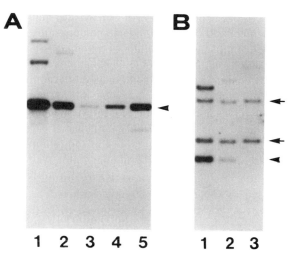

Figure 1. Identification of transgenic mice by Southern blot hybridization analysis. (A) DNA (10 μg) isolated from the tails of transgenic mice carrying the human immunodeficiency virus (HIV) long terminal repeat (LTR)–*tat* gene was digested with *Bgl*II, fractionated on an agarose gel, transferred onto a nitrocellulose membrane, and hybridized with a transgene-specific probe. Lane 1, animal C4; lane 2, animal F2; lanes 3 to 5, titration of the unit-length injected transgene fragment (arrowhead) representing one, three, and nine copies, respectively. (B) DNA (10 μg) isolated from the tails of transgenic animals carrying the metallothionein–*ras* transgene was cut with *Hind*III and hybridized with a *ras*-specific probe. The transgenic DNA construct, when cut with *Hind*III, releases a 900-bp internal fragment (arrowhead) that is easily distinguishable from the two bands derived from the endogenous gene (arrows). Lane 1, animal D1; lane 2, animal G2; lane 3, animal H10.

are interested in defining the copy number and the orientation of a transgene that has no sequence homology with any endogenous gene. To do so, we have selected a restriction enzyme that cuts only once within the transgene. If this one-cut enzyme cleaves more to one side of the molecule, it will facilitate our defining the orientation of the transgene at a particular integration site. Had the tandem copies been arranged in a head-to-head/tail-to-tail fashion, then cleavage with an asymmetric one-cut enzyme would generate both a larger and a smaller than unit-length molecule. If, however, the integrated copies are arranged in a head-to-tail array, then cleavage by the same enzyme would generate unit-length molecules (arrowhead), as is the case for the two animals represented in lanes 1 and 2 (Fig. 1A). In this case, apart from defining the orientation of the integrated copies, the analysis also provides an indication that the transgene is more or less intact.

The size of the unit-length molecule is defined by including, in the same gel, the injected DNA fragment that has been previously mixed with nontransgenic genomic DNA which has been digested with the same enzyme (Fig. 1A, lanes 3–5). By using known concentrations of the injected DNA fragment, for example, equivalent to one, three, and nine copies per diploid genome, one should be able to estimate the copy number in each of the founder animals. In the present example, the two transgenic animals (Fig. 1A, lanes 1 and 2) have 10 and 20 copies of the transgene per cell. It should be pointed out that, although we have defined the number of copies per cell, this experiment does not tell us how many integration sites exist in each of the animals, nor does it tell us whether the animals are mosaic for any particular integration site.

The analysis of transgenes that have a related copy in the host is somewhat more complicated. In that case, one can make use of restriction enzymes that can differentially distinguish between them. Figure 1B shows the analysis of three transgenic mice that carry a cDNA copy of an endogenous gene. Choice of an enzyme that would cleave not only within the coding sequence but also within one or more of the introns which are unique to the genomic copy would allow the assignment of DNA fragments in a Southern blot as host-derived (arrows) or transgene-derived. If the enzyme chosen cuts out a large contiguous stretch of the transgene (arrowhead), it will further provide information as to the intactness of the transgene. The benefit of having an endogenous counterpart is that it may serve as an internal control for copy number analysis. For example, lane 1 (Fig. 1B) contains DNA from an animal with approximately 2–3 copies of the transgene per cell, lane 2 has less than 1 copy, suggesting that it is likely to be a genetic mosaic, and lane 3 definitely does not have an intact copy of the transgene.

From a single screening by Southern blot analysis, it can be determined that the transgene is present, intact, in how many copies, and in what orientation. Neither of the other commonly used techniques, slot-blot hybridization or polymerase chain reaction (PCR), yield nearly the same amount of information. Slot-blot hybridization analysis can confirm the presence of the transgene and, if performed accurately, can be informative as to the approximate copy number, but it cannot be used to obtain information with regard to the intactness or orientation of the transgene. The PCR usually amplifies an internal DNA segment of a gene and is used to confirm the presence of the transgene. It is at best only semiquantitative and provides little information as to copy number or intactness. Despite these limitations, slot-blot hybridization and PCR analysis are not without value for the analysis of transgenic animals.

B. Polymerase Chain Reaction

The polymerase chain reaction is useful for detecting low copy number genetic target sequences by repeated rounds of amplification. The technique is relatively simple, quick, and sensitive, and thus can accommodate many more samples than the Southern blot hybridization procedure. The PCR has been applied to transgenic screening using DNA from a variety of tissues including tail biopsies (Lin *et al.*, 1989; Hanley and Merlie, 1991), ear clippings (Chen and Evans, 1990), and whole blood (Chen and Evans, 1990; Skalnik and Orkin, 1990). Briefly, the tissue is protease-digested, boiled, and used as a crude extract in the PCR reaction.

Figure 2A shows an example of screening by PCR of potential transgenic animals using crude extracts of ear clippings. All mice except one (Fig. 2A, lane 1)

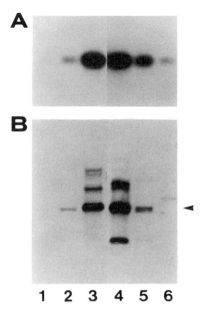

Figure 2. Comparison between polymerase chain reaction and Southern blot hybridization analysis for the identification of transgenic mice. (A) PCR amplification of a 173-bp DNA fragment of the HIV LTR–*nef* transgene, using crude extracts of ear clippings. The PCR products were denatured, hybridized with a transgene-specific probe, electrophoresed through a polyacrylamide gel, vacuum-dried, and exposed to X-ray film. Lanes 1–6 contain DNA samples from potential transgenic mice. (B) Southern blot hybridization analysis of DNA obtained from tail biopsies from the same set of mice as in (A), using the same transgene-specific probe. Intact copies of the transgene are evident by the appearance of a 900-bp component (arrowhead).

were shown to have the transgene. Although there is a suggestion that these mice have varying numbers of copies of the transgene, it is difficult to quantify the copy number in each of these mice owing to the fact that PCR amplification gives rise to exponential increases in signal strength. Depending on where the reaction is on the amplification curve, minor differences may become grossly exaggerated and major differences may become less significant. In addition, PCR analysis cannot provide information as to either the intactness or orientation of the transgene. Furthermore, in whatever application of the PCR technique, there is associated with its high sensitivity an inherent risk of detecting contamination, which will lead to the identification of false positives. Despite its limitations, the PCR can be useful as a preliminary screening procedure.

In situations where one has large numbers of animals to analyze, as when one expects the transgene to induce early lethality, the use of PCR would allow the identification of potential transgenic animals and reduce the number of samples that have to be analyzed by Southern blot hybridization to determine transgene intactness and copy number. Figure 2B depicts the same set of samples as in Fig. 2A but analyzed by Southern blot hybridization, which, first of all, confirms that the DNA from lanes 2–6, but not lane 1, was derived from transgenic animals. Unlike the PCR, the Southern blot further reveals that, except for the DNA in lane 6, all of the transgenic animals have intact copies of the transgene (Fig. 2B, arrowhead), albeit at different copy numbers. Additionally, many of the transgenic animals also show other hybridizing components, which may represent partial genetic rearrangements and/or junctional sequences at one or more integration sites. Each pattern will serve as a unique profile to identify an individual transgenic mouse and will be useful in the subsequent establishment of independent transgenic lines.

C. Slot-Blot Hybridization

Similar to the PCR, the slot-blot hybridization technique also has its limitations and strengths. The technique involves the direct application of the DNA sample to a solid support such as a nitrocellulose membrane and subsequent incubation of the membrane with a transgene-specific probe. Compared to Southern blot hybridization, slot-blot analysis does not require cleavage of the input DNA with a restriction enzyme and, for that reason, does not require extensive purification of the DNA samples. As a result, the procedure will significantly reduce processing time and allow handling of many more samples. However, concomitant with the use of undigested DNA, which has a much higher viscosity, one can expect a higher background hybridization signal and, as a consequence, an increased possibility of detecting false positives. Although there are clear advantages for the use of this technique, it cannot provide any information on either the orientation or the intactness of a transgene.

An example of transgenic screening by slot-blot hybridization is shown in Fig. 3A. Genomic DNA from the F1 offspring of a founder mouse is screened using a transgene-specific probe. Slots 2, 4, and 6 (Fig. 3A) show positive signals which vary in intensities. If an equivalent amount of DNA is added to each slot, the result would imply that the transgenic offspring do not all contain the same number of copies of the transgene. The DNA in slot 2 must have at least 4 times as many copies as the DNA in slots 4 or 6. If this is correct, then the founder mouse must have more than one site of integration. The latter concern can be resolved by subjecting the relevant DNA samples to Southern blot hybridization analysis (Fig. 3B) using the same DNA probe.

With the restriction enzyme selected, all three F1 mice show three hybridizing components (Fig. 3B, arrowheads); the relative molar ratios between the three components remain constant among the animals, confirming that they are internal fragments derived from the transgene. By densitometric tracing, one estimates that the high copy number animal (Fig. 3B, lane 1) has about 4 times as many copies as the other two animals (lanes 2 and 3). In addition to the common components, the high copy number animal has extra bands that are likely to be junctional fragments and are derived from an integration site which is different from the other two F1 littermates. The uncertainty suggested by the slot-blot technique is now explained by the

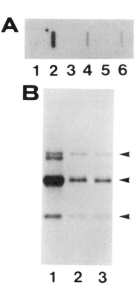

Figure 3. Comparison of slot-blot and Southern blot hybridization analyses for the identification of transgenic mice carrying the HIV LTR–*nef* transgene. (A) Tail DNA (5 μg) was vacuum-applied to a nitrocellulose membrane, air-dried, and hybridized to a transgene-specific probe. Hybridization was detected in slots 2, 4, and 6, albeit at different signal intensities. (B) Southern blot hybridization analysis of DNA from samples shown in (A). Lanes 1 to 3 contain DNA samples that are the same as those in slots 2, 4, and 6, respectively. Whereas all three samples are confirmed for the presence of the transgene (arrowheads), minor differences in the restriction pattern between the DNA in lane 1 and those in lanes 2 and 3 suggest the existence of two different sites of integration.

finding with Southern blot analysis that the founder mouse has more than one integration site. What is left unresolved is whether there are other sites of integration not identified in this small litter of mice.

V. ESTABLISHING A TRANSGENIC LINE

Ideally, integration occurs at a single site in the genome during the one-cell stage of embryogenesis. In this case, assuming that the transgene does not have a deleterious effect on gametogenesis that induces transmission ratio distortion (Palmiter *et al.*, 1984), half of the F1 offspring should inherit the transgene. In practice, integration may occur at several different sites within the genome. These sites may be linked in close proximity on the same chromosome or can be on completely different chromosomes. The distance between the integration sites will dictate the frequency of meiotic recombination which in turn determines how the multiple integration sites carrying the transgenes will segregate.

To complicate matters further, integration at a single site or at multiple genomic sites may occur later than the one-cell stage of development, resulting in the generation of a genetically mosaic animal (Palmiter *et al.*, 1984; Wilkie *et al.*, 1986). A mosaic founder animal with a single integration site will transmit the transgene to less than 50% of its offspring, and the exact percentage will be dictated by the stage at which integration occurs and the contribution of the recipient blastomere to the formation of germ cells during development. If multiple integration events occur, the timing of each event may be independent; that is, one event may occur at the one-cell stage, while another occurs at a later stage. In this scenario, the first integration site will be transmitted to 50% of the F1 animals, whereas a lower percentage will inherit the second integration site. A single founder may thus be mosaic for one, but not for another integration site.

What are the indications that multiple integration sites exist? One obvious sign is the inheritance of a transgene by greater than 50% of F1 generation animals. For example, in the case of a founder carrying two nonmosaic integration sites, 75% of the F1 animals will be transgenic. A second indication would be a variation in the transgene copy number among F1 offspring. A third sign would be variations in the pattern of so-called flanking bands among F1 animals.

A Southern blot analysis of offspring derived from a founder mouse carrying two integration sites is shown in Fig. 4. DNA from the founder animal (F0) contains five major hybridizing components. The most intense band represents complete copies of the transgene, whereas the remaining four are indicative of partial copies or flanking bands. The remaining 10 lanes (Fig. 4) contain DNA from individual F1 mice sired by the founder. Of the 10 animals, 8 have inherited the transgene. Three of the offspring represented in lanes 1, 8, and 10 have inherited a single copy of the complete transgene and a higher molecular weight flanking band which

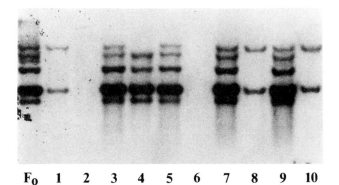

Figure 4. Southern blot hybridization analysis of mice sired by a founder with multiple integration sites. DNA (10 μg) isolated from the offspring of a founder animal carrying an HIV LTR–*tat* transgene was digested with *Bgl*II and hybridized with a *tat*-specific probe. F0, founder; lanes 1 to 10, F1 offspring.

together constitute profile "A." The animal represented in lane 4 has inherited approximately 10 copies of the complete transgene and three other hybridizing components which constitute profile "B." The remaining four transgenic animals, represented in lanes 3, 5, 7, and 9 (Fig. 4), display a similar profile as the founder, showing all five hybridizing components, indicating that they have inherited both profiles A and B. It should be noted that the single copy complete transgene component from profile A is masked by the higher copy number complete transgene component present in profile B. However, because the A profile also includes a unique high molecular weight flanking band, it is possible to readily identify animals that have inherited both A and B. If profile A did not include a unique flanking band, it would be considerably more difficult to determine which animals had inherited both A and B.

Not all genomic restriction enzyme digests of transgenic DNA will reveal unique flanking bands for a given integration site. However, by screening transgenic founders with a panel of restriction enzymes, it is usually possible to find one that is useful for this purpose. Frequently, the flanking bands can be of high molecular weight and may contain only a fraction of the transgene, making their detection by Southern analysis difficult. For this reason, it is essential that Southern analysis of transgenic animals be carried out at the highest possible sensitivity.

VI. DERIVATION OF HOMOZYGOUS MICE

Once the number and arrangement of transgene integration sites have been well-defined, one can consider whether it is appropriate to derive homozygous mice by establishing transgenic brother–sister breeding pairs. In general, the most efficient approach is to first determine which lines express the transgene, then breed a selected few to homozygosity if necessary, since the amount of time and resources involved in generating homozygotes is significant and success is not guaranteed. Although hemizygous mice are usually the starting point in expression analyses, there may be situations where having homozygotes is essential. For example, if the level of transgene expression is anticipated to be near the threshold for detection, then increasing the gene dosage by achieving homozygosity may be important. Alternatively, if a very small tissue is to be assayed and samples from many animals must be pooled, then deriving homozygotes early on may be useful.

The most straightforward method of deriving animals that are homozygous for a transgene is to establish breeding pairs consisting of hemizygous transgenic brothers and sisters that have been carefully analyzed and shown to carry a single integration site. Although it is also possible to backcross a transgenic F1 to the founder, there is a greater risk of picking up a second integration site that may exist in a mosaic form in the founder.

Homozygous animals can be easily identified in a Southern blot analysis that

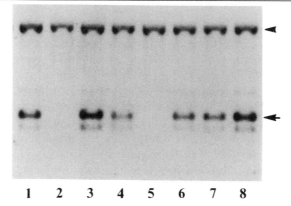

Figure 5. Southern blot hybridization analysis to detect homozygous transgenic mice. DNA (10 μg) isolated from eight offspring of a hemizygous brother–sister breeding pair carrying an SL3-3 virus LTR/HIV *tat* transgene was digested with *EcoRI* and hybridized with two probes simultaneously. One probe was a 4.1-kb *KpnI*–*ApaI* fragment from the upstream region of the *Hox1.4* gene which detects a 6-kb *EcoRI* fragment (arrowhead) that is endogenous and invariant among the offspring; the second probe was specific for the HIV *tat* gene and detects a 1.6-kb band that represents the transgene (arrow).

simultaneously employs two different probes: one specific for the transgene and one specific for an invariant endogenous gene. An example of such an analysis is shown in Fig. 5. Lanes 1–8 contain DNA isolated from offspring of a brother–sister transgenic breeding pair. Assuming Mendelian inheritance, this cross should generate 25% homozygous, 50% hemizygous, and 25% nontransgenic offspring. The lower molecular weight hybridizing component (arrow) represents the transgene locus. The upper band (arrowhead) represents an endogenous gene locus that should be conserved and constant in gene dosage among all the offspring. Comparing the ratio of the transgene hybridization signal to the endogenous gene hybridization signal within each lane, it is clear that lanes 3 and 8 (Fig. 5) contain DNA from animals that are good candidates for homozygosity since their ratios are about twice those in lanes 1, 4, 6, and 7, which most likely contain DNA from hemizygous animals. It is possible, though usually not necessary, to quantitate the ratios of hybridization signals by scanning densitometry.

It is also feasible to identify homozygous animals in a Southern blot analysis using a single probe specific for the transgene. In this case, great care must be taken to ensure that equivalent amounts of DNA, digested to completion with restriction enzyme, are loaded into each lane on the gel. Using this strategy, one can compare the level of fluorescence in each lane from a photograph of an ethidium bromide-stained gel to the level of hybridization signal seen in each lane on a Southern blot autoradiogram, assuming that the transfer of DNA from the gel to the nitrocellulose membrane was complete for each lane.

Because neither of the approaches outlined above are foolproof for detecting homozygous animals, it is important, though laborious, to carry out a backcross analysis whenever possible. A true homozygote, when crossed to a nontransgenic animal, should transmit the transgene to 100% of its offspring. Failure to transmit to all offspring could be an indication of hemizygosity, or it may indicate that the transgene locus is unstable and can be lost at some point during germ cell development.

VII. SUMMARY

The success of any transgenic project hinges on the ability to identify and subsequently characterize the mice that contain the integrated transgene. Therefore, the methods for the identification of the presence and intactness of a transgene must be considered during the early phase of a project. Knowing what aspects of a transgene to exploit, one can design the transgenic construct accordingly. Having identified the transgenic animals by Southern blot hybridization analysis, one has to propagate each of the founder animals in order to generate independent lines. Maintenance of a transgenic line may be achieved by the use of more rapid and simple methods, such as the polymerase chain reaction or slot-blot hybridization analysis, to identify the presence or absence of the transgene in progeny mice. In summary, early anticipation of problems associated with the identification of the transgene and the correct choice of methods for tracking the presence of the transgene are the key to success in deriving transgenic animals.

ACKNOWLEDGMENTS

We thank Dr. Lian-Sheng Chen and Mr. Mario Cepeda for generous contribution of figures. We also thank Drs. Robert A. Pollock and Frank M. LaFerla for critical review of the manuscript.

REFERENCES

Bieberich, C., Scangos, G., Tanaka, K., and Jay, G. (1986). Regulated expression of a murine class I gene in transgenic mice. *Mol. Cell. Biol.* **6,** 1339–1342.

Brinster, R. L., Braun, R. E., Lo, D., Avarbock, M. R., Oram, F., and Palmiter, R. D. (1989). Targeted correction of a major histocompatibility class II E_α gene by DNA microinjected into mouse eggs. *Proc. Natl. Acad. Sci. U.S.A.* **86,** 7087–7091.

Chen, S., and Evans, G. A. (1990). A simple screening method for transgenic mice using the polymerase chain reaction. *BioTechniques* **8,** 32–33.

Covarrubias, L., Nishida, Y., and Mintz, B. (1986). Early postimplantation embryo lethality due to DNA rearrangements in a transgenic mouse strain. *Proc. Natl. Acad. Sci. U.S.A.* **83,** 6020–6024.

Einat, P., Bergman, Y., Yaffe, D., and Shani, M. (1987). Expression in transgenic mice of two genes of different tissue specificity integrated into a single chromosomal site. *Genes Dev.* **1,** 1075–1084.

Feigenbaum, L., Hinrichs, S. H., and Jay, G. (1992). JC virus and simian virus 40 enhancers and transforming proteins: Role in determining tissue specificity and pathogenicity in transgenic mice. *J. Virol.* **66,** 1176–1182.

Fuchs, E., Esteves, R. A., and Coulombe, P. A. (1992). Transgenic mice expressing a mutant keratin 10 gene reveal the likely genetic basis for epidermolytic hyperkeratosis. *Proc. Natl. Acad. Sci. U.S.A.* **89,** 6906–6910.

Goring, D. R., Rossant, J., Clapoff, S., Breitman, M. L., and Tsui, L.-C. (1987). *In situ* detection of β-galactosidase in lenses of transgenic mice with a γ-crystallin/*lacZ* gene. *Science* **235,** 456–458.

Greaves, D. R., Wilson, F. D., Lang, G., and Kioussis, D. (1989). Human CD2 3'-flanking sequences confer high-level, T cell-specific, position-independent gene expression in transgenic mice. *Cell (Cambridge, Mass.)* **56,** 979–986.

Grosveld, F., van Assendelft, G. B., Greaves, D. R., and Kollias, G. (1987). Position-independent, high-level expression of the human β-globin gene in transgenic mice. *Cell (Cambridge, Mass.)* **51,** 975–985.

Hammer, R. E., Krumlauf, R., Camper, S. A., Brinster, R. L., and Tilghman, S. M. (1987). Diversity of α-fetoprotein gene expression in mice is generated by a combination of separate enhancer elements. *Science* **235,** 53–58.

Hanley, T., and Merlie, J. P. (1991). Transgene detection in unpurified mouse tail DNA by polymerase chain reaction. *BioTechniques* **10,** 56.

Hogan, B., Costantini, F., and Lacy, E. (1986). "Manipulating the Mouse Embryo: A Laboratory Manual." Cold Spring Harbor Laboratory, Cold Spring Harbor, New York.

Lin, C. S., Magnuson, T., and Samols, D. (1989). Laboratory methods: A rapid procedure to identify newborn transgenic mice. *DNA* **8,** 297–299.

Lo, D., Burkly, L. C., Widera, G., Cowing, C., Flavell, R. A., Palmiter, R. D., and Brinster, R. L. (1988). Diabetes and tolerance in transgenic mice expressing class II MHC molecules in pancreatic beta cells. *Cell (Cambridge, Mass.)* **53,** 159–168.

Mahon, K. A., Overbeek, P. A., and Westphal, H. (1988). Prenatal lethality in a transgenic mouse line is the result of a chromosomal translocation. *Proc. Natl. Acad. Sci. U.S.A.* **85,** 1165–1168.

Overbeek, P. A., Lai, S.-P., Van Quill, K. R., and Westphal, H. (1986). Tissue-specific expression in transgenic mice of a fused gene containing RSV terminal sequences. *Science* **231,** 1574–1577.

Palmiter, R. D., Chen, H. Y., and Brinster, R. L. (1982). Differential regulation of metallothionein–thymidine kinase fusion genes in transgenic mice and their offspring. *Cell (Cambridge, Mass.)* **29,** 701–710.

Palmiter, R. D., Wilkie, T. M., Chen, H. Y., and Brinster, R. L. (1984). Transmission distortion and mosaicism in an unusual transgenic mouse pedigree. *Cell (Cambridge, Mass.)* **36,** 869–877.

Schedl, A., Beerman, F., Thies, E., Montoliu, L., Kelsey, G., and Schütz, G. (1992). Transgenic mice generated by pronuclear injection of a yeast artificial chromosome. *Nucleic Acids Res.* **20,** 3073–3077.

Skalnik, D. G., and Orkin, S. (1990). A rapid method for characterizing transgenic mice. *BioTechniques* **8,** 31.

Stewart, T. A., Pattengale, P. K., and Leder, P. (1984). Spontaneous mammary adenocarcinomas in transgenic mice that carry and express MTV/*myc* fusion genes. *Cell (Cambridge, Mass.)* **38,** 627–637.

Townes, T. M., Lingrel, J. B., Chen, H. Y., Brinster, R. L., and Palmiter, R. D. (1985). Erythroid-specific expression of human β-globin genes in transgenic mice. *EMBO J.* **4,** 1715–1723.

Wilkie, T. M., Brinster, R. L., and Palmiter, R. D. (1986). Germline and somatic mosaicism in transgenic mice. *Dev. Biol.* **118,** 9–18.

Yokoyama, T., Silversides, D. W., Waymire, K. G., Kwon, B. S., Takeuchi, T., and Overbeek, P. A. (1990). Conserved cysteine to serine mutation in tyrosinase is responsible for the classical albino mutation in laboratory mice. *Nucleic Acids Res.* **18,** 7293–7298.

8

Molecular Approaches Involved in Mammalian Gene Transfer: Evaluation of Transgene Expression

Charles J. Bieberich, Lien Ngo, and Gilbert Jay

Department of Virology
Jerome H. Holland Laboratory
American Red Cross
Rockville, Maryland 20855

I. INTRODUCTION

The purpose of this chapter is to introduce the techniques available to assay transgene expression, to critically evaluate each approach to bring out its merits and pitfalls, and to provide an example of the results one can expect to achieve with each method. Our intent is not to present actual protocols for carrying out assays for gene expression; most protocols have already been compiled in several excellent and widely available volumes. However, in those cases where protocols are not available in a reference-type format, the reader will be directed to an appropriate source in the original literature.

By and large, the techniques we discuss are extant in most molecular biology laboratories. Many of the circumstances to be considered by the investigator assaying transgenic animals for gene expression are the same as those to be considered by one who is transfecting genes into cells in culture. For example, how does one distinguish an endogenous transcript or protein from its transgene-derived counterpart? How does one compare the relative levels of transgene expression between cells? How does one recognize the use of alternate transcriptional start sites in different cell types? The complexities of measuring transgene expression in the whole animal often pose unique challenges, and we evaluate each approach in light of this fact.

A. Design of the Transgene to Allow Expression Analysis

Before launching into a transgenic project, in fact even before the actual construction of the transgene, it is critical to consider how the problem of analyzing transgene expression will be approached. In this regard, there are two broad categories of transgenes: those that have an endogenous counterpart already present in the genome of the animal and those that do not. Those that do not are in general straightforward to assay for expression since they will produce a novel transcript that can be detected by a unique nucleic acid probe, or they will produce a novel protein or enzyme activity that can be detected using antisera and/or histochemical methods.

Those that do have an endogenous counterpart are more problematic, especially if the tissue distribution of endogenous and transgene expression overlap. If overlapping expression is anticipated, then one must include in the transgene features that will allow one to distinguish the transgene-derived mRNA or protein from the endogenous. The most facile approach is to alter the nucleotide sequence of one or more exons in the transgene. For example, one can delete a portion of the 5′ or 3′ noncoding region of the gene and distinguish between the two on the basis of size. By the same token, one can insert a piece of foreign DNA to serve as a molecular "tag." The latter approach has the advantage of providing a unique sequence in the transgene mRNA that can be detected by a variety of methods. The extent and nature of each distinguishing feature designed into the transgene will be dictated by individual experimental circumstances and by the preference of the investigator in terms of how transgene expression will be detected. The most important consideration, though, is whether altering the transgenic mRNA will perturb gene expression. In other words, will including extra sequences or deleting sequences alter the half-life of the transcript or change its translation efficiency? Because these issues are usually difficult to address, as a rule, the minimal change that allows one to distinguish between and yet detect both transcripts is preferred.

B. Considerations in the Analysis of Transgene Expression

Generating transgenic animals may seem to be a formidable task, and, indeed, it does pose a considerable technical challenge. However, it is most often the case that the subsequent analyses, namely, identifying founders and characterizing the expression of the new genetic material, prove to be the real challenges.

Characterizing the expression of a transgene is a crucial step in all transgenic projects. In some cases it may represent the crux of the whole experiment, for example, if one is using transgenic animals to define the specificity of a new putative *cis* regulatory element. In other experiments it may be only the first step in a long process of assessing a disease phenotype. Indeed, in the latter case, a careful definition of transgene expression may prompt the investigator to anticipate the phenotypic consequences of transgene expression.

Having derived a set of transgenic founder animals, the first decision to be made with regard to analysis of transgene expression is whether to analyze the founder animals themselves or to carry out the analysis on the transgenic offspring. An important consideration in making this decision is whether it is possible to assess expression in a tissue that can readily be biopsied without endangering the health or reproductive capacity of the founder. For example, one can easily use a tail biopsy to look for expression in skin, muscle, and bone. One can also safely remove a full thickness of skin biopsy from other regions of the body. Expression in liver, spleen, kidney, adrenal gland, lymph node, testis, ovary, and pancreas can be assessed by partial or complete removal of these organs without any long-term deleterious effects on the health of the animal. Blood can also be safely removed by retroorbital or tail bleeding. Procedures for performing many of these surgeries in rats have been compiled (Waynforth, 1980), and most can be readily adapted to mice with minor modifications. With practice, they can be safely carried out and generally do not require specialized equipment.

Although in general, it is desirable to avoid any invasive procedures that may impose a risk of losing a founder animal before deriving F1 offspring, there is still considerable merit to analyzing expression in the founder animals. Knowing ahead of time which founders to breed to derive an F1 generation can save a significant amount of time and resources that would otherwise go to analyzing offspring of nonexpressing families. Implicit in this discussion is the assumption that the offspring of a founder animal that does not express a transgene will themselves also be nonexpressors.

In general, analyses of transgene expression are carried out on material derived from F1 generation offspring of founder animals. Any of several methods may be used to determine which offspring have inherited the transgene, and these are described in detail in Chapter 7. However, before an accurate assessment of expression can be carried out, the pattern of transgene inheritance must be well defined. For this purpose, a Southern blot analysis will yield the most information with regard to transgene copy number and arrangement. To complicate matters, integration at a single site or at multiple genomic sites may occur later than the one-cell stage of development, resulting in the generation of a genetically mosaic animal. It is important to be absolutely certain that expression is analyzed from a single integration site.

Once the number and arrangement of transgene integration sites have been well defined, one can consider whether it is appropriate to homozygose the transgene by establishing transgenic brother–sister breeding pairs. In general, the most efficient approach is to first determine which lines express the transgene, then breed a selected few to homozygosity if necessary, since the amount of time and resources involved in generating homozygotes is significant and success is not guaranteed. Although hemizygous mice are usually the starting point in expression analyses, there may be situations where having homozygotes is essential. For example, if the level of transgene expression is anticipated to be near the threshold for detection, then increasing the gene dosage by achieving homozygosity may be important. Alternatively, if a very small tissue is to be assayed and samples from many animals must be pooled, then deriving homozygotes early on may be useful.

II. ANALYSES OF STEADY-STATE LEVELS OF TRANSGENIC RNA TRANSCRIPTS

Frequently, the initial analysis to evaluate transgene expression involves a broad survey of a variety of tissues at the mRNA level. The number of different samples taken from an individual mouse is limited primarily by the dissecting skill of the investigator but most often includes at least portions of all major organ systems: brain, salivary gland, thymus, heart, lung, liver, spleen, kidney, small intestine, large intestine, pancreas, testis or ovary, muscle, bone, lymph nodes, and skin.

A. *Isolation of RNA from Tissues*

Isolation of intact RNA, free of protein and DNA, is the critical first step in analyzing steady-state levels of transgene mRNA. The most effective method for inactivating endogenous nucleases and deproteinizing RNA is by rapid homogenization in the presence of guanidinium thiocyanate or guanidine hydrochloride (MacDonald *et al.*, 1987). A strong reductant, either 2-mercaptoethanol or dithiothreitol, is also included to disrupt disulfide bonds. Rapid and thorough homogenization is essential for isolating intact RNA, especially in the case of tissues with relatively high levels of endogenous RNase activity, for example, pancreas and spleen. The use of a high-speed homogenizer, for example, a Tissumizer (Tekmar, Cincinnati, OH) or Poly-tron (Janke & Kunkel AG, Bresigan, Germany), is a virtual necessity. Physical separation of RNA from DNA, protein, and other tissue components can be achieved by ultracentrifugation through cesium chloride (MacDonald *et al.*, 1987), by phenol extraction under acidic conditions (Chomczynski and Sacchi, 1987), or by differential precipitation (Cathala *et al.*, 1983).

B. Slot-Blot Hybridization Analysis

One approach to analyzing a large number of RNA samples in a single experiment is by slot-blot analysis (Sambrook *et al.*, 1989). In this method, denatured total or poly(A)-enriched RNA is applied to a solid support (typically a nitrocellulose membrane) with the aid of a vacuum applied through slots in a plexiglass manifold. The bound RNA is then hybridized with a radiolabeled nucleic acid probe, and a signal is detected by autoradiography. Figure 1A shows an autoradiogram of a typical slot-blot filter containing RNA from eight different tissues of a transgenic animal, and Fig. 1B shows the results of a densitometric tracing of the same autoradiogram. RNA samples from five of the tissues show strong hybridization to the probe, and the relative strength of the autoradiographic signal is indicated. This method has the advantages of being relatively simple, having a high sample throughput, and being amenable to quantitation. For most accurate quantitation, the transgene probe can be stripped off and the support hybridized a second time with a different probe to control for the amount of RNA loaded into each slot. The signal seen in the autoradiogram using the transgene probe can then be normalized with respect to the signal seen with a control probe. The difficulty with such a normalization process is finding an appropriate control probe that is applicable to all of the tissues.

A major disadvantage of the slot blot method is that no information with regard to the size of the transgene-derived transcript or transcripts is obtained. This problem can be circumvented by carrying out a Northern blot analysis which is an equally effective method for measuring the steady-state level of mRNA (Selden, 1989). In this technique, total or poly(A)-enriched RNA is separated by electrophoresis in a denaturing agarose gel, and transferred, typically by capillary action, to a solid membrane support. The support is then hybridized to a radioactive probe to detect the transgenic mRNA.

C. Northern Blot Hybridization Analysis

Although it is more laborious than the slot-blot method, Northern analysis has the added benefit of allowing the investigator to determine the size of the transcript of interest and can reveal the presence of multiple transcripts that differ from one another by greater than 50–100 bases. If, however, multiple transcripts are de-

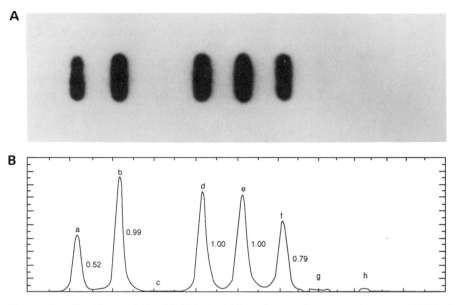

Figure 1. Slot-blot analysis of RNA isolated from tissues of a transgenic mouse. (A) Whole RNA (15 µg) isolated from various tissues of a transgenic mouse carrying an SL3-3 virus long terminal repeat (LTR)/human immunodeficiency virus (HIV) *tat* transgene was applied to a nitrocellulose membrane and probed with an HIV *tat*-specific probe. Slot a, brain; b, bone; c, spleen; d, heart; e, muscle; f, kidney; g, salivary gland; h, intestine. (B) The autoradiogram in (A) was densitometrically scanned, and expression was normalized with respect to the slot e which showed the strongest signal.

tected, a Northern analysis will usually not reveal the basis of the size difference [e.g., alternate transcription start sites, alternate splicing, or alternate use of poly(A) sites] unless a specific probe is available to address each possibility. Northern analysis also shares with the slot-blot method the advantages of high sample throughput and quantitation. Another advantage of the Northern technique is that it often allows one to measure the transgenic mRNA and its endogenous counterpart in the same sample, often in the same experiment.

As an example of the results of a typical Northern blot analysis, Fig. 2A shows an autoradiogram of liver RNA extracted from three independent strains of mice carrying a metallothionein I (MT) major histocompatibility complex (MHC) D^d class I transgene. Liver RNA was extracted from mice that were either not treated or treated with heavy metals administered through the drinking water, to induce transcription from the metallothionein promoter. The ratio of normal to induced transgenic mRNA levels, as quantitated by densitometry, is indicated. The probe used to detect the transgene was a ^{32}P end-labeled 17-base oligonucleotide specific for the transgenic D^d class I gene. Figure 2B shows an autoradiogram of the same Northern blot membrane that has been stripped and hybridized a second time with

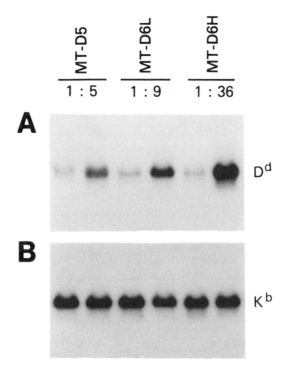

Figure 2. Northern blot analysis of RNA isolated from livers of transgenic mice. Poly(A) $^+$ / RNA (5μg) from three independent transgenic lines carrying an MT I-D^d transgene (see text) was separated on a denaturing formaldehyde agarose gel and transferred to nitrocellulose. Mice were maintained on either normal drinking water or water supplemented with zinc sulfate to induce MT promoter activity prior to RNA isolation. (A) Hybridization was carried out with a D^d-specific oligonucleotide probe. The ratios of uninduced to induced levels of expression are indicated for each line. (B) The same filter shown in (A) was stripped and hybridized again with a K^b-specific probe to detect expression of the endogenous K^b MHC class I gene.

an oligonucleotide probe to detect the endogenous K^b class I gene. The latter is not affected by treatment with Zn^{2+} and serves as an internal control.

D. Nuclease Protection Assay

Another approach to measuring the steady-state level of the transgenic mRNA is by the nuclease protection assay (Sambrook *et al.*, 1989; Greene and Struhl, 1987; Gilman, 1987). In this procedure, a radiolabeled nucleic acid probe of defined length is incubated in solution with an RNA sample. The probe is typically in one of three forms: end-labeled DNA, uniformly labeled DNA, or uniformly labeled synthetic RNA. Following the hybridization step, a nuclease or combination of nucleases is used to destroy any part of the probe that does not hybridize to the specific mRNA. The sample is then analyzed by electrophoresis through a denaturing polyacrylamide gel, and the signal is detected autoradiographically.

Although the sample throughput in nuclease protection assays is somewhat lower than slot-blot or Northern blot analyses owing to the number of manipulations involved, it offers several advantages over these methods. Because both the probe and the mRNA are in solution, hybridization follows second-order kinetics, compared to first-order kinetics for filter hybridization where the mRNA is immobilized. Rapid and quantitative hybridization can, therefore, be achieved at comparatively low probe concentrations. That translates into a higher signal-to-noise ratio for nuclease protection assays compared to Northern or slot-blot analysis. Furthermore, specific information with regard to transcription initiation or termination can be obtained depending on what region of the mRNA is covered by the probe. This is a particularly important consideration in analyzing gene expression in transgenic mice. Because transgenes often consist of heterologous pieces of DNA juxtaposed by virtue of cloning rather than by evolutionary selection, cryptic initiation, splice, or termination sites may come into play. Some of these may be tissue specific, whereas others may appear more globally.

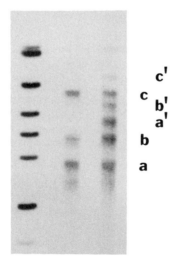

Figure 3. Ribonuclease protection assay to detect transgene expression. Whole RNA (10 μg) was isolated at 12.5 days of gestation from normal (lane 1) or *Hox3.1/β*-gal transgenic mouse embryos (Bieberich *et al.*, 1990) and hybridized with a synthetic RNA probe. Bands a, b, and c are common to both RNA samples, whereas bands a', b', and c' are found only in the RNA sample from the transgenic embryo (see text for explanation). The lengths of the size markers are, from top to bottom, in bases, 242, 238, 217, 201, 190, 180, 160, and 147.

Figure 3 shows the results of an RNase protection experiment designed to detect expression of a transgene consisting of a mouse homeobox (*Hox 3.1*) promoter driving the β-galactosidase (*β-gal*) reporter gene. The probe covers the 5' end of the *Hox 3.1* transcription unit and detects three major transcriptional start sites evidenced by the protected fragments labeled a, b, and c. Lane 1 (Fig. 3) contains RNA from a normal midgestation mouse embryo. The probe also contains 25 bases near its 3' end that are present in the transgenic *Hox 3.1/β-gal* transcript but not present in the endogenous *Hox 3.1* mRNA. When RNA from a transgenic embryo (Fig. 3, lane 2) is hybridized to the same probe, six protected fragments are observed. The three novel protected fragments, labeled a', b', and c', are each 25 bases longer than their endogenous counterparts (a, b, and c) and indicate the presence of the transgene-derived mRNA species. Because the same probe is used to detect both sets of transcripts, one can directly compare the steady-state level of endogenous versus transgenic mRNA in the same sample. This example demonstrates that the transgene-derived mRNAs are initiated at the same three major start sites as the endogenous *Hox 3.1* transcripts. However, the level of transgenic mRNA initiated at start site c is lower than that of the endogenous transcript, as evidenced by the relatively low signal observed for the c' protected fragment.

E. Reverse Transcriptase–Polymerase Chain Reaction Analysis

The most sensitive method for detecting transgene expression is by the reverse transcriptase–polymerase chain reaction (RT-PCR) method (Veres *et al.*, 1987; Kawasaki, 1990). To carry out this assay, a specific "antisense" primer is allowed to anneal in solution to an RNA sample, and reverse transcriptase enzyme is used to convert a region of the mRNA of interest to cDNA. A second "sense" primer is then added to the sample along with a thermostable DNA polymerase. The sample is then heated to denature the mRNA–cDNA hybrid, cooled to permit the "sense" primer to anneal to the cDNA, and incubated to allow the DNA polymerase to convert the single-stranded cDNAs to complete double-stranded molecules. The regimen of denaturation, annealing, and repair, often referred to as a "cycle," is then repeated for a specified number of times. With each cycle, the number of double-stranded products theoretically doubles, resulting in an exponential accumulation of the specific molecule derived initially from the mRNA of interest.

When the process is completed, each sample is assayed for the predicted, specific amplification product. In many cases, the products of an RT–PCR amplification can be examined by agarose gel electrophoresis in conjunction with ethidium bromide staining. In theory, the products of every RT–PCR amplification could be assayed in this fashion, since one molecule of a 200-bp cDNA amplified during 40 cycles of the PCR should yield nearly 250 ng of product, an amount that is easily detectable by ethidium bromide staining. In practice, not all reactions approach that level of efficiency, even after many more cycles than what is theoretically required. The factors that can affect the efficiency of amplification are many, and the dynamics of the so-called anemic PCR amplifications are not well understood (Mullis, 1991).

Many PCR amplifications also result in the formation of multiple products, including the molecule of interest. Sorting out the desired product from the background components by electrophoresis and ethidium bromide staining alone can be difficult. To circumvent the problems created by either weak specific amplification or efficient specific plus nonspecific amplification, it is often desirable to increase the sensitivity of product detection by using a radiolabeled oligonucleotide probe. The probe can be used in a Southern blot analysis or by performing a solution hybridization experiment directly with the PCR product before analysis on a nondenaturing polyacrylamide gel (Yoshioka *et al.*, 1991). Not only is sensitivity increased, but a confirmation of specificity can also be achieved by using a probe that does not overlap with the PCR primers. If information with regard to the absolute or relative size of the amplified product is not required, the products can be simply assayed by slot-blot analysis.

The merits of using RT–PCR to study transgene expression are manifold. It is indisputably the most sensitive method now available to detect the presence of a specific transcript; a single mRNA molecule can be converted to cDNA and amplified to a detectable level. The procedure is technically straightforward and can accommodate many samples in a single experiment. Importantly, for transgenic work, it can be used to detect very small structural differences between endogenous and transgenic transcripts. In addition, RNA isolated from formalin-fixed and paraffin-embedded tissues can be analyzed by this method, allowing for retrospective studies of transgene expression (Rupp and Locker, 1988).

Along with those advantages come significant pitfalls. In most cases, only a small region of the mRNA is amplified, so no information on overall size of the transcript is obtained. Also, amplification of genomic DNA that virtually always copurifies to some degree with RNA can give rise to false-positive signals. This is of particular concern when analyzing tissues from a transgenic mouse with many copies of a transgene and hence many DNA-based target sites for the PCR primers. In some cases it is possible to derive PCR primers that span an intron, thereby allowing one to distinguish between genomic DNA and cDNA-based products on the basis of size. In other instances where transcripts are derived from intronless transgenes, more rigorous steps to eliminate trace amounts of DNA from the RNA sample may be required. For example, one can include a step to degrade the DNA enzymatically, or one can perform differential precipitation using lithium chloride to separate DNA from RNA (Cathala *et al.*, 1983). An additional precaution one can take in the PCR experiment is to include a "no reverse transcriptase" control for each RNA sample by simply omitting the RT step. The great care that must be taken to avoid contamination of RNA samples with "exogenous" DNA, such as products of previous amplification reactions, also encumbers this technique.

Another significant shortfall with the RT–PCR method is the fact that it is difficult to use as a quantitative technique for measuring mRNA levels. The difficulty stems from the exponential nature of the amplification process. The amount of product (N) in a PCR amplification is dependent on the number of input molecules (N_0), the number of "rounds" of amplification (n), and the efficiency (eff) of the reaction:

$$N = N_0(1 + eff)^n$$

Clearly, small differences in the efficiency of each reaction brought about by, for example, differing levels of RNA purity can have a significant impact on the number of molecules produced. A 10% difference in the efficiency of a reaction between two samples with the same number of input molecules can lead to a 5-fold difference in the amount of product produced after 30 rounds of amplification. Despite these difficulties, several semiquantitative (Frye *et al.*, 1989; Brenner *et al.*, 1989; Rappolee *et al.*, 1988) and quantitative (Robinson and Simon, 1991; Wang and Mark, 1990; Gilliland *et al.*, 1990; Becker-Andre and Hahlbrock, 1989; Chelly *et al.*, 1988) approaches for measuring mRNA levels using RT–PCR have been devised.

An example of a semiquantitative analysis of gene expression using RT–PCR is shown in Fig. 4. In this experiment, RNA samples from various tissues of a transgenic mouse carrying a transgene in which the myelin basic protein (MBP) promoter drives the expression of the MHC K^b class I gene were analyzed (Yoshioka *et al.*, 1991). The analysis of transgene expression in the mice was compli-

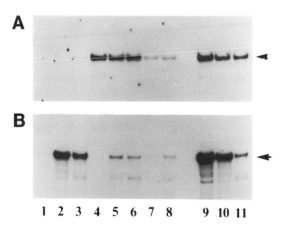

Figure 4. RT–PCR analysis of expression of an MBP–K^b transgene. (A) Whole RNA (100 ng) was analyzed by RT–PCR to detect expression of the endogenous K^b class I MHC gene (arrowhead) (Yoshioka *et al.*, 1991). (B) A different aliquot of the same RNA samples was analyzed using primers to detect expression of the MBP–K^b transgene. Liver RNA was titrated in lane 9 (300 ng), lane 10 (100 ng), and lane 11 (33 ng) to demonstrate that the level of signal was dependent on the RNA concentration. Lanes 1, No RNA; lanes 2, brain; lanes 3, spinal cord; lanes 4, thymus; lanes 5, lung; lanes 6, liver; lanes 7, kidney; lanes 8, testis.

cated by the fact that the endogenous K^b class I gene is expressed in virtually all tissues. The transgenic mRNA differed only slightly from the endogenous mRNA; the first 12 nucleotides of the endogenous K^b transcript were replaced by 36 nucleotides immediately downstream of the MBP transcription start site.

The results of RT–PCR with primers to detect either the endogenous K^b mRNA (Fig. 4A) or the transgenic MBP–K^b mRNA (Fig. 4B) are shown. After amplification, the reaction products were denatured and hybridized in solution to a ^{32}P end-labeled oligonucleotide probe common to both of the predicted PCR products. The hybridized products were then separated from free probe by electrophoresis in a nondenaturing polyacrylamide gel. The solution hybridization step serves three purposes: it increases the sensitivity of the assay by several orders of magnitude, it confirms the specificity of the amplified product, and it allows for quantitation across a broad range of signals. For each pair of primers, a titration of input RNA was carried out simultaneously (Fig. 4, lanes 9–11) to demonstrate that the amount of product produced was dependent on the number of input target mRNA molecules. A marked difference in the pattern of expression of the endogenous K^b gene and the transgenic MBP–K^b gene is apparent. Most notably, expression of the endogenous gene is barely detectable in the brain and spinal cord, where the transgenic mRNA is abundant. In contrast, the endogenous mRNA is abundant in thymus, lung, liver, and kidney, whereas the transgenic mRNA was either low or below detection.

F. In Situ *Hybridization*

All of the approaches to determine transgene expression described thus far suffer from a common shortcoming: they all depend on homogenization of tissues to extract mRNA. As a result, no information with regard to cell type specificity of expression can be obtained without employing painstaking cell separation techniques prior to grinding up the sample. The signal that is observed is necessarily an average over the entire population of cells in the homogenate, and no information as to the number of cells expressing the transgene or their relative levels of expression is gained. To achieve a higher level of resolution of gene expression, one must turn to *in situ* methods.

In situ hybridization is an effective method for analyzing transgene expression at the cellular level (Awgulewitsch and Utset, 1991; Bandtlow *et al.*, 1987; Young, 1991). In this technique, microtome sections of either frozen or paraffin-embedded tissues are fixed to a solid support, typically a glass microscope slide. Following a series of treatments to fix, dehydrate, and permeabilize the tissue sections, a radiolabeled nucleic acid probe is allowed to hybridize to the mRNA *in situ*. After washing to remove unbound probe, the slide is either exposed directly to X-ray film, exposed to a photographic emulsion-coated coverslip, or dipped directly into emulsion to detect the signal. Sulfur-35 is the radionuclide of choice for *in situ* analyses since its relatively low-energy β particle provides adequate resolution with reasonable exposure times. Although nonisotopic methods of labeling both DNA and RNA for use as *in situ* hybridization probes to detect mRNA have been developed, they have failed to replace isotopic methods for use in mouse tissues. The main reason for the lack of popularity for these seemingly attractive alternatives is that, in the present forms, they are less sensitive than isotopic methods.

The high degree of resolution of *in situ* hybridization comes at a price; the technique is time-consuming and technically demanding, requires specialized equipment, and cannot accommodate a large number of samples in a single experiment. Although it is certainly the most difficult of the methods described thus far, it yields the most information with regard to localization of gene expression. Subsets of transgene-expressing cells within an organ can readily be identified. In fact, if fixation and hybridization conditions are optimal, the level of resolution can approach the single cell.

The "sensitivity" of *in situ* hybridization depends somewhat on the pattern of expression of the gene under study. For example, if the gene of interest is expressed in all cells of an organ or embryo and its mRNA accumulates to a relatively low steady-state level, then any filter or solution hybridization method of detection would likely yield a relatively strong signal. On the other hand, an *in situ* hybridization analysis would show a weak and diffuse signal everywhere that would be difficult to interpret. However, if the gene of interest is expressed in only a small subset of cells within a tissue or embryo, but its mRNA is relatively abundant within that small group of cells, then the *in situ* analysis would give an unequivocally positive signal on a few of the cells in the section, whereas other detection methods (with the exception of RT–PCR) could give a negative result.

Figure 5 shows the results of an *in situ* hybridization experiment using an RNA probe to detect expression of the *Hox 3.1* homeobox gene. Figure 5A shows an autoradiogram of a frozen near-midsagittal section of a newborn mouse. Specific signal is apparent primarily in the cervical region of the spinal cord (indicated by the arrow in Fig. 5A). Exposing hybridized sections to standard X-ray film is a quick method for assessing the success of an experiment. Although the level of resolution of the film is low, in many cases it is sufficient to allow one to identify the major sites of specific gene expression. Higher resolution imaging of *in situ* hybridization signal can be achieved by coating the slides with a photographic emulsion. Figure 5B,C shows a near-midsaggital section of only the cervical region of a newborn mouse hybridized with a *Hox 3.1*-specific probe. The slide has been coated with emulsion, exposed, developed, and counterstained. Figure 5B is a bright-field image of the section. Clearly, much greater information with regard to tissue structure is gained compared to the film exposure in Fig. 5A. Figure 5C is a dark-field photomicrograph of the same section as in Fig. 5B showing strong hybridization signal in the ventral portion of the cervical spinal cord. The hybridization signal appears white as the silver grains in the emulsion scatter light. The advantage of the emulsion-coating procedure is that, in addition to higher silver

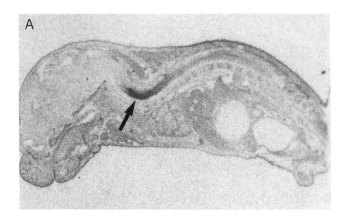

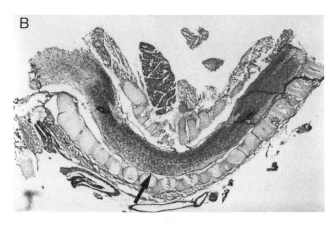

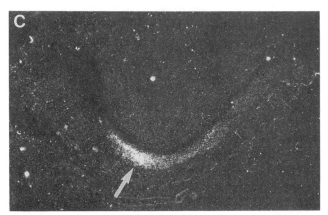

Figure 5. *In situ* hybridization analysis of gene expression. (A) X-Ray film (X-Omat AR, Kodak, Rochester, NY) autoradiograph of a midsagittal section of a newborn mouse embryo hybridized with a *Hox 3.1*-specific RNA probe. Specific hybridization signal in the spinal cord is indicated by the arrow. (B) Bright-field image of a midsagittal section through the dissected cervical region of a newborn mouse embryo stained with hematoxylin and eosin. (C) Dark-field image of the tissue section shown in (B) after hybridization with the *Hox 3.1*-specific RNA probe. Specific hybridization signal is evident primarily in the ventral portion of the spinal cord (arrow).

grain density compared to film, one can counterstain the slides using standard histological reagents to reveal tissue and cellular architecture.

III. ANALYSES OF STEADY-STATE LEVELS OF TRANSGENIC PROTEIN PRODUCTS

Although the analysis of RNA is the usual first step for determining transgene expression, it is by no means a given in all situations. For genes that are expressed at very low levels and whose gene products have high enzymatic activities which can be easily detected without significant background interference, analysis of the protein product may prove more sensitive and direct. Furthermore, detection of RNA does not necessarily imply that a protein product will be made and that it will be active. Efficient translation of certain mRNAs, such as the ferritin transcript (Koeller *et al.*, 1989), requires specific trans-acting factors that may be restricted in tissue distribution. Alternatively, gene products may have to undergo tissue-specific posttranslational modifications, such as proteolytic cleavage or glycosylation, to become functional. In each of these cases, detection of transgenic mRNA in a particular tissue does not imply the expression of a functional protein product. Equally important is the recognition that the site of synthesis of the product of a transgene need not be its site of action. The protein may be secreted into the circulation and may have one or more target tissues. Given any of these circumstances, the analysis of the protein product of the transgene may prove more informative and rewarding. The choice of whether to analyze the expression of the mRNA, the protein product, or both would depend on the specific transgene and its expected function.

As with RNA analyses that take advantage of the availability of a unique DNA fragment with high specificity as a detection probe, protein analyses frequently make use of the availability of an antibody with exquisite specificity. Each of the available methods has its strengths and weaknesses, and the method of choice frequently depends on the characteristics of the particular antibody, the properties of the antigen, and the type of tissues to be analyzed.

A. Radioimmunoprecipitation Assay of Tissue Extracts

Polyacrylamide gel electrophoretic (PAGE) analysis of immunoprecipitates obtained with the use of a specific antibody depends on the ability to first radiolabel the antigen to be detected. With transgenic animals, this can best be done by intraperitoneal injection of radiolabeled amino acids, such as [^{35}S]methionine and/or [^{35}S]cysteine (Kress *et al.*, 1983). In general, radioimmunoprecipitation has the advantage of providing size confirmation of the protein product that is specifically detected. It may also reveal whether the transgenic gene product is properly modified, by using procedures such as two-dimensional gel electrophoresis (Bonifacino, 1991) or tryptic peptide analysis of the immunoprecipitate (Morrison *et al.*, 1983). Unfortunately, successful use of this technique to detect small amounts of protein is predicated on the ability to radiolabel the protein product to a sufficiently high specific activity. *In vivo* radiolabeling of proteins by either intraperitoneal, intravenous, or intramuscular injections of amino acid precursors invariably yields proteins with relatively low specific activities. Of even more concern is the fact that the rate of uptake of the radiolabeled amino acids by the various tissues is unavoidably different, making comparisons of the relative accumulation of a particular protein impossible.

As an alternative to *in vivo* labeling, one can consider radiolabeled amino acid uptake by organ or tissue cultures derived from a transgenic mouse. However, this approach is not applicable to all tissues. Of even greater concern is the fact that expression of a particular gene by cells in culture may not accurately reflect the natural state of affairs within an animal. As a result, the detection of artifacts cannot easily be excluded. Given these constraints, the method of radioimmunoprecipitation to detect the protein product of a transgene in various tissues is rarely used, and it is feasible only in unique situations.

B. Western Immunoblot Analysis of Tissue Extracts

The Western immunoblot analysis approach involves separating total protein in different tissue extracts by sodium dodecyl sulfate (SDS)–PAGE analysis, transferring the protein from the polyacrylamide gel to a solid support (typically a nitrocellulose membrane) with the aid of an electric field, and incubating the membrane with the appropriate antibody (Bonifacino, 1991). The immune complex formed between the antibody and its specific protein antigen on the membrane is then detected by further incubation with either ^{125}I-labeled *Staphylococcus aureus* protein A or an ^{125}I-labeled second antibody, followed by autoradiography. Nonisotopic methods may also be used to detect the immune complex, such as incubation with a peroxidase-conjugated second antibody and subsequent reaction with any of several peroxidase substrates.

An example of the use of this approach is shown in Fig. 6. In this study, the v-H-*ras* gene was placed under the control of the metallothionein I (MT) promoter, which targets expression to the liver and intestine. Western immunoblot analysis using a panreactive Ras antibody of extracts from the intestine of a control mouse (Fig. 6, lane 1), a negative littermate (lane 3), and a transgenic mouse (lane 5) revealed the presence of the activated Ras protein only in the latter; the observed doublet represents the phosphorylated and nonphosphorylated forms of the protein. The endogenous Ras protein, also appearing as a doublet, is detected in the brains

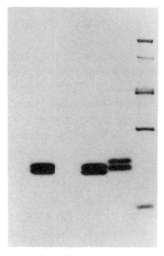

1   2   3   4   5   M$_r$

Figure 6. Western immunoblot analysis of the expression of the Ras protein in the MT– v-*ras* transgenic mice. Extracts from the intestines of a control mouse (lane 1), a normal littermate (lane 3), and its transgenic littermate (lane 5), together with extracts from the brains of the same normal (lane 2) and transgenic (lane 4) littermates, were fractionated by electrophoresis in an SDS–polyacrylamide gel, transferred to a nitrocellulose membrane, and incubated first with a panreactive anti-Ras antibody and then with ^{125}I-labeled protein A. The immune complexes were subsequently visualized by exposure of the nitrocellulose membrane to an X-ray film. The apparent molecular mass ratios (M$_r$) of the markers used were 92,500, 68,000, 45,000, 30,000, and 12,500.

of both the negative littermate (Fig. 6, lane 2) and the transgenic mouse (lane 4). The fact that the doublet detected in the intestine migrates distinctly behind that observed in the brain confirms that it was derived from the activated *ras* gene which was introduced under the control of the MT promoter.

Western immunoblot analysis allows relative quantitation of the product of the transgene in virtually all tissues that can be dissected from an animal. Because it involves the analysis of tissue extracts, it does not provide information as to whether individual cells within the tissue express similar or differing levels of the protein, or whether only a subset of the cells within the tissue is expressing the transgene. Furthermore, not all antibodies are suitable for this purpose. Crude antisera may give a high overall or specific background, and monoclonal antibodies that detect SDS-sensitive epitopes will not work. In general, the limitation of the method is dependent on how much of the protein extract can be loaded onto a gel for the analysis. If needed, the sample may first be enriched for the specific protein by performing an immunoprecipitation reaction with the antibody and *S. aureus* protein A prior to the Western immunoblot analysis. This will greatly increase sample input and enhance the detection of the transgenic protein product.

C. In Situ *Immunohistochemical Staining* of Tissue Sections

A procedure that allows single-cell analysis of transgenic protein products is immunohistochemical staining of tissue sections (Beltz and Burd, 1989). Either frozen, fixed and frozen, or fixed and paraffin-embedded tissues, depending on the particular antibody to be used, are sectioned in a microtome and mounted onto glass microscope slides. Following a series of treatments to fix, dehydrate, and permeabilize the tissue sections, an appropriate concentration of antibody is added to allow binding to the target protein *in situ*. The immune complex is then visualized by a second reaction which, depending on the sensitivity needed, may involve the use of a peroxidase-conjugated second antibody and a peroxidase substrate. After immunochemical staining, the sections may be further treated with a light counterstain, such as hematoxylin and/or eosin, to facilitate histological identification of tissue architecture.

An example of the use of this method is shown in Fig. 7, where a fixed and paraffin-embedded section from the liver of a transgenic mouse carrying the *HBx* gene of the hepatitis B virus (Kim *et al.*, 1991) has been immunostained with an antipeptide antibody specific to the HBx protein. The immunostaining is not detected in all cells in the liver, but rather is restricted to groups of cells surrounding the blood vessels. The staining is predominantly cytoplasmic. The counterstain by hematoxylin allows identification of the nuclei of both HBx-positive and HBx-negative cells.

This procedure permits evaluation of the percentage of cells within the tissue expressing the transgenic protein product as well as the relative level of expression among the cells. Admittedly, these estimates are at best subjective and crude. With appropriate morphological evaluation, such as the use of different cytochemical stains or histochemical markers, one may also define the specific cell type involved. Success in this procedure is predicated on the availability of an adequate antibody with minimal background cross-reactivities. Antibodies that can be used on fixed and paraffin-embedded sections usually are more revealing than those which work only on frozen sections. At times, antibodies which detect different epitopes on the same protein may stain different subsets of cells within the same tissue section. Such information can only come from procedures which allow single-cell analysis.

D. *Immunofluorescent Staining* of Isolated Cells

For circulating cells and those that can be derived from tissues as single cells, direct or indirect immunostaining of live or fixed cells, either in suspension or as a monolayer, can be performed. Using fluorescent antibodies, including those which are conjugated with fluorescein or rhodamine, single cells may be treated and analyzed

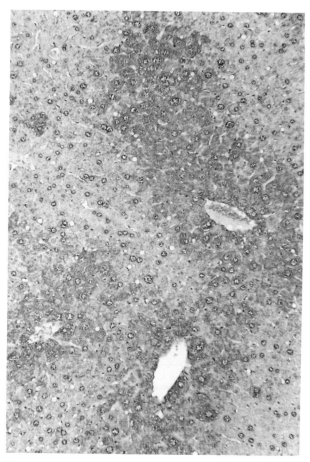

Figure 7. Immunohistochemical staining of a liver biopsy from an *HBx* transgenic mouse. A 10-μm fixed and paraffin-embedded section of liver from a 6-month-old mouse with early focal lesions of altered hepatocytes was deparaffinized and reacted with a rabbit antiserum against a C-terminal peptide (amino acids 139–154) of the HBx protein. The immune reaction was then visualized by the avidin–biotin complex (ABC) method, which involved sequential incubations with biotinylated goat anti-rabbit immunoglobulin G and a mixture of avidin and biotinylated horseradish peroxidase, followed by a further incubation with the peroxidase substrate. The slide was subsequently counterstained with hematoxylin.

either under a fluorescence microscope or through a fluorescence-activated cell sorter (FACS) (Jaffe and Raffeld, 1991).

An example of the use of this method for the analysis of transgenic mice carrying the MT promoter driving the MHC D^d class I gene is shown in Figs. 8 and 9. When live epithelial cells, removed by hyaluronidase treatment of the intestine from transgenic mice that were either untreated (Fig. 8A) or zinc treated (Fig. 9B), were incubated with anti-D^d antibody and subsequently incubated with a phycoerythrin-

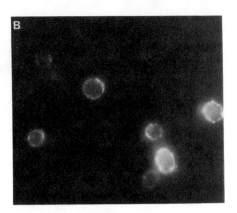

Figure 8. Fluorescence microscopic analysis of intestinal epithelial cells from the MT−D^d transgenic mice. Live cells obtained from the intestines of transgenic mice that were either untreated (A) or zinc treated (B) were stained with an anti-D^d monoclonal antibody and a rhodamine−phycoerythrin-conjugated second antibody. The stained cells were photographed under an epifluorescence microscope.

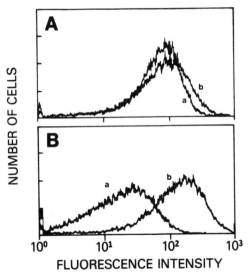

Figure 9. Flow microfluorometric analysis of intestinal epithelial cells from the MT−D^d transgenic mice. Intestinal epithelial cells stained with either an anti-$K^b L^b$ or an anti-D^d monoclonal antibody, followed by a R-phycoerythrin-conjugated second antibody, were analyzed in an EPICS 752 flow cytometer. (A) Cells from the untreated (trace a) and zinc-treated mice (trace b) showed similar levels of fluorescence for the endogenous class I antigen when stained with the anti-$K^b L^b$ monoclonal antibody. (B) Cells from the zinc-treated mouse (trace b) showed a much higher level of fluorescence than those from the untreated mouse (trace a) when stained for the transgenic class I antigen using the anti-D^d monoclonal antibody.

conjugated second antibody, specific cell surface immunofluorescence was detected using an epifluorescence microscope only in cells from the treated mouse. The extent of fluorescence was variable among the individual cells, with some cells staining significantly more intensely than others. When the same cells after immunostaining were subjected to FACS analysis, a fluorescence intensity profile for each cell population was obtained (Fig. 9). In this study, one can deduce not only the mean fluorescence of each population but also the heterogeneity among the individual cells within a population.

Although such methods allow direct quantitation and comparison, one has to assume that the metabolic state of the isolated cells was not too different from that within the tissues of an animal, and that the observed gene expression or the lack thereof was not an artifact resulting from an altered state of metabolism of the isolated cells. At times, this may be difficult to control. The main advantage of this procedure, however, lies in the fact that one can select for subsets of cells, such as those with a specific cell surface marker, for analysis. This greatly facilitates identification of cells that express the transgene.

In situations where antibodies are not available, one has to resort to specific properties of the gene product of interest to facilitate its detection. For example, knowledge of its isoelectric point, state of phosphorylation, or ability to bind specific cofactors may allow the detection of the gene product. In the case of enzymes, substrates can be designed that will permit direct detection and quantitation.

E. Nonimmunologic Reporter Gene Analysis

There is no doubt that the transgenic mouse is the ultimate proving ground for measuring the regulatory potential of DNA elements that act in *cis* to direct gene expression. In a typical experiment, potential *cis*-regulatory sequences are used to direct expression of a so-called reporter gene. Any transcription unit can be used as a reporter provided its product, transcriptional or translational, is innocuous and can be detected and distinguished from any endogenous counterpart. Detection of reporter gene expression can be based on the presence of unique target sequences for a probe within the transcript, altered transcript length, the presence of unique epitopes on a protein product, or, if the reporter gene encodes a functional enzyme, the presence of a novel enzyme activity. The use of enzymes as reporters is particularly attractive because they provide an opportunity to study *cis*-acting elements that are either weakly active or are restricted to a very small subset of cells.

Genes for three enzymes have been employed as reporters in transgenic mice: a bacterial gene that encodes chloramphenicol acetyltranfersase (CAT), a bacterial gene that encodes β-galactosidase (*lacZ*), and a firefly gene that encodes the enzyme luciferase. CAT has been extensively used as a reporter gene in transgenic mice. To detect CAT activity, a protein extract is made from a population of cells or from a whole tissue, and the extract is subsequently incubated with the substrates [^{14}C]chloramphenicol and acetyl-CoA (Gorman *et al.*, 1982). The CAT enzyme

A   **B**

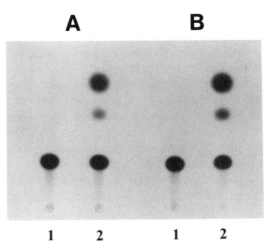

1   2   1   2

Figure 10.   CAT assay to detect reporter transgene expression. Protein extracted from the tails of normal (lanes 1) or transgenic mice (lanes 2) carrying the HTLV-1LTR driving expression of the CAT reporter gene was analyzed for activity by thin-layer chromatography.

will catalyze the acetylation of chloramphenicol, and the acetylated form can easily be separated from the unacetylated form by either thin-layer chromatography (TLC) or high-performance liquid chromatography (HPLC). The results of a typical CAT assay with TLC analysis are shown in Fig. 10. In this experiment, protein was extracted from the tails of two mice from independent lines (Fig. 10A, B) carrying the CAT reporter driven by the long terminal repeat (LTR) of the human T-lymphotropic virus type 1 (HTLV-1). CAT activity is readily detectable in the tail extracts from both lines, and it appears that the extract from the mouse represented in Fig. 10B has greater specific activity than the extract represented in Fig. 10A (cf. lanes 2), when equivalent amounts of cell extracts were analyzed. Lanes 1 in Fig. 10 represent the activity of extracts of nontransgenic littermates from each line.

What kind of information with regard to gene regulation can be garnered from this type of analysis? Clearly, in the example shown in Fig. 10 one can safely conclude that the *cis*-acting element driving the CAT reporter gene is active in the tails of the transgenic mice. The absence of background CAT activity in this or any other mouse tissue makes this assay unambiguous. One might also conclude that the extract in Fig. 10B has a higher specific activity than that in Fig. 10A, a point that can be substantiated by densitometric scanning of the autoradiogram or by cutting out the spots on the chromatogram and counting the eluate by liquid scintillation methods. One problem with this sort of quantitation is that it is difficult to control precisely for the amount of protein loaded and the relative quality of each sample. Unlike nucleic acid hybridization studies where a third-party probe can serve as an internal control within each sample, normalization of CAT assays is more problematic.

The most significant drawback to using CAT assays to define promoter strength and specificity is that no information with regard to cell-type specificity or relative level of expression between cells is obtained. For example, the activity observed from the tail homogenates in Fig. 10 could arise from a low level of CAT enzyme in many cell types or a relatively high level of activity coming from only a specific cell type. This drawback can be partially circumvented in that certain cell populations can be purified from tissues by a variety of separation techniques. However, these techniques are generally cumbersome, result in only partial purification, and often require prolonged incubation of the dissociated tissue, allowing sufficient time for changes in gene expression to occur.

The luciferase gene isolated from the common North American firefly *Photinas pyralis* has also been employed as a reporter in transgenic mice (DiLella *et al.,* 1988). The luciferase enzyme catalyzes the production of light in the presence of ATP, molecular oxygen, and a heterocyclic carboxylic acid substrate termed luciferin (De Wet *et al.,* 1987). The emission of light can be detected with a luminometer. As a reporter gene, luciferase shares with CAT the advantages of sensitivity, quantitation, and low background. Luciferase has the added advantage of being detectable using nonisotopic methods. However, it also shares the disadvantage of being an "averaging" technique since the assays are carried out on tissue homogenates.

The most versatile enzyme reporter is *Escherichia coli* β-galactosidase. This 100-kDa protein, encoded by the *lacZ* gene, catalyzes the cleavage of lactose into glucose and galactose. Importantly, several chromogenic and flurogenic substrates have been developed to detect β-galactosidase activity. The two most widely used are the chromogenic molecule 5-bromo-4-chloro-3-indolyl-β-D-galactopyranoside (X-Gal) and the fluorogenic substrate fluorescein di-β-D-galactopyranoside (FDG). The availability of several different substrates allows for considerable flexibility in terms of detecting β-galactosidase activity.

The use of FDG as a substrate in combination with flow cytometric analysis provides the most sensitive assay for β-galactosidase activity (Nolan *et al.,* 1988). The FDG assay can also be carried out on live cells, providing the opportunity to enrich for cells expressing the reporter gene by fluorescence-activated cell sorting. Although this technique, called FACS–FDG, is easily applied to hematopoietic cells, its general utility is limited to tissues that can be readily dissociated into single-cell populations suitable for flow cytometric analyses.

X-Gal is the only β-galactosidase substrate that has been routinely used to detect transgene activity *in situ.* When X-Gal is cleaved by the β-galactosidase enzyme, a striking blue halogenated indolyl derivative is formed, pinpointing the site of reporter gene expression. X-Gal is particularly useful in that it can be used to detect β-galactosidase activity in whole mounts of embryos as well as whole adult organs or razor blade sections of adult organs with minimal tissue preparation time (Sanes *et al.,* 1986). Essentially, tissues or embryos are dissected, fixed, and incubated directly in the staining cocktail. In the presence of detergents, X-Gal can readily penetrate through several millimeters of most tissues. A notable exception is skin, which can act as an effective barrier to penetration. After incubation with X-Gal, the whole mounts can be processed by standard paraffin embedding techniques and counterstained to facilitate histological identification of expressing cells. However, one must beware that some counterstains, for example, hematoxylin and Giemsa, can interfere with detection of weak X-Gal staining because these stains also yield a bluish color. It may be advisable then, in some cases, to counterstain only alternate sections.

Some mouse tissues, for example, salivary gland, kidney, and intestine, have considerable amounts of endogenous lysosomal β-galactosidase activity that can interfere with interpretation of results of reporter gene expression studies. To minimize background activity, it is important to perform the FDG and X-Gal assays at a neutral pH to favor the activity of the bacterial β-galactosidase. Minimizing staining time is also critical for keeping background low.

As an alternative to whole-mount staining to detect β-galactosidase activity, tissues can be cryosectioned prior to incubation with X-Gal. Although it is more time-consuming, this approach alleviates any potential problems of substrate penetration. Tissues can also be fixed and cryoprotected to preserve morphology before sectioning without inhibiting β-galactosidase activity (Mucke *et al.*, 1991).

The advantages of analyzing reporter gene activity *in situ* are obvious. One can rapidly determine not only which tissues show reporter gene activity, but also whether all or only a subpopulation of cells within a given tissue are capable of supporting transgene expression. An X-Gal staining analysis of three 10.5-day gestation mouse embryos that carry a homeobox/*lacZ* transgene is shown in Fig. 11. Figure 11A shows a dark-field micrograph of a whole mount after fixation and staining with X-Gal. β-galactosidase activity is evident in a broad posterior region of the embryo, as well as in a small patch of skin overlying the hindbrain. Figure 11B shows a cross section through the neural tube of a 10.5-day gestation embryo. Strong β-galactosidase activity is evident in two groups of cells that flank the neural tube. Weaker activity is detected in a central band of neuronal precursor cells within the neural tube. Figure 11C shows a parasagittal section through a third embryo. In this section it is possible to distinguish single blue cells on a large background of cells that show no β-galactosidase activity. The photomicrographs in Fig. 11B,C were produced using differential interference contrast optics without counterstaining on paraffin-embedded sections of embryos that had been stained with X-Gal as whole mounts.

Recently, the *Drosophila* alcohol dehydrogenase (ADH) gene has been used as a reporter gene in transgenic mouse embryos (Nielsen and Pederson, 1991). *Drosophila* ADH activity was detected *in situ* in preimplantation mouse embryos by a simple histochemical assay. The assay takes advantage of the fact that *Drosophila* ADH preferentially uses 2-butanol, a substrate not usually used by mammalian forms of ADH. Although this gene shows great promise as a single-cell resolution

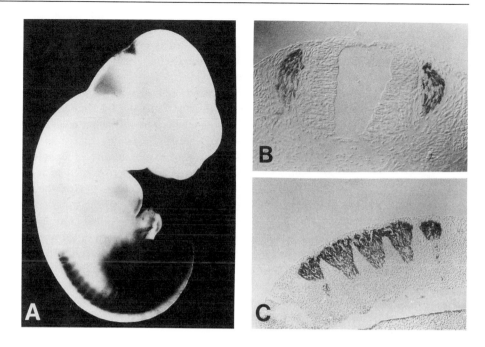

Figure 11. Analysis of β-galactosidase activity in transgenic embryos. Transgenic embryos carrying the *lacZ* gene under the control of a homeobox gene promoter were analyzed for β-galactosidase activity using the X-Gal histochemical assay. (A) Whole-mount embryo shows a broad posterior domain of dark staining indicative of β-galactosidase activity. (B) Cross section through the neural tube of a 10.5-day β-galactosidase-expressing embryo reveals activity in the developing spinal ganglia flanking the neural tube. (C) Parasagittal section through a 10.5-day transgenic embryo also reveals strong β-galactosidase activity in spinal ganglia. Dorsal is toward the top in (B) and (C).

reporter in early embryos, its utility in later embryos and adult tissues remains to be demonstrated.

IV. CONCLUSIONS

Once a pattern of transgene expression has been determined by any of the techniques described here, it is critical to confirm the pattern in a second independent transgenic strain, given that the site of integration may strongly influence gene expression. In cases where two strains of mice are examined and their expression patterns do not corroborate one another, then more strains must be analyzed to reveal a consistent pattern.

Obviously, no one particular method is any better than another for the evaluation of gene expression in transgenic animals. The choice of procedures will depend on a combination of factors, including the following: availability of reagents; number of tissues to be analyzed; sites of suspected gene expression; requirement for absolute quantitation of the level of expression; whether information on heterogeneity of expression among cells is required; the need to identify expressing cell types; the ability to distinguish between the products of the transgene and the endogenous gene; the need to confirm the size of the transcriptional or translational product; the ability to determine the transcriptional start sites, splice sites, or termination sites; the need to confirm the posttranslational modifications of the product; and whether information on the subcellular localization of the gene product is required. In short, the method of choice will be dependent on the specific gene of interest and the type of information to be obtained. The fact that each procedure has its own potentials and limitations is axiomatic. By considering the advantages and drawbacks of each method *before* deriving animals and by carefully designing the transgene to facilitate expression analyses, one can expedite the timely completion of many trangenic mouse experiments.

REFERENCES

Awgulewitsch, A., and Utset, M. F. (1991). Detection of specific RNA sequences in tissue sections by *in situ* hybridization. *In* "Methods in Nucleic Acids Research" (L. Chao, J. Karam, and G. Warr, eds.), pp. 359–375. CRC Press, Boca Raton.

Bandtlow, C. E., Heumann, R., Schwab, M. E., and Thoenen, H. (1987). Cellular localization of nerve growth factor synthesis by *in situ* hybridization. *EMBO J* **6,** 891–899.

Becker-André, M., and Hahlbrock, K. (1989). Absolute mRNA quantification using the polymerase chain reaction (PCR). A novel approach by a PCR aided transcript titration assay (PATTY). *Nucleic Acids Res.* **17,** 9437–9446.

Beltz, B. S., and Burd, G. D. (1989). "Immunocytochemical Techniques: Principles and Practice." Blackwell, Cambridge, Massachusetts.

Bieberich, C. J., Utset, M. F., Awgulewitsch, A., and Ruddle, F. H. (1990). Evidence for positive and negative regulation of the *Hox 3.1* gene. *Proc. Natl. Acad. Sci. U.S.A.* **87,** 8462–8466.

Bonifacino, J. S. (1991). Isolation and analysis of proteins. *In* "Current Protocols in Immunology" (J. E. Coligan, A. M. Kruisbeek, D. H. Margulies, E. M. Shevach, and W. Strober, eds.), Vol. 1, pp. 8.0.1.–8.12.9. Green and Wiley (Interscience), New York.

Brenner, C. A., Tam, A. W., Nelson, P. A., Engleman, E. G., Suzuki, N., Fry, K. E., and Larrick, J. W. (1989). Message amplification phenotyping (MAPPing): A technique to simultaneously measure multiple mRNAs from small numbers of cells. *BioTechniques* **7,** 1096–1103.

Cathala, G., Savouret, J. F., Mandez, B., West, B. L., Karin, M., Martial, J. A., and Baxter, J. D. (1983). A method for isolation of intact, translationally active ribonucleic acid. *DNA* **2,** 329–335.

Chelly, J., Kaplan, J. C., Maire, P., Gautron, S., and Kahn, A. (1988). Transcription of the dystrophin gene in human muscle and non-muscle tissues. *Nature (London)* **333**, 858–860.

Chomczynski, P., and Sacchi, N. (1987). Single-step method of RNA isolation by acid guanidinium thiocyanate–phenol–chloroform extraction. *Anal. Biochem.* **162**, 156–159.

De Wet, J. R., Wood, K. V., DeLuca, M., Helinski, D. R., and Subramani, S. (1987). Firefly luciferase gene: Structure and expression in mammalian cells. *Mol. Cell. Biol.* **7**, 725–737.

DiLella, A. G., Hope, D. A., Chen, H., Trumbauer, M., Schwartz, R. J., and Smith, R. G. (1988). Utility of firefly luciferase as a reporter gene for promoter activity in transgenic mice. *Nucleic Acids Res.* **16**, 4159.

Frye, R. A., Benz, C. C., and Liu, E. (1989). Detection of amplified oncogenes by differential polymerase chain reaction. *Oncogene* **4**, 1153–1157.

Gilliland, G., Perrin, S., and Bunn, H. F. (1990). Competitive PCR for quantitation of mRNA. *In* "PCR Protocols: A Guide to Methods and Applications" (M. A. Innis, D. H. Gelfand, J. J. Sninsky, and T. J. White, eds.), pp. 60–69. Academic Press, San Diego.

Gilman, M. (1987). Ribonuclease protection assay. *In* "Current Protocols in Molecular Biology" (F. M. Ausubel, R. Brent, R. E. Kingston, D. D. Moore, J. G. Seidman, J. A. Smith, and K. Struhl, eds.), pp. 4.7.1–4.7.8. Wiley, New York.

Gorman, C. M., Moffat, L. F., and Howard, B. H. (1982). Recombinant genomes which express chloramphenicol acetyltransferase in mammalian cells. *Mol. Cell. Biol.* **2**, 1044–1051.

Greene, M. J., and Struhl, K. (1987). Analysis of RNA structure and synthesis. *In* "Current Protocols in Molecular Biology" (F. M. Ausubel, R. Brent, R. E. Kingston, D. D. Moore, J. G. Seidman, J. A. Smith, and K. Struhl, eds.), pp. 4.6.1–4.6.13. Wiley, New York.

Jaffe, E. S., and Raffeld, M. (1991). Immunofluorescence and cell sorting. *In* "Current Protocols in Immunology" (J. E. Coligan, A. M. Kruisbeek, D. H. Margulies, E. M. Shevach, and W. Strober, eds.), Vol. 1, pp. 5.0.1.–5.8.8. Green and Wiley (Interscience), New York.

Kawasaki, E. S. (1990). Amplification of RNA. *In* "PCR Protocols: A Guide to Methods and Applications" (M. A. Innis, D. H. Gelfand, J. J. Sninsky, and T. J. White, eds.), pp. 21–27. Academic Press, San Diego.

Kim, C.-M., Koike, K., Saito, I., Miyamura, T., and Jay, G. (1991). HBx gene of hepatitis B virus induces liver cancer in transgenic mice. *Nature (London)* **351**, 317–320.

Koeller, D. M., Casey, J. L., Hentze, M. W., Gerhardt, E. M., Chan, L. N. L., Klausner, R. D., and Harford, J. B. (1989). A cytosolic protein binds to structural elements within the iron regulatory region of the transferrin receptor mRNA. *Proc. Natl. Acad. Sci. U.S.A.* **86**, 3574–3578.

Kress, M., Cosman, D., Khoury, G., and Jay, G. (1983). Secretion of a transplantation-related antigen. *Cell (Cambridge, Mass.)* **34**, 189–196.

MacDonald, R. J., Swift, G. H., Przybyla, A. E., and Chirgwin, J. M. (1987). Isolation of RNA using guanidinium salts. *In* "Methods in Enzymology" (S. L. Berger and A. R. Kimmel, eds.), Vol. 152, pp. 219–227. Academic Press, San Diego.

Morrison, B., Kress, M., Khoury, G., and Jay, G. (1983). SV40 tumor antigen: Isolation of the origin-specific DNA binding domain. *J. Virol.* **47**, 106–114.

Mucke, L., Oldstone, M. B. A., Morris, J. C., and Nirenberg, M. I. (1991). Rapid activation of astrocyte-specific expression of GFAP–LacZ transgene by focal injury. *New Biol.* **3**, 465–474.

Mullis, K. B. (1991). The polymerase chain reaction in an anemic mode: How to avoid cold oligodeoxyribonuclear fusion. *PCR Methods Appl.* **1**, 1–4.

Nielsen, L. L., and Pederson, R. A. (1991). *Drosophilia* alcohol dehydrogenase: A novel reporter gene for use in mammalian embryos. *J. Exp. Zool.* **257**, 128–133.

Nolan, G. P., Fiering, S., Nicolas, J.-F., and Herzenberg, L. A. (1988). Fluorescence-activated cell analysis and sorting of viable mammalian cells based on β-D-galactosidase activity after transduction of *Escherichia coli lacZ*. *Proc. Natl. Acad. Sci. U.S.A.* **85**, 2603–2607.

Rappolee, D. A., Mark, D., Banda, M. J., and Werb, Z. (1988). Wound macrophages express TGF-α and other growth factors *in vivo*: Analysis by mRNA phenotyping. *Science* **241**, 708–712.

Robinson, M. O., and Simon, M. I. (1991). Determining transcript number using the polymerase chain reaction: Pgk-2, mP2, and PGK-2 transgene mRNA levels during spermatogenesis. *Nucleic Acids Res.* **19,** 1557–1562.

Rupp, G. M., and Locker, J. (1988). Purification and analysis of RNA from paraffin-embedded tissues. *BioTechniques* **6,** 56–60.

Sambrook, J., Fritsch, E. F., and Maniatis, T. (1989). "Molecular Cloning: A Laboratory Manual," 2nd Ed. Cold Spring Harbor Laboratory, Cold Spring Harbor, New York.

Sanes, J. R., Rubenstein, J. L. R., and Nicolas, J.-F. (1986). Use of a recombinant retrovirus to study post-implantation cell lineage in mouse embryos. *EMBO J.* **5,** 3133–3142.

Selden, R. F. (1989). Analysis of RNA by Northern hybridization. *In* "Current Protocols in Molecular Biology" (F. M. Ausubel, R. Brent, R. E. Kingston, D. D. Moore, J. G. Seidman, J. A. Smith, and K. Struhl, eds.), pp. 4.9.1–4.9.8. Wiley, New York.

Veres, G., Gibbs, R. A., Scherer, S. E., and Caskey, C. T. (1987). The molecular basis of the sparse fur mouse mutation. *Science* **237,** 415–417.

Wang, A. M., and Mark, D. F. (1990). Quantitative PCR. *In* "PCR Protocols: A Guide to Methods and Applications" (M. A. Innis, D. H. Gelfand, J. J. Sninsky, and T. J. White, eds.), pp. 70–75. Academic Press, San Diego.

Waynforth, H. B. (1980). "Experimental and Surgical Technique in the Rat." Academic Press, New York.

Yoshioka, T., Feigenbaum, L., and Jay, G. (1991). Transgenic mouse model for central nervous system demyelination. *Mol. Cell. Biol.* **11,** 5479–5486.

Young, W. S. (1991). *In situ* hybridization and Northern analyses of the expression of protein kinase C genes using oligodeoxyribonucleotide probes. *Focus* **13,** 46–49.

PART III

Production of Transgenic Laboratory and Domestic Animal Species

Production of Transgenic Rats and Rabbits

James M. Robl
Department of Veterinary and Animal Sciences
University of Massachusetts
Amherst, Massachusetts 01003

Jan K. Heideman
Transgenic Animal Facility
University of Wisconsin
Madison, Wisconsin 53706

I. Introduction
II. Materials and Methods
    A. Generation of Transgenic Rats
    B. Generation of Transgenic Rabbits
References

I. INTRODUCTION

The methodology used in the generation of transgenic mice is well established and has become routine in many laboratories. Transgenic mice have proved to be valuable research tools. Although there are many reasons why mice are popular laboratory animals, there are a variety of research and commercial areas in which alternative species are more suitable models.

Some experiments require the use of an animal larger than a mouse. For example, biochemistry often involves isolation and purification procedures for which the final yield of product is low. A larger animal is also desirable in fields such as endocrinology where repeated blood samples must be assayed to monitor hormone levels, or where cannulation of blood vessels or localization of specific nerves is necessary. Rats and rabbits have the advantage over other large laboratory species, such as the guinea pig, in that they have a short gestation period and yield large numbers of embryos. In comparison to hamsters, rat and rabbit embryos are not particularly sensitive to *in vitro* handling.

Rats have historically been used more than mice in many types of research, including studies of neuroanatomy and heptacarcinogenesis (Pitot *et al.*, 1988).

Recently, a transgenic rat model was made in order to study the multistage nature of hepatocarcinogenesis. Rats carrying the alb–SV transgene (12 kb of the 5′ flanking region of the mouse albumin gene fused to the coding sequence of SV40 T-antigen) develop focal lesions similar to altered hepatic foci in their livers, with an occasional small heptocellular carcinoma exhibiting histological characteristics of hepatoblastoma (Heideman *et al.*, 1991).

Rats are also the model of choice for human mammary tumors. Most murine mammary tumors are associated with mouse mammary tumor virus, whereas in humans and rats breast cancer is not typically of viral origin. Also, genetic variability in the susceptibility of various rat strains to chemically induced mammary tumors has been described (Gould *et al.*, 1986), which provides a model for the study of human genetic variability in the susceptibility to breast cancer. It would be desirable to be able to study the effects of mammary tumor enhancer and suppressor genes found in different strains of rats on the development of tumors caused by expression of an oncogenic transgene in the mammary glands of rats.

The first two published papers describing transgenic rats both address human diseases for which development of mouse models was not feasible. Mullins *et al.* (1990) produced outbred Sprague-Dawley × WKY rats with hypertension caused by expression of a mouse *Ren-2* renin transgene. Hammer *et al.* (1990) describe spontaneous inflammatory disease in inbred Fischer and Lewis rats expressing human class I major histocompatibility allele HLA-B27 and human β_2-microglobulin transgenes. Both investigations were preceded by the failure to develop human disease-like symptoms in transgenic mice (Mullins *et al.*, 1989; Taurog *et al.*, 1988) expressing the same transgenes that successfully caused the desired symptoms in transgenic rats.

Of the many reports on the production of transgenic rabbits, most have been directed toward using the rabbit as a model for large domestic animals or as a basic biological model. As an example, the first report of the production of transgenic rabbits investigated the possibility of enhancing growth rate by the incorporation of a metallothionein–human growth hormone fusion gene (Hammer *et al.*, 1985). Several other reports along this line have followed (Brem *et al.*, 1985; Gazaryan *et al.*, 1988, 1989; Enikolopov *et al.*, 1988).

Rabbits were the first animal model of atherosclerosis (Clarkson *et al.*, 1974). It was found that feeding rabbits a diet high in cholesterol resulted in lesions of the arteries having some resemblance to fatty streaks in humans. The Watanabe heritable hyperlipidemic (WHHL) rabbit, which spontaneously develops hypercholesterolemia, has proved to be an invaluable resource in the study of atherosclerosis in humans (Brown and Goldstein, 1986). The WHHL rabbit has a mutation in the low-density lipoprotein receptor gene that is similar to the class 2 mutations in human familial hypercholesterolemia (FH). The application of transgenic technology on the WHHL rabbit could provide insight into the efficacy of gene therapy for FH in humans. A transgenic rabbit model for atherogenic disturbances in lipid metabolism has recently been produced containing a human apolipoprotein A-I antisense gene (Perevozchikov *et al.*, 1990). In one line the synthesis of high-density lipoproteins was suppressed.

Another area in which rabbits have traditionally been the species of choice is for the study of the humoral response to antigens or simply as a production system for polyclonal antibodies against an antigen of interest (Cohen and Tissot, 1974). The rabbit is still used in immunology research, and one report of transgenic rabbits concerned the production of a line containing the rabbit immunoglobulin heavy chain enhancer fused to the rabbit c-*myc* oncogene (Knight *et al.*, 1988; Suter *et al.*, 1990). This line had a high incidence of malignant lymphocytic leukemia and may be useful as a source of transformed B cells for a variety of studies. Currently, widespread production and use of murine monoclonal antibodies has shifted emphasis from the rabbit to the mouse as a model for studying the humoral immune system.

In the future, with the aid of transgenic technology, the rabbit may be an ideal small laboratory model for acquired immunodeficiency syndrome. The rabbit is the only small animal that is susceptible to experimental infection with human immunodeficiency virus type 1 (Gardner and Luciw, 1989). Insertion of an appropriate human gene may enhance its susceptibility and response to the virus.

The transgenic rabbit may also be an important bioreactor for the production of various pharmaceutical proteins. Massoud *et al.* (1990) successfully expressed human α_1-antitrypsin in the blood of transgenic rabbits at levels of 1 mg/ml. Buehler *et al.* (1990) have successfully expressed human interleukin-2 in the milk of transgenic rabbits using the β-casein promoter. Rabbits have an advantage in this system over larger animals because transgenic animals can be made quickly and inexpensively. Furthermore, for harvesting the product from milk, rabbits have the advantages of nursing only one time per day, having a high protein content in the milk (12.3%), and giving reasonable volumes of milk (170–220 g/day).

II. MATERIALS AND METHODS

In the following sections differences between rats, rabbits, and mice in techniques and species characteristics will be emphasized. In general, the rat is very similar to the mouse in anatomy, general reproduction, gestation, and parturition; consequently, these species characteristics will not be discussed in detail. The rabbit, on the other hand, is quite different and warrants a more detailed discussion.

A. Generation of Transgenic Rats

1. Superovulation and Embryo Recovery

Rats, like mice, can be induced to ovulate using PMSG (pregnant mare's serum gonadotropin; Sigma, St. Louis, MO) and HCG (human chorionic gonadotropin; Sigma), but this technique is successful only with female rats in a narrow range of age and weight. Embryos are obtained from prepubescent (4 weeks old, 70–80 g)

Sprague-Dawley females maintained on a 4 a.m. on–4 p.m. off lighting schedule. They are given an intraperitoneal injection of 15 IU PMSG between 1 and 3 p.m. followed 48 hr later by 5 IU HCG and then are caged with mature fertile males overnight. Mating is confirmed by a plug or sperm in the vaginas of the female rats.

Donors are sacrificed by anesthetic overdose at approximately 4 p.m. the afternoon after mating. Oviducts are removed through flank incisions, transfered to M2 culture medium (Hogan *et al.*, 1986) and flushed to collect the cumulus–embryo masses. Cumulus cells are removed with hyaluronidase, and embryos are washed and inspected. The yield of normal-appearing embryos varies between 5 and 20 per donor, with an average of 10.

Other researchers (Mullins *et al.*, 1990; Hammer *et al.*, 1990) who have generated transgenic rats used a method described by Armstrong and Opavsky (1988) to induce superovulation in embryo donors. Osmotic minipumps dispensing 1 IU/day of purified follicle-stimulating hormone (FSH; Burnes Biotech, Omaha, NE) over a 60-hr period prior to mating resulted in a yield of about 65 fertilized ova per donor, 80% of which were normal and competent to develop, as judged by the ability to form morulae and blastocysts. This yield was significantly better than that resulting from PMSG and HCG injections given by the same investigators. It is likely that minipumps dispensing FSH will become the method of choice for supplying embryos for the generation of transgenic rats.

2. Rat Embryo Microinjection

The membranes of rat embryos are more resistant to penetration than those of mouse embryos. Injection pipettes are prepared by chipping the ends against the holding pipette, using micromanipulators, to produce sharp tips of freshly broken glass. Figure 1 shows an injected pronuclear rat embryo.

3. Embryo Transfer

A pool of about 30 mature Sprague-Dawley females (at least 13 weeks old) are maintained as embryo transfer recipients. Two to four randomly cycling rats with vaginal smears indicating estrus the afternoon on which the donor females are given hCG are selected and placed with vasectomized Copenhagen males. Mating behavior usually begins within a few minutes, and the abundance of leukocytes in vaginal smears the following morning confirms pseudopregnancy.

The procedures used for oviductal embryo transfer in the rat are similar to standard mouse techniques, with the following exceptions:

• To minimize losses resulting from time in culture, rat embryos are always transferred on the day that they are microinjected.

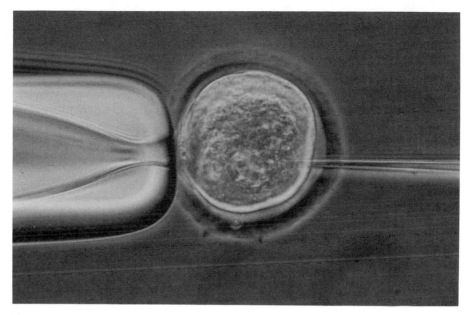

Figure 1. Injection of a pronuclear rat embryo. The pronucleus is swollen from the injected solution. Magnification, 400 ×.

• Because rats respond inconsistently to injectable anesthetics, anesthesia is induced with halothane U.S.P. (2-bromo-2-chloro-1,1,1-trifluoroethene) and maintained with Metofane (methoxyfluorane) (both inhalants).

• The ovarian fat pad is always large in the adult rat, and it must be carefully arranged to permit access to the oviduct.

• The rat ovarian bursa is highly vascular, so that some bleeding is almost unavoidable; however, clotting occurs rapidly.

• The infundibulum of the rat oviduct is usually oriented horizontally, buried between the oviduct and ovary. It is easier to insert the transfer pipette if about 4 mm of the tip is bent at a 30° angle.

About 90% of the transfers that included carrier embryos resulted in the birth of live pups.

4. Carrier Embryos

The survival rate of microinjected mouse embryos can be up to 90%, but the more aggressive technique required to inject rat embryos results in a survival rate of about 60%. In addition, roughly half as many rat as mouse embryos can be injected in a given time period, and rat superovulation and fertilization are sometimes inconsistent. To ensure sufficient litter sizes for normal pregnancy, parturition and lactation, coat-color marked carrier embryos are used. Microinjected embryos are from donors mated with Sprague-Dawley males. The number of embryos surviving microinjection is divided by 10 to estimate the number of resulting pups. Sufficient noninjected embryos from donors mated with Copenhagen males are transferred along with the injected embryos to give expected litter sizes of 6–7. About 80% of carrier embryos develop into live pups. Techniques used for sampling and identification of mice are suitable for rats as well (see Chapter 7).

B. Generation of Transgenic Rabbits

1. General Reproduction in the Rabbit

The rabbit is unique among the mammalian species that have been used most commonly in embryo research in at least two aspects. First, it is a reflex ovulator. Ovulation and the formation of a functional corpus luteum are induced by mating, contact with other females, cervical stimulation, or hormonal treatment. However, does will not necessarily always accept a male. Domestic rabbits exhibit a cyclicity of 4 to 6 days in sexual receptivity which corresponds to changes in blood estrogen levels (Hagan, 1974). Ovulation occurs 10 to 13 hr after mating. During the first 4 days following ovulation the embryos, or unfertilized oocytes, are transported through the oviduct and to the uterotubal junction. A second unique aspect of the rabbit is that the embryos become layered with a thick mucopolysaccharide coat. This mucopolysaccharide, or mucin, coat is secreted by the isthmic epithelial cells in response to estrogen. Following entry into the uterus the mucin coat thins and eventually disintegrates. Implantation occurs on days 6 and 7, and pregnancy lasts for a total of 30 to 31 days. If the mating was not fertile a pseudopregnancy results, and the corpora lutea secrete progesterone for about 16 to 17 days (Hagan, 1974).

2. Superovulation, Synchronization, and Mating

Does can be superovulated with six consecutive subcutaneous injections of 0.3 mg FSH given 12 hr apart. This is followed by an intravenous injection of 75 IU of HCG 12 hr after the last FSH injection. Intravenous injections are given in the marginal ear vein using a 26-gauge needle. With this procedure it is possible to recover an average of 20 embryos from each female, with a range of 6 to 70. A single injection of PMSG at the time of the first FSH injection (150 IU) can be used instead of the FSH sequence. In our hands this gives a less consistent and lower overall response than FSH. Embryo recipients are induced to ovulate using a single injection of HCG (75 IU) and/or mating to a vasectomized buck at the time of donor mating. Mating of the recipient doe is not necessary, and we have had pregnancies in does which would not accept a buck; however, we prefer to use recipients that are receptive over those that are not. In general, superovulated does readily accept a buck and mate without assistance. Recipient does often are not receptive but can be mated by restraining the doe with one hand over the neck and shoulders and lifting the rump and tail with the other.

3. Culture and Manipulation Media

As with embryos from any species, it is always beneficial to minimize the amount of time an embryo is handled *in vitro*. Rabbit embryos, however, are more resistant to the harmful effects of *in vitro* culture than are embryos from many species. A simple Krebs–Ringer bicarbonate medium supplemented with 20% (v/v) fetal calf serum is sufficient to support development of a high percentage of embryos to the blastocyst stage. Because fetal calf serum is beneficial for culture, we supplement a Dulbecco's phosphate-buffered saline (PBS) solution with 20% fetal calf serum for flushing, recovering, and manipulating embryos.

Although a simple medium with fetal calf serum will support development to the blastocyst stage, the number of cells is considerably fewer than comparable embryos to the blastocyst stage. Because fetal calf serum is beneficial for culture, we supplement a Dulbecco's phosphate-buffered saline (PBS) solution with 20% fetal calf serum for flushing, recovering, and manipulating embryos. processed by a quick centrifugation at 15,000 g, sterilization by filtration (0.22-μm pore size) directly into a culture dish (50-μl drops), and overlaying with equilibrated mineral oil. Embryos are cultured in a humidified atmosphere of 7% CO_2 in air at 39°C.

4. General Surgical Procedures

Surgical recovery of embryos or transfer of embryos into a recipient is done via midventral laparotomy under general anesthesia. Rabbits are first given an intramuscular injection of xylazine (11 mg/kg; Lloyd Laboratories, Shenandoah, IA). This is followed 5 min later by an intramuscular injection of ketamine (37 mg/kg; Aeveco Co., Fort Dodge, IA). Anesthesia is induced within a few minutes of the last injection and lasts for approximately 45 min. The abdominal region is clipped and scrubbed with a surgical soap solution such as Betadine (Purdue Frederick Co., Norwalk, CT). An incision is made through the skin and body wall beginning just caudal to the naval and extending approximately 7 cm. Following embryo recovery or transfer, the body wall is sutured using a continuous stitch, and the skin is closed with an interrupted mattress stitch. The suture material is a nonabsorbable polyamide suture (Butler, Westfield, MA, size 3-0). It is important in suturing the skin to avoid pulling the sutures excessively tight. This may result in a skin irritation and the rabbit's pulling the stitches out.

5. Embryo Recovery

Pronuclear embryos are recovered at 20 hr post-HCG by first inserting the flared end of a plastic cannula into the fimbriated end of the oviduct approximately 2 cm (Fig. 2A, B). The opposite end of the cannula is placed in a collection tube. The

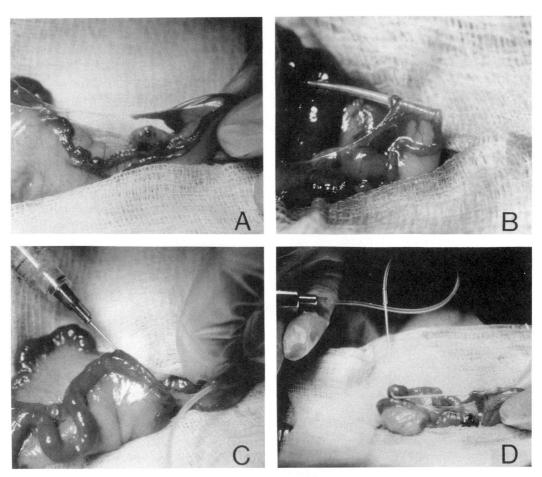

Figure 2. Embryo recovery and transfer procedure in the rabbit. (A) Rabbit oviduct in relation to ovary. (B) Insertion of flared tubing into the fimbriated end of the oviduct. (C) Insertion of a blunt 20-gauge needle into the tip of the uterine horn and through the tubouterine junction for flushing. (D) Transfer of embryos into the oviduct using a fire-polished glass cannula.

cannula is made from a 3-cm length of polyethylene tubing (i.d. 1.19 mm, o.d. 1.70 mm; Becton Dickinson Co., Parsippany, NJ), which is flared by passing it over a flame. Five milliliters of medium is flushed through the oviduct via a 20-gauge blunted needle inserted into the uterus 1 cm below the uterotubal junction and threaded up 2 cm into the isthmus of the oviduct (Fig. 2C). Following flushing, the embryos are recovered using a dissecting microscope and placed in culture until microinjection. Because embryo donors are used a second time and it is not possible to be certain that all embryos are recovered, the does are given an intramuscular injection of prostaglandin $F_{2\alpha}$ ($PGF_{2\alpha}$) (4 mg Lutalyse; Upjohn, Kalamazoo, MI) 4 to 6 days following surgery to induce luteolysis. They are then allowed a 3-week period for recovery.

6. Microinjection

Procedures for the manufacture of micropipettes and setting up the micromanipulator are the same as with mice and have been described in detail in Chapter 2 and elsewhere (Hogan *et al.*, 1986). Briefly, the injection routine involves placing approximately 30 embryos (no more than can be injected in 30–45 min) at one time in a 100-μl drop of DPBS plus 20% FCS, overlaid with mineral oil, in a 100-mm petri dish. A 10- to 30-μl drop of DNA solution is placed adjacent to the embryo manipulation drop. The injection pipette is front loaded after chipping the tip against the holding pipette. Rabbit embryos are larger than mouse embryos (160 versus 110 μm from outside of zona pellucida to outside of zona pellucida) and slightly more opaque; however, the nuclei can be readily seen under phase-contrast or interference contrast microscopy (Fig. 3). The zona pellucida is also considerably thicker and at 20 hr post HCG may contain a very thin mucin coat. Neither the zona pellucida nor the mucin coat are difficult penetrate with the injection pipette. The vitelline membrane, however, is more difficult to penetrate, and gentle tapping on the stage of the microscope is sometimes beneficial. Slow penetration of the membrane and slow removal of the pipette improve survival. Rabbit embryos are generally more resistant to lysis than mouse embryos. The volume of solution injected, as in the mouse, is judged by the swelling of the pronucleus. Following injection, embryos are placed into culture until the time of transfer.

7. Embryo Transfer

For embryo transfer the oviducts of recipient females are exposed surgically as described previously. The transfer is done using a coagulation capillary (1 mm; Fisher Scientific, Medford, MA) bent at a 90° angle. The ends of the glass tubing are fire polished, and one end is connected to a screw pipettor with a 30-cm length of Silastic tubing. The loading procedure involves first pulling a 2-cm length of medium into and around the bend in the pipette. The medium is simply used to

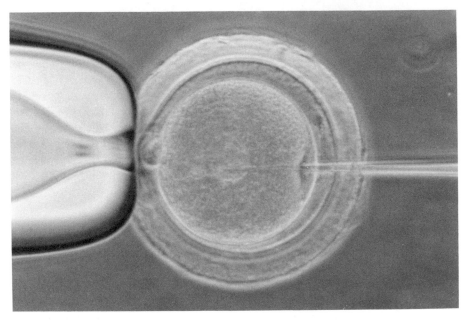

Figure 3. Injection of a pronuclear rabbit embryo. The pronucleus is swollen from the injected solution. Magnification, 400 ×.

monitor the movement of the fluid during the expulsion of the embryos. The tip of the transfer pipette is then loaded by first pulling a 2-cm length of medium into the pipette followed by a 0.5-cm air bubble. The embryos are then pulled into the pipette in as small a volume of medium as possible. The embryo-containing drop is followed by another air bubble and a 1-cm length of medium. The transfer pipette is inserted into the fimbriated end of the oviduct to the first kink, or about 2 to 3 cm (Fig. 2D). The medium containing embryos is gently expelled, and the transfer pipette is then slowly removed from the oviduct with further expulsion of medium. During the transfer process the oviduct is not handled.

8. Gestation and Parturition

Pregnancy can be assessed at 2 weeks following transfer by abdominal palpation. Fetuses can be detected by allowing the uterine horns to slip between the fingers. Approximately 5 days before parturition a nest box containing bedding material should be placed in the cage. The recipient should be monitored daily for nest building. The nest is quite obvious and consists of a large mound of hair. Parturition will then occur within a few hours. The exception to this is that some does will undergo a false labor one or more weeks before the normal time of parturition and will pull a small amount of hair. Following nest building the recipient should be monitored closely. Because of the variable and often low survival rates of gene-injected embryos, litters of one, two, or three offspring are common. Frequently with small litters one or all of the offspring are excessively large and may be unable to pass through the cervix, preventing normal parturition. If pregnancy of the recipient has been confirmed by palpation and yet the recipient has not delivered by day 32, a cesarean delivery should be considered. With either cesarean-delivered offspring or litters of one or two, cross-fostering to large litters, if possible, improves survival. Cross-fostered offspring are readily accepted if they are of similar age and are rubbed with nest material of the surrogate mother. To ensure that a surrogate mother is available, a female can be mated synchronously with a group of gene transfer recipients.

9. Sampling and Identification

Offspring may be sampled as soon as the litter is well established, which is approximately 1 week after birth, or sampling may be done at the time of sexing, which can be done about 2 weeks after birth. A 22-mm section can be taken from the tip of the tail for assessment of gene integration. The tail is first dipped in an antiseptic solution, then clipped with scissors and pinched to stop any bleeding. At this age it is not possible to tattoo the ears of the offspring, which is the most common method of permanent identification for rabbits. Ear tags could be used but may interfere with the marginal ear vein, which is the primary vein used for blood collection and intravenous administration of compounds. A simple method of marking the young rabbits temporarily is by shaving the hair in various locations until they are old enough to ear-tattoo.

ACKNOWLEDGMENTS

J.M.R. thanks Lisa Korpiewski for assistance with preparation of the manuscript and John Balise for technical assistance. Work with rabbits was supported in part by a grant from TSI Inc. to J.M.R.

REFERENCES

Armstrong, D. T., and Opavsky, M. A. (1988). Superovulation of immature rats by continuous infusion of follicle-stimulating hormone. *Biol. Reprod.* **39**, 511–518.
Brem, G., Brenig, B., Goodman, H. M., Selden, R. C., Graf, F., Kruff, B., Springman, K., Hondele, J., Meyer, J., Winnacker, E. L., and Krausslich, H. (1985). Production of transgenic mice, rabbits and pigs by microinjection into pronuclei. *Zuchthygiene* **20**(5), 251–252.

Brown, M. S., and Goldstein, J. L. (1986). A receptor-mediated pathway for cholesterol homeostasis. *Science* **232,** 34–47.

Bühler, T. A., Bruyere, T., Went, D. F., Stranzinger, G., and Bürki, K. (1990). Rabbit β-casein promoter directs secretion of human interleukin-2 into the milk of transgenic rabbits. *Biotechnology* **6,** 140–143.

Clarkson, T. B., Lehner, N. D. M., and Bullock, B. C. (1974). Specialized research applications: I. Arteriosclerosis research. *In* "The Biology of the Laboratory Rabbit" (S. H. Weisbroth, R. E. Flatt, and A. L. Kraus, eds.), pp. 155–165. Academic Press, New York.

Cohen, C., and Tissot, R. G. (1974). Specialized research applications: II. Serological genetics. *In* "The Biology of the Laboratory Rabbit" (S. H. Weisbroth, R. E. Flatt, and A. L. Kraus, eds.), pp. 167–177. Academic Press, New York.

Collas, P., Duby, R. T., and Robl, J. M. (1991). *In vitro* development of rabbit pronuclear embryos in rabbit peritoneal fluid. *Biol. Reprod.* **44,** 1100–1107.

Enikolopov, G. N., Zakharchenko, V. I., Grashchuk, M. A., Suraeva, N. N., Georgiev, G. P., Tinyaeva, E. A., Rubtsov, P. M., Skryabin, K. G., Baev, A. A., and Ernst, L. K. (1988). Obtaining transgenic rabbits containing and expressing the human somatotropin gene. *Dokl. Akad. Nauk SSSR* **299,** 1246–1249.

Gardner, M. B., and Luciw, P. A. (1989). Animal models of AIDS. *FASEB J.* **3,** 2593–2606.

Gazaryan, K. G., Andreeva, L. E., Serova, I. A., Tarantul, V. Z., Kuznetsova, E. D., Khaidarova, N. V., Gening, L. V., Kuznetsov, Yu. M., and Gazaryan, T. G. (1988). Production of transgenic rabbits and mice that contain bovine growth hormone gene. *Mol. Genet. Mikrobiol. Virusol.* **10,** 23–26.

Gazaryan, K. G., Andreeva, L. E., Kuznetsova, E. D., Khamidov, D. Kh., Koval, T. Yu, and Tarantul, V. Z. (1989). Obtaining a transgenic rabbit bearing the transcribed gene of human growth hormone releasing factor. *Dokl. Akad. Nauk SSSR* **305,** 726–728.

Gould, M. N. (1986). Inheritance and site of expression of genes controlling susceptibility to mammary cancer in an inbred rat model. *Cancer Res.* **46,** 1199–1202.

Hagen, K. W. (1974). Colony husbandry. *In* "The Biology of the Laboratory Rabbit" (S. H. Weisbroth, R. E. Flatt, and A. L. Kraus, eds.), Academic Press, New York.

Hammer, R. E., Pursel, V. G., Rexroad, C. E., Jr., Wall, R. J., Bolt, D. J., Ebert, K. M., Palmiter, R. D., and Brinster, R. L. (1985). Production of transgenic rabbits, sheep and pigs by microinjection. *Nature (London)* **315,** 680–683.

Hammer, R. E., Malka, S. D., Richardson, J. A., Tang, J.-P., and Taurog, J. D. (1990). Spontaneous inflammatory disease in transgenic rats expressing HLA-B27 and human β₂m: An animal model of HLA-B27-associated human disorders. *Cell (Cambridge, Mass.)* **63,** 1099–1112.

Heideman, J., Su, Y., Jully, J. H., Moser, A., Griep, A. E., Neveu, M., and Pitot, H. C. (1991). Transgenic hepatocarcinogenesis in the rat. *J. Cell. Biochem.* **15A** (Suppl.), 194 (abstract).

Hogan, B., Costantini, F., and Lacy, E. (1986). "Manipulating the Mouse Embryo." Cold Spring Harbor Laboratory, Cold Spring Harbor, New York.

Knight, K. L., Spieker-Polet, H., Kazdin, D. S., and Oi, V. T. (1988). Transgenic rabbits with lymphocytic leukemia induced by the c-*myc* oncogene fused with immunoglobulin heavy chain enhancer. *Proc. Natl. Acad. Sci. U.S.A.* **85,** 3130–3134.

Massoud, M., Bischoff, R., Dalemans, W., Pointu, H., Attal, J., Schultz, H., and Clesse, D. (1990). Production of human proteins in the blood of transgenic animals. *C.R. Acad. Sci. Ser. 3* **311,** 275–280.

Mullins, J. J., Sigmund, C. D., Kane-Haas, C., and Gross, K. W. (1989). Expression of the DBA/2J *Ren-2* gene in the adrenal gland of transgenic mice. *EMBO J.* **8,** 4065–4072.

Mullins, J. J., Peters, J., and Ganten, D. (1990). Fulminant hypertension in transgenic rats harboring the mouse *Ren-2* gene. *Nature (London)* **344,** 541–544.

Perevozchikov, A. P., Vaisman, B. L., Dozortsev, D. I., Sorokin, A. V., Orlov, S. V., Denisenko, A. D., Dyban, A. P., and Klimov, A. N. (1990). The transgenic rabbits containing a human apolipoprotein A-I antisense gene as a model for the genetic correction of atherogenic disturbances of lipid metabolism. *Biopolim. Kletka* **6**(2), 17–24.

Pitot, H. C., Beer, D., and Hendrich, S. (1988). Multistage carcinogenesis: The phenomenon underlying the theories. *In* "Theories of Carcinogenesis" (O. Iversen, ed.), pp. 159–177. Hemisphere Press, New York.

Suter, M., Becker, R. S., and Knight, K. L. (1990). Rearrangement of VHa1-encoding Ig gene segment to the a2 chromosome in an a^1/a^2 heterozygous rabbit: Evidence for *trans* recombination. *J Immunol.* **144**, 1997–2000.

Taurog, J. D., Lowen, L., Forman, J., and Hammer, R. E. (1988). HLA-B27 in inbred and non-inbred transgenic mice. *J. Immunol.* **141**, 4020–4023.

10

Production of Transgenic Poultry and Fish

Linda C. Cioffi
Progenitor Incorporated
Athens, Ohio 45701

Howard Y. Chen
Merck, Sharp and Dohme Research Laboratories
Rahway, New Jersey 07065

John J. Kopchick
Edison Animal Biotechnology Center
Ohio University
Athens, Ohio 45701

I. INTRODUCTION

The genetic selection of poultry and fish with improved growth and health characteristics is a relatively slow process. This is particularly true in the fish industry, where fish farming or aquaculture techniques are very new. In the poultry field, birds with impressive growth phenotypes have been selected. However, feed efficiency and viral and parasitic diseases are still important husbandry issues. Therefore, to address these issues, investigators are presently introducing genes into poultry and fish using molecular genetics and gene transfer approaches. This technology of manipulating genes to increase growth rate, improve feed efficiency, and import disease resistant traits would benefit both the poultry and aquaculture industries. This chapter reviews the methodology for the production of transgenic fish and poultry.

II. TRANSGENIC POULTRY

A. Background

The total poultry market in the United States is estimated to be approximately $10 billion. Genetic traits relating to growth and health of poultry have been selected classically by poultry breeders. The ability to introduce traits (genes) into the germ line of poultry would certainly enhance the ability of breeders to selectively modify a breeding stock and would be important economically. The major lines of research related to transgenic poultry can be found in Table 1.

TABLE 1
Major Lines of Research Related to Transgenic Poultry

Methods for gene insertion
    Direct DNA transfer or "transfection" into chicken follicles
    Replication-competent avian retroviral vectors
    Replication-defective avian retroviral vectors
    Genetic manipulation of precursor germ cells
    Embryo manipulation, artificial hatching, and embryonic stem cells
Identification of economically important traits
    Growth-related genes
        Growth hormopne
        Growth hormone-releasing factor
        Insulin-like growth factor
        Growth hormone receptor
        Insulin
        Thyroid-stimulating hormone
    Disease-related genes
        Avian leukosis *env* gene—intracellular immunity
        Influenza
        Bacterial resistance
        Coccidiosis
        Immunomodulators

Currently, meat and egg products from transgenic poultry are not available. In fact, the ability to produce transgenic birds has lagged behind the means by which transgenic mammals are generated. In part, this is due to the novel methods required for production of transgenic poultry. In the literature, one can find only a few accounts of production of transgenic chickens. However, two of those accounts dramatically forecast the exciting possibilities related to growth and disease resistance traits that can be introduced into poultry.

A major obstacle stands directly in the path of production of transgenic birds. Methods whereby genes can be introduced safely, easily, and reproducibly into the germ line must be established. To date, avian retroviruses seem to be the method of choice. Paradoxically, avian retroviruses are pathogenic to poultry. Genetic manipulation of these viruses, it is hoped, will eliminate the pathogenic potential of these agents while still allowing them to be used as carriers of genetic information. Issues relating to methods of introduction of DNA into poultry either directly or by use of retroviruses are reviewed. Also, manipulation of early chicken embryos and embryonic cells is discussed.

The purpose of this section is to document methods by which transgenic poultry can be produced. Poultry can be defined as domestic fowl, such as chickens, turkeys, ducks, and geese, that are raised for meat or eggs. Exogenous genetic material (DNA) inserted into somatic or germinal cells of poultry defines transgenic poultry. The enabling technology to create these animals is developing. Products from transgenic poultry have not advanced to the marketplace.

A conventional method used to introduce recombinant DNA into animals is via microinjection of DNA into the male pronuclei of fertilized eggs. This procedure has been used to produce transgenic animals in other agriculturally important species such as fish, pigs, rabbits, and sheep (Palmiter and Brinster, 1986; Hammer et al., 1985). Other methods for the production of transgenic mammals have been developed, including use of replication-defective mammalian retroviruses and the use of embryonic stem cells.

Direct microinjection of recombinant DNA into the pronuclei of embryonic cells of poultry is technically impossible. Avian ova are telolecithal and are normally fertilized approximately 30 min following ovulation. Cell division occurs in the oviduct for 20 hr before oviposition. At this time, the embryo is composed of approximately 60,000 pluripotent cells (Spratt and Haas, 1961), which are collectively called the blastoderm (Kochav et al., 1980; Eyal-Giladi and Kochav, 1976; and Mindur et al., 1985). The presence of a large yolk and multiple pronuclei makes direct microinjection of recombinant DNA impractical.

The data presented in this chapter show attempts to produce transgenic poultry using the following strategies: introduction of recombinant DNA into unfertilized chicken ova; transfection of recombinant DNA into day 1 blastoderms; the use of avian retroviruses as genetic transducing vectors; and manipulation of chicken blastodermal cells or primordial germ cells for the production of chimeric organisms. The current methods for the production of transgenic poultry involve four approaches: (1) insertion of recombinant DNA into cells of an unincubated embryonic blastoderm; (2) insertion of DNA into the chicken germ line using replication-competent retroviral vectors; (3) insertion of DNA into the chicken germ line using replication-defective retroviral vectors; and (4) sperm-mediated gene transfer into poultry. Results to date relative to these procedures are presented below.

B. Direct DNA Transfer into Chicken Follicles and Day 1 Blastoderms

Recombinant DNA, in particular, the bovine growth hormone (bGH) gene linked to the Rous sarcoma virus (RSV) long terminal repeat (LTR), has been successfully introduced into chicken blastodermal cells. However, injection of the DNA into unfertilized ova of surgically exposed follicles did not result in the development of chickens containing the bGH gene (Kopchick et al., 1990). The procedures for this type of DNA transfection experiment are listed below (see Fig. 1).

1. Methods

a. Recombinant DNA Used   DNA used for transfection of day 1 blastoderms and unfertilized ova contained the bGH gene ligated to the RSV LTR (pBGH-4) and has previously been described (Kopchick et al., 1985). This plasmid DNA directs expression of bGH in cultured mammalian and avian cells. pBGH-4 DNA is introduced into day 1 chicken blastoderm as either covalently closed circular super-coils or linear molecules in the presence or absence of DEAE-dextran (200 μg/ml)

Feathers removed from dorsolateral thoracic region

↓

Disinfect with Chlorheximide

↓

Incision made with #22 surgical blade

↓

Exposure of chicken follicles

↓

Introduction of DNA near the day 1 blastoderm

↓

Musculature of body cavity closed and bacitracin applied

Figure 1.   Schematic of direct DNA transfer into chickens.

or calcium phosphate precipitates (Kopchick *et al.,* 1985). The amount of DNA used is 5, 50, or 100 μg. Volumes of injections range from 10 to 50 μl in buffer (10 mM Tris, 0.1 mM EDTA, pH 7.5). The DNA is placed near or under the blastoderm by injection using a glass capillary pipette.

Also, day 1 embryos have been transfected with the firefly luciferase gene (pRSV-L) by a liposome-mediated gene transfer technique. This DNA encodes the firefly luciferase gene linked to RSV LTR (obtained from S. Subramani, University of California, San Diego, La Jolla, CA). The liposomes, either Lipofectin or TransfectAce, are obtained from Bethesda Research Laboratories (Gaithersburg, MD). DNA and lipids are added to Dulbecco's modified Eagle's medium separately, and then the two solutions are mixed gently. The injection of the DNA/liposome solution into the subgerminal cavity has been described previously. The injected embryos are analyzed at day 3 or 8 of incubation.

b. Exposure of Chicken Follicles To expose the chicken ovary for subsequent manipulation, the following protocol is employed. Feathers are removed in the dorsolateral thoracic region ventrally from the backbone to the tip of the ribs. The area is disinfected with Chlorhexidine, and an incision is made with a No. 22 surgical blade. The intercostal muscles are separated from the sixth and seventh vertebral ribs. A Weitlaner retractor (4.5 in., 2 × 3 prongs) is used to open the body cavity. A probe is used to puncture the air sac and expose the avian follicles. Individual maturing ovarian follicles are manipulated and injected with 1 to 25 μg pBGH-4 DNA as near to the oocyte as possible using a 30-gauge, 1 in. needle.

The musculature of the body cavity is closed using 8 to 10 sutures, and zinc bacitracin is applied to the entire area. Following surgery, Combiotic (trademark of Pfizer, 2 ml/bird) is injected for 3 days postsurgery to prevent infection and promote the healing process. This procedure does not affect subsequent fertilization or egg laying.

c. Transfection of Day 1 Blastoderms Fertilized eggs (day 1 blastoderms) are collected immediately after laying and stored at 15°C. A portion of the shell (3 × 3 mm) is removed without disturbing the membrane. After localizing the blastoderm, 1 to 100 μg of pBGH-4 DNA is injected in the immediate area of the blastoderm using glass capillary pipettes or a 27-gauge needle attached to a 100-μl micropipette. Following this manipulation, the opening in the shell is covered with cellophane tape and sealed with paraffin. Eggs are incubated at 38°C until hatching. This protocol does not affect hatchability.

A variation of the above procedure is to inject DNA directly into the subgerminal cavity of the embryo. Eggs are stored on their sides (major axis horizontal) overnight at 15°C to allow the germinal disk to rotate to a new position for ease of injection. The side of egg facing up is marked and then wiped with 70% (v/v) 2-propanol before opening the eggshell. A square of approximately 5 mm is etched using a Foredom Tool (series RB) employing a Dremel cut-off wheel (No. 409).

The shell and inner membrane are removed with forceps to expose the germinal disk. Four microliters of DNA solution is injected into the subgerminal cavity by employing a glass needle mounted on a Drummond 25-μl microdispenser. This in turn is attached to a Narishige (Greenvale, NY) micromanipulator. The glass needles are pulled with a Kopf (Tujunga, CA) micropipette puller to an outer diameter of approximately 25 μm. After injection, the eggs are sealed with 3M Durapore surgical tape and incubated as previously described. The hatchability is significantly reduced by this procedure.

d. Detection of Recombinant DNA and Bovine Growth Hormone in Adult Chickens   Chicken genomic DNA, derived from various tissues, is prepared by homogenizing 50 mg of tissue or 50 μl of packed red blood cells (RBCs) in 3 ml of a solution containing proteinase K (0.5 mg/ml), 1% (w/v) sodium dodecyl sulfate (SDS), 5 mM EDTA, and 10 mM Tris, pH 7.5. The solution is incubated overnight at 42°C. Following two extractions with equal volumes of phenol and chloroform, DNA is precipitated with ethanol at -20°C. Precipitated DNA is collected by centrifugation, washed with 70% (v/v) ethanol, and suspended in 2.0 ml sterile water.

For detection of the bGH gene in chicken genomic DNA, a *Pvu*II DNA fragment from pBGH-4 is used. Probes with specific activities of at least 1×10^9 cpm/μg are used in the analyses. Detection of bGH in chicken serum is by radioimmunoassay (RIA) using rabbit anti-bGH sera (Kopchick *et al.*, 1985). The antibody used to detect bGH in culture fluids or in serum does not react with endogenous chicken growth hormone.

e. Restriction Endonuclease Digestion and DNA Hybridization   Genomic DNA (10 μg) is treated with *Hind*III or *Pvu*II (50 units) overnight at 37°C and resolved by 1% agarose gel electrophoresis. Following electroblotting to Gene-Screen Plus membranes (DuPont/NEN, Boston, MA) at 42 V in 0.33$\times$ TAE 0.04 m Tris-Acetate, 0.001 m EDTA, membranes are exposed to a ^{32}P-labeled DNA hybridization probe (6×10^5 cpm/ml of hybridization solution per the manufacturer's suggestions).

2. Results

a. Introduction of pBGH-4 DNA into Unfertilized Ova   Following surgery which resulted in exposure of the chicken ovary, pBGH-4 was inserted via capillary pipette injections into the largest (5–10) ova. Between 1 and 100 μg of linear pBGH-4 DNA was injected either in the presence of DEAE-dextran or as calcium phosphate precipitates. Approximately 100 ova were injected in 100 artificially inseminated hens. Of the 400 offspring produced and analyzed in this experiment, none were found to possess pBGH-4 DNA in RBC genomic DNA as determined by blot hybridization analyses, and none expressed detectable bGH in serum as assayed by RIA. Attempts to introduce recombinant DNA into unfertilized chicken follicles were unsuccessful. Although the surgical procedures and follicle manipulation did not affect the ovulation and fertilization capacity of the egg, no recombinant DNA positive progeny were obtained by this protocol. One possible explanation of these negative results is that the inserted DNA is degraded in the unfertilized egg. When DNA is exposed to this type of tissue and incubated for 1 hr at 37°C *in vitro,* the DNA is cleaved.

b. Transfection of pBGH-4 DNA into Day 1 Blastoderms A second protocol for the production of transgenic chickens was to insert or transfect recombinant DNA (pBGH-4) into the day 1 fertilized embryo. At this time in development, the embryo is composed of approximately 60,000 cells (Spratt and Haas, 1961). Also, primordial germ cells are not detectable at this time and are indistinguishable from somatic cells (Ginsburg and Eyal-Giladi, 1986, 1987). Therefore, in all probability, positive animals generated by this procedure would be mosaic relative to the cells which take up the DNA and also mosaic relative to the site(s) of DNA integration.

Of 1500 animals generated via this protocol, seven animals possessed elevated levels of serum bGH (between 20 and 40 ng/ml). pBGH-4 DNA was not found in RBC genomic DNA of the seven animals, as determined by hybridization blot analyses. These animals were grown to maturity and mated with control animals. Offspring (200) derived from the crosses were negative for serum bGH and also negative for pBGH-4 DNA in RBC genomic DNA. The seven founder chickens were sacrificed, then liver, muscle, brain, kidney, and intestine tissue were removed. Following hybridization blot analysis of the DNA preparations, three chickens were found to possess DNA which hybridized with the *Pvu*II fragment of bGH. One animal possessed over 30 copies of bGH DNA in the muscle preparation and 1–2 copies in the kidney and liver DNA preparations. A second chicken possessed bGH DNA (>30 copies) in DNA derived from the brain tissue. A third animal contained bGH DNA in genomic DNA derived from intestinal tissue (>30 copies) and kidney tissue (2 copies).

It appears that different cells in the blastoderm can take up DNA, since a variety of tissues in positive animals possessed the DNA. It must be pointed out that, in these analyses, only animals which secreted bGH into serum at detectable levels (>10 ng/ml) or animals which possessed bGH DNA in blood cells would be classified as positive for transgene expression. Animals which possessed integrated DNA in cells other than blood or which did not express and secrete bGH would be classified as negative. To establish whether precursor germ cells of the blastoderm could be transfected by this procedure, all of the 1500 resultant animals generated in this experiment should have been tested by analysis of progeny. However, in general, facilities are not available for such an ambitious experiment.

c. Expression of Firefly Luciferase in Chicken Embryos Following Liposome-Mediated Gene Transfer Following liposome-mediated gene transfer or day 1 blastoderm with pRSV-L, embryos were collected at day 3 or 8 of incubation and assayed for luciferase activity (C. Rosenblum and H. Y. Chen, unpublished results, 1990). Tissues were homogenized in a buffer containing 100 μM dithiothreitol using a Teflon or glass tissue homogenizer. The homogenate was centrifuged at 12,000 g at 4°C for 30 min. The pelleted material was retained for isolation of DNA, and the supernatant was used in a luciferase assay. Samples were measured in a Berthold luminometer (Berthold Systems, Inc., Pittsburgh, PA). Results showed that luciferase expression levels varied greatly among different embryos. This indicates that most of the expression is transient in nature, as can be expected from a transfection experiment.

The possibility of permanent integration and germ line transmission in adult chickens requires further studies. Recent development of cationic liposomes as efficient transfecting agents hold some promise in this approach. We have tested two of the liposomes, Lipofectin and TransfectAce, in transfecting embryonic cells *in ovo*. Results indicated that both liposomes are capable of transfecting the embryos. TransfectAce is the choice between the two because it can work efficiently in the presence of proteins, which interfere with Lipofectin. Although high levels of expression of a luciferase reporter gene can be detected in day 3 embryos, most of the activity appeared transient in nature. The efficiency of permanent integration and germ line transmission awaits further studies in adult animals.

3. Conclusions

In summary, cells of the blastoderm can be transfected with recombinant DNA; however, the frequency of transfections of germ line precursor cells was rare. Therefore, direct DNA injection by these methods is not a viable approach unless a much more efficient method of transfection can be found.

C. Replication-Competent Avian Retroviral Vectors

Five groups have reported successful gene transfer into chickens by use of replication-competent retroviral vectors (Souza *et al.*, 1989; Salter *et al.*, 1987; Hippenmeyer *et al.*, 1988; Chen *et al.*, 1990; Kopchick *et al.*, 1990). Two groups (Salter *et al.*, 1987; Kopchick *et al.*, 1990) have shown successful insertion of viral genes into the germ lines of chickens. In the case of Salter *et al.* (1987), an endogenous viral LTR was used in the retroviral vector, whereas Kopchick *et al.* (1990) used an exogenous viral LTR. Avian leukosis virus (ALV) containing exogenous LTRs have been shown to be more oncogenic than those with endogenous LTRs. Hippenmeyer *et al.* (1988) have shown successful insertions of exogenous genes in somatic tissue of 20-day chicken embryos. Chen *et al.* (1990) have reported successful insertion and expression of the exogenous gene (bGH) in adult animals. The protocols used for these experiments were similar; therefore, the methods from Kopchick *et al.* are used to describe the generation of transgenic chickens using replication-competent retroviruses.

1. Methods

a. Cell Culture and Virus Propagation   Primary chicken embryo fibroblasts (CEFs) are derived from SPAFAS (Storrs, CT) C/E day-10 embryos. CEFs are established by trypsinization followed by plating of cells in Ham's F10 medium 199 (50/50) containing 5% (v/v) tryptose phosphate broth and 4% newborn calf serum.

Avian leukosis virus is propagated following transfection of CEFs with cloned viral DNA encoding a transformation-defective Schmidt-Ruppin A strain of RSV, obtained from Steve Hughes of the National Cancer Institute (Frederick, MD) (Chen *et al.*, 1990). Most, if not all, of the replication-competent retroviruses used by all groups in these studies are obtained from Dr. Steve Hughes. Again, although the vectors differ slightly, the basic protocols are similar. The *src* gene in the virus is deleted and replaced by a unique *Cla*I restriction enzyme cleavage site (Chen *et al.*, 1990). Virus is propagated in CEFs at 37°C in an atmosphere of 5% (v/v) CO_2 for 5 days, at which time the supernatants from the lysed cells are assayed for viral p27 antigen by enzyme-linked immunosorbent assay (ELISA). Rabbit anti-p27 and rabbit anti-p27 conjugated to horseradish peroxidase used in the assay are obtained from SPAFAS. Virus titers in the culture medium are calculated as described (Lannett and Schmidt, 1980).

b. Virus Assays on Chicken Blood, Meconium, and Cloacal Swabs   Virus isolations on day-old chicks are performed by expressing meconium from the chick into a tube containing 1.5 ml of cell culture medium with 5 times the normal amount of antibiotics. Samples are frozen at -70°C. Samples are thawed and centrifuged (10,000 g), and 0.4 ml of supernatant is inoculated into primary CEF cultures made from SPAFAS C/E embryos. Cultures are incubated at 37°C in an atmosphere of 5% (v/v) CO_2 for 5 days, passaged, and grown for an additional 5 days; supernatants from lysed cells are assayed for p27 antigen by ELISA.

Older birds are screened for virus by using either blood samples or cloacal swabs. Cloacal swabs are placed in 1.5 ml of cold cell culture medium containing $5\times$ antibiotics and frozen at -70°C. Both blood and cloacal swab samples are inoculated on primary CEF cultures treated as described above.

c. Blastoderm Injection with Retrovirus   Most of the groups use the following protocol for introduction of virus into the day 1 egg. Eggs to be injected are collected and stored at 15°C (large end up) for 1–2 days to allow the air cell to form and for the blastoderm to rotate to the large end of the eggs. The air cell is outlined in pencil and a small section (1 cm²) of shell cut and removed. Eggs are examined, and all eggs in which the blastoderm could be seen through the membrane are saved for injection.

Approximately $10^5 - 10^6$ infectious units of virus are injected adjacent to and slightly below the blastoderm. Volumes of $10-50$ μl are injected using glass pipettes or repeating micropipettes equipped with a 27-gauge needle. All injection procedures are carried out using biological safety cabinets. After injection, eggs are sealed with cellophane tape and the large end dipped in melted paraffin. Alternatively, virus can be directly injected into the subgerminal cavity of the blastoderm as previously described. Eggs are set in an incubator, and, at hatching, chicks are screened for virus by analysis of meconium samples. All virus-positive chicks are saved for DNA hybridization assays. For detecting exogenous or endogenous proviral DNA sequences in chicken genomic DNA, a 250-bp fragment derived from the U_3 region of an exogenous viral clone, SR-RSV-A, or a 146-bp fragment derived from the U_3 region of an endogenous viral clone, RAV-O, is used. The exogenous and endogenous cloned viral DNA was kindly provided by L. Crittendon (U.S. Department of Agriculture, East Lansing, MI).

d. Genes Used for Insertion   Chen *et al.* (1990) have used the bGH gene which had been incorporated into the Schmidt-Ruppin A strain of RSV. In a similar fashion, Hippenmeyer *et al.* (1988) have incorporated a bacterial neomycin phosphotransferase gene (NPT-II) into a Schmidt-Ruppin A strain of RSV. Kopchick *et al.* (1990) and Salter *et al.* (1987) have incorporated no exogenous genes into these vectors but have used the viral genes as markers for gene insertion.

2. Results

Exogenous virus ($10^5 - 10^6$ per 50 μl) was introduced into day 1 chicken blastoderms. First generation (G_1) viremic males were selected by serum assayed for infectious virus. Also, chicken RBC DNA was analyzed by hybridization blots or polymerase chain reaction (PCR). Data from Kopchick *et al.* (1990) will be discussed, although similar results were obtained by Salter *et al.* (1987) using a viral vector containing an endogenous viral LTR. The production frequency of the G_1 males ranged between 7 and 12% (Table 2). Dot-blot hybridization analyses of the G_1 males revealed that less than one exogenous provirus was present per RBC genome (data not shown). This result is consistent with the hypothesis that G_1 males are mosaic relative to exogenous retroviral DNA sequences; that is, the virus infected some but not all cells of the blastoderm.

Fourteen exogenous virus positive males were mated individually with exogenous virus negative females. Blot analysis of RBC DNA derived from the progeny animals employed exogenous and endogenous virus-specific DNA hybridization probes. A total of 316 G_2 animals were generated by the crosses, and RBC DNA was analyzed by DNA blot hybridization (Chen *et al.*, 1990). Generally, positive animals were shown to possess one to two copies of the exogenous viral DNA per cell. Transmission frequencies of exogenous retroviral DNA sequences from G_1

TABLE 2
Production Frequencies of Viremic Females (Shedders) and G_1 Viremic Males

	Number positive/total	Production efficiency (%)
G_1 viremic males and viremic SPAFUS females	28/240	11.66
G_1 viremic males and viremic Hubbard line 139 females	62/797	7.78

males to G_2 individuals ranged from 0 to 40% (Table 3). Transmission frequencies of G_1 exogenous virus positive individuals to subsequent generations (G_3 and G_4) were approximately 50% (data not shown).

a. Southern Blot Analysis of DNA from Transgenic Chickens   To insert foreign DNA into the chicken germ line, day 1 blastoderms were exposed to virus. If germ cell precursors are infected by the virus, offspring (G_1) should be generated that are capable of passing the viral DNA to the next generation (G_2). A pedigree study using Southern blot analysis would verify this prediction and also reveal valu-

TABLE 3
Transmission Frequencies of Viral DNA by Viremic G_1 Males to G_2 Progeny

G_1 Males	Nonviremic female	G_2 Progeny Number positive/total	G_2 Progeny Transmission frequency (%)
95-289	39	2/20	10.0
95-286	43	1/14	7.1
95-294	50	3/28	10.7
96-259	36	4/17	23.5
96-260	38	3/13	23.0
96-246	42	7/28	25.0
96-273	44	0/23	0
96-268	49	14/39	35.9
97-211	46	0/29	0
90-627	47	0/25	0
91-258	48	3/22	13.6
91-218	40	2/5	40.0
91-181	41	1/28	3.4
98-232	37	1/21	4.8

able information concerning the timing of viral infection during early embryonic development of the G_1 chickens and the frequency of transmission of viral DNA to the progeny (G_2). For this purpose, DNA was isolated from a G_1 viremic male. This male was also mated to a nonviremic female. DNA from the G_2 viremic offspring resulting from the cross was isolated and subjected to Southern blot analysis (Chen *et al.*, 1990).

No discrete bands or smears of DNA were seen in the nonviremic female chicken, indicating that the circulating RBCs were not infected by the exogenous virus. A smear of DNA can be detected in the G_1 chicken, indicating heterogeneous viral DNA integration sites in RBC DNA. All 10 viremic G_2 progenies showed discrete bands of DNA, indicating all of the RBCs of a given chicken had the same viral integration site and therefore confirming the predicted germ line transmission. In these studies, the presence of two bands after hybridization analysis is indicative of one viral integration event. Although most of the G_2 viremic chickens had one integration site, one chicken had two integration sites. This was confirmed by restriction enzymatic analysis of RBC DNA using other enzymes. Furthermore, none of the viremic G_2 chickens showed the same integration pattern, indicating that each chicken was derived from a different germ cell precursor. These results showed that although a significant portion of the germ cell precursors can be infected by the virus, the infection took place at a relatively late stage of germ cell differentiation. Evidence for this is the fact that no identical DNA integration pattern can be detected in the offspring. A pedigree study involving three other viremic G_1 males and viremic G_2 progenies showed similar results (Chen *et al.*, 1990).

To gain further information concerning viral DNA distribution and integration patterns in various tissues, a G_1 viremic bird and its G_2 viremic offspring were sacrificed. DNA isolated from various tissues was subjected to Southern blot analysis. Results showed all tissues from a G_2 viremic chicken had an identical DNA pattern, indicating all tissues were derived from the same embryonic cell which contained integrated provirus, as can be expected with G_1 to G_2 germ line transmission (Chen *et al.*, 1990). However, no discrete bands can be detected in the DNA of various tissues of the G_1 viremic chicken. Finally, comparison of Southern blot DNA patterns of G_2 and G_3 transgenic chickens in three different lines showed identical patterns between parent and offspring (Chen *et al.*, 1990), indicating stable transmission of the integrated genes to offspring.

b. Resistance of Transgenic Chickens to Challenge by Subgroup A Rous Sarcoma Virus   Results of Salter *et al.* (1987) and Kopchick *et al.* (1990) show that viral DNA (transformation-defective, Schmidt-Ruppin A strain of RSV) was inserted into the germ line of chickens. These transgenic chickens should be refractory to subsequent infection by ALV because they produce the envelope gene product, which interferes with the ability of the virus to infect chicken cells. To test this possibility, transgenic and nontransgenic chickens were challenged with subgroup A and subgroup B strains of RSV. This virus induces tumors in the chicken

TABLE 4
Comparison of Tumor Development in Transgenic and Control Chickens
Challenged with Subgroup A and Subgroup B Viruses

| | | | Tumor incidence | | | |
| | | | Subgroup A | | Subgroup B | |
	Number of birds	Viremic	2 weeks	3 weeks	2 weeks	3 weeks
Transgenic	24	+	0/24	1/23	19/24	18/23
Control	11	−	9/11	9/11	9/11	8/11

within 2–4 weeks following injection. The subgroup designation refers to a viral gene (*env*) encoded by the viral genome. Recall that the exogenous virus used to produce the transgenic chickens was a subgroup A retrovirus that did not cause this type of tumor. A comparison of the development of tumors in transgenic and non-transgenic chickens was determined (Table 4). Results indicated that transgenic chickens possessing subgroup A virus in the genome are resistant to subsequent infection with subgroup A virus. These chickens were also challenged with subgroup B virus. Transgenic chickens that were resistant to subgroup A virus were susceptible to subgroup B virus (Table 4).

The transgenic chickens produce virus and are specifically resistant to subgroup A virus. This indicates that protection occurs by an interference phenomenon caused by the envelope gene product. Salter and Crittenden (1989) have found a similar phenomenon in transgenic birds that express the subgroup A envelope gene. In these experiments, transgenic lines (which did not produce infectious virus but which did produce the *env* gene product) were found to be resistant to infection by exogenous laboratory and field strains of subgroup A virus (Salter and Crittenden, 1989; Crittenden *et al.*, 1989). These types of transgenic birds would be more practical in the field because they do not produce infectious virus. These results are important in that the transgenic chickens are resistant to a virus which causes avian leukosis. A similar approach can be applied to control other viral diseases that are economically important to chicken producers.

c. Gene Expression in Founder Chickens Hippenmeyer *et al.* (1988) have successfully inserted the NPT-II gene into fertilized, day 1 eggs (SPAFAS line 11) using similar protocols as described above. Approximately 12% of the embryos possessed the NPT-II gene. Expression of the NPT-II gene varied as a function of tissue type, with the highest activity found in muscle and foot tissue.

Chen *et al.* (1990) have successfully inserted the bGH gene into day 1 chicken embryos and obtained two adult transgenic chickens in which high levels of bGH expression resulted in larger birds. Because only two birds were generated that possessed elevated serum bGH and a corresponding enhanced growth performance, more animals must be analyzed in order to evaluate the statistical significance of these findings. However, these results are certainly encouraging for breeders interested in increasing the growth performance of chickens.

3. Conclusions

Schmidt-Ruppin A replication-competent retrovirus vectors were used in these studies. The virus will infect precursor germ cells in chickens. Exogenous virus positive G_1 males, when mated to virus negative females, generate offspring (G_2), some of which are transgenic.

Transmission of the exogenous viral genes from G_1 males to G_2 individuals was confirmed by Southern hybridization blot analysis. DNA derived from G_2 viremic males showed a heterogeneous mixture of molecules that hybridize to the exogenous probe. This observation is consistent with many different infectious events occurring in the embryonic cells of the developing G_2 individual. DNA derived from G_2 offspring revealed exogenous viral DNA in all of the virus positive individuals. These results demonstrate that the viral DNA sequences were passed via the germ line from G_1 sires to G_2 progeny. This result also confirms the mosaic nature of viral integration into the germ cells of G_1 males. Presumably, many different precursor germ cells are infected with the virus. In general, identical banding patterns have not been encountered for any G_2 progeny derived from any one G_1 male. Also, in rare instances, one precursor germ cell can be infected by two viruses.

If the G_2 exogenous virus-specific animals were produced by way of the germ cell of a sire which possessed an integrated copy of an exogenous retrovirus, all cells derived from a G_2 positive animal should possess the same viral integration pattern. This has been shown by Chen *et al.* (1990). Also, G_2 positive animals produced in this manner should pass the exogenous viral DNA to approximately 50% of progeny, and the Southern hybridization pattern in the positive offspring should be identical to that in the sire. This was confirmed by analyses of families where G_2 sires were mated with nonviremic females and G_3 individuals analyzed. Approximately 50% of offspring possessed exogenous viral DNA. Also, the Southern hybridization banding pattern was identical between the three G_2 sires and respective G_3 positive offspring (Chen *et al.*, 1990).

Thus, exogenous replication-competent subgroup A retrovirus can infect germ cells or precursor germ cells in a developing embryo. Apparently, the virus infects the precursor germ cells at a relatively late stage in development. The efficiency of producing G_1 males that pass genetic information to offspring via the germ line is relatively high. Therefore, the opportunity to manipulate the germ line of chickens with replication-competent avian retroviruses seems practical. Encouraging results also have been presented concerning disease resistance and growth parameters of transgenic chickens produced via this method.

D. *Replication-Defective Avian Retroviral Vectors: Vector/Helper Cell Systems*

Replication-defective or incomplete avian retroviral vector/helper cell systems have been developed (Savatier *et al.*, 1989; Stoker and Bissell, 1988; Watanabe and Temin, 1983) and have been used successfully in the production of transgenic chickens (Bosselman *et al.*, 1989a,b). In general, the viral envelope gene, *env*, and other viral genes (*gag–pol*) are expressed in a population of cells (helper cells). These gene products produce virus particles but have been engineered such that they will not package or transmit viral genetic information (viral RNA). Vector DNA or DNA that encodes RNA which can be packaged is subsequently introduced into the cells. The emerging virus will possess RNA from the vector that will ultimately infect fresh chicken cells. Using these systems, a complete viral genome is not generated; therefore, the virus is replication defective. A potential problem associated with these vector/helper cell systems is that of recombination, in which replication-competent virus is produced (Hu *et al.*, 1987).

A replication-defective reticuloendotheliosis virus (REV) vector has been developed (Watanabe and Temin, 1983; Emerman and Temin, 1984) and has been shown to infect chicken embryonic blastodermal cells (Bosselman *et al.*, 1989a). This vector system is important because it does not contain genetic information found in the germ line of chickens, whereas the other systems contain viral genes that are found in the chicken genome (endogenous viral DNA sequences). Approximately 8% of male birds generated by use of this viral vector contained the REV DNA in their semen and were able to pass vector sequences to offspring. Replication-competent helper viruses were detected in about 10% of the birds derived from embryos infected with the vectors (Bosselman *et al.*, 1989b). Virus negative birds were used for progeny testing.

Two other helper cell/vector systems have been developed and tested in cultured chicken embryonic cells. Savatier *et al.* (1989) used an avian leukosis virus (RAV-1) vector that contains the viral *gag–pol* and *env* genes but lacks the viral RNA packaging site. A quail cell line was generated that expresses the viral genes. DNA was inserted into these cells that encodes the bacterial neomycin resistance gene coupled to the avian retroviral LTRs and packaging signal. Use of this system resulted in production of replication-defective virus at high titers ($10^4 - 10^5$ virus/ml). Importantly, this system does not produce replication-competent virus particles.

In a similar system, Stoker and Bissell (1988) have developed an avian retroviral vector/helper cell packaging system. In this system, the viral DNA packaging signal has been removed and helper quail cell lines generated that express the viral *gag–pol* and *env* genes. These cells were analyzed for the ability to produce defective virus that carry the neomycin-resistance gene (as described above). High titers of replication-defective virus ($10^5 - 10^6$ virus/ml) were reported. Importantly, production of very low levels of replication-competent virus were found with the system.

In all three vector/helper cell systems described above, production of replication-competent helper virus is still a concern. Before use of the vector/helper cell systems in a commercial setting, more work must be performed to ensure the safety of the systems relative to the production of infectious virus. Encouragingly, one of the systems, namely, that of Bosselman *et al.* (1989a) has been shown to infect chicken embryonic cells at a relatively high efficiency. However, because REV infects human cells, it would not be suitable as a vector for producing transgenic chickens for human consumption. It may be impossible to produce high-titer vectors that have zero probability of producing replication-competent virus.

E. Sperm-Mediated Gene Transfer in Chickens

It has been reported that gene transfer in chickens has been achieved through the use of sperm cells as vectors (Grunebaum *et al.*, 1991). An ERS buffer was developed [which is a modified extender containing NaCl, KCl, $CaCl_2$, $NaHCO_3$, $MgSO_4$, bovine serum albumin (BSA), lecithin, egg yolk, and glycerol] and shown to maintain sperm motility much better, and it can be used for cryopreservation of sperm. Gruenbaum *et al.* reported success in producing transgenic chickens by insemination of hens with chicken sperm that was previously incubated with DNA constructs in ERS buffer for 1 hr at room temperature. The success rate was 30–60% but resulted in mostly mosaic animals. The transmission of the gene to the progenies was similar to that of mosaic animals produced by the retroviral method (i.e., 1–10%). It is theoretically possible to use sperm as a vector for gene transfer, and from other laboratories it has been found that chicken sperm binds DNA (H. Y. Chen, unpublished data, 1990). Owing to the fact that the use of retroviruses has its limitations, this method of sperm-mediated gene transfer is very exciting and may hold great promise in the future production of transgenic chickens.

F. Genetic Manipulation of Embryonic Cells and Embryos

1. Manipulation of Primordial Germ Cells

Infection of chicken primordial germ cells (PGC) with defective retrovirus and transfer of the cells to developing embryos have been reported by Simkiss *et al.* (1990). Similar procedures were used by Rosenblum and Chen to express the firefly luciferase gene in the gonads of day 17 embryos (C. Rosenblum and H. Y. Chen, unpublished results, 1990). Primordial germ cells can be obtained from either blood samples taken from stage 15 embryos (~55 hr of incubation) or the germinal crescent of stage 12 embryos (~48 hr of incubation). Blood samples can be removed from either heart or peripheral circulation using a glass needle. The pooled blood samples are infected with recombinant retrovirus before being transferred into the veins or the heart of a similarly aged recipient embryo. To obtain primordial germ cells from the germinal crescent, incubated eggs are removed from the eggshell and submerged in phosphate-buffered saline (PBS) in a small bowl. The embryonated region is cut out from the surface of the yolk with a pair of small scissors and transferred to a petri dish in order to trim excess tissues and yolk. The germinal crescents are incubated in Ca^{2+}- and Mg^{2+}-free saline with 0.02% (w/v) EDTA to break up the tissue. The cells are infected with a defective RSV vector containing a firefly luciferase reporter gene and then washed before transferring to a recipient embryo. Analysis of gonads of day 17 embryos showed detectable levels of luciferase activity, indicating successful transfer of infected primordial germ cells to the recipient embryos.

2. Embryo Manipulation

As stated above, the embryo found in the day 1 egg can be directly transfected with DNA or infected with replication-competent or replication-defective retroviruses. These manipulations (e.g., removal of the eggshell and injection of DNA or virus) do not dramatically affect the hatchability. Also, follicles within the hen can be exposed and manipulated without affecting subsequent fertilization and hatchability.

3. Artificial Hatching

An early chicken embryo has been successfully grown *in vitro* (Perry, 1987). This result is very exciting because it opens the possibility of infecting the fertilized eggs at a stage earlier than the day 1 blastoderm. Therefore, this should increase the efficiency of producing first-generation animals that have the transgene in germinal tissue. However, this approach is unlikely to be of importance commercially. Hatchability of these animals is at best 7%. To obtain one embryo, a hen must be killed. Because commercial broiler parental lines are expensive, this is currently not a viable alternative.

4. Embryonic Stem Cells

Currently there is no documentation of chicken embryonic stem cells similar to mouse lines which are used for production of chimeric mice. However, Petitte *et al.* (1990) have isolated day 1 embryonic cells from the chicken blastoderm (Barred

Plymouth Rock chickens) and introduced these cells into the blastoderm of Dwarf White Leghorn chickens. Approximately 11% of the resulting embryos were phenotypically chimeric with respect to feather color. One animal developed to hatching. The single surviving male was a germ line chimera, as it produced both Barred Plymouth Rock and Dwarf White Leghorn progeny. Therefore, with this protocol, one may be able to generate transgenic chickens in a manner similar to the embryonic stem cell approach used for the production of transgenic mice. However, the early embryonic cells from the blastoderm are not true embryonic stem cells but are a mixed population of cells. It is therefore necessary to develop a procedure to isolate chicken embryonic stem cells and to establish conditions to culture these cells *in vitro* genetic manipulations.

G. Identification of Economically Important Traits

Genetic traits that would influence the economics of poultry production are those associated with growth and health of the birds. Two examples of transgenic chickens possessing genes related to growth or disease resistance have been presented. In one case, bGH gene has been inserted and, when expressed in adult birds, appears to increase the mass of the animals (Chen *et al.*, 1990). In the other case, the *env* gene of an avian leukemia virus has been placed in the germ line of the chicken. Cells that express this gene are resistant to subsequent infection with the same strain of virus (Salter and Crittenden, 1989; Kopchick *et al.*, 1990). Disease resistance genes that should be used in the future include those which combat viral, bacterial, or parasitic infections. Also, immunomodulators such as interferons and interleukins may be important genetic targets.

H. Current State of the Art

The method of choice for the production of transgenic poultry appears to be utilization of replication-defective avian retroviral vectors with corresponding helper cells. However, more vigorous characterization of the various versions of these genetic delivery systems must be established before any one method is eventually used. Issues concerning safety, gene delivery, and recombination in founder animals and offspring must be addressed relative to use of these systems. Also, methods that enhance the efficiency of infecting precursor germ cells are needed. Toward that end, the exciting results concerning culturing and hatching of chickens using artificial systems, the ability to transfer blastodermal cells from one individual to another, and the possibility of sperm-mediated gene transfer are important steps toward the overall goal of transgenic poultry production.

III. TRANSGENIC FISH

A. Background

The genetic selection of fish that have improved growth and other characteristics is a slow process, and, therefore, many investigators are presently introducing genes into fish using molecular genetics and gene transfer techniques. The technology of manipulating genes to increase growth rate and temperature tolerance, improve feed efficiency, and impart the acquisition of disease resistance has great potential applications in the aquaculture industry as well as for the study of vertebrate embryology and development.

Methods for introducing foreign DNA into the germ line of fish are reviewed in this section. Genes currently being targeted to generate transgenic fish are discussed. Also, there is a need and responsibility to implement safeguards to protect the environment against the introduction of unwanted transgenes into the native fish population, and possible solutions to this problem are examined.

Improved growth characteristics in fish have been achieved using natural selection. Unfortunately, this approach is a slow process. Dramatic effects on growth have been produced by injection of growth hormone (GH) into fish. Initial studies involved injecting piscine growth hormone into hypophysectomized killifish (Pickford, 1954). In coho salmon, increased feed efficiency as well as an enhanced growth rate was found on injection of GH (Markert *et al.*, 1975). However, GH gene injection is a laborious procedure and one in which repeated injections are necessary. Recently, the use of a synthetic enkephalin analog has been found to significantly increase growth, food conversion efficiency, body weight, and specific growth rate in tilapia (Chang and Lin, 1991).

Several methods of gene transfer in fish have been successful. These include microinjection, electroporation, and sperm-mediated gene transfer. The use of microinjection to transfer genes in fish has produced significantly enhanced growth rates. The first report by Zhu *et al.* (1985) involved the microinjection of the human GH gene into goldfish. GH was found to be expressed and the gene transmitted to offspring in the goldfish. Also, a significant increase in growth rate was detected in the GH transgenic fish (Maclean *et al.*, 1987). Other reports on GH gene transfer into fish include studies involving rainbow trout (Chourrout *et al.*, 1986, 1988; Guise *et al.*, 1991), medaka (Ozata *et al.*, 1986; Inoue *et al.*, 1990), loach (Zhu *et al.*, 1986; Benyumov *et al.*, 1989; Xie *et al.*, 1992), salmon (McEvoy *et al.*, 1988; Rokkones *et al.*, 1989), tilapia (Brem *et al.*, 1988), catfish (Dunham and Eash, 1987; Powers *et al.*, 1991), pike (Guise *et al.*, 1991), goldfish (Zhu *et al.*, 1985; Yoon *et al.*, 1988), and carp (Zhang *et al.*, 1990; Powers *et al.*, 1991; Hernandez *et al.*, 1991; Xie *et al.*, 1992).

Genes that have been successfully introduced into fish can be found in Table 5. These genes include growth hormone genes from human, rat, rainbow trout, and

TABLE 5
Summary of Successful Gene Transfer in Fish

Species	Construct	Ref.
Goldfish	mMT/hGH	Zhu et al. (1985)
Loach	mMT/hGH	Zhu et al. (1986)
Medaka	SV40/cCr	Ozata et al. (1986)
Trout	SV40/hGH	Chourrout et al. (1986)
Catfish	mMT/hGH	Dunham and Eash (1987)
Trout	mMT/rGH	Maclean et al. (1987)
Salmon	fAFP/fAFP	Fletcher et al. (1988)
Zebrafish	SV40/ Hygro	Stuart et al. (1988)
Salmon	mMT/rGH	McEvoy et al. (1988)
Salmon	mMT/β-Gal	McEvoy et al. (1988)
Tilapia	mMT/hGH	Brem et al. (1988)
Zebrafish	SV40/CAT	Stuart et al. (1988)
Goldfish	SV40/Neo	Yoon et al. (1988)
Salmon	mMT/hGH	Rokkones et al. (1989)
Loach	mMT/hGH	Benyumov et al. (1989)
Medaka	mMT/rtGH	Inoue et al. (1990)
Medaka	fLus/fLuc	Tamiya et al. (1990)
Trout	cG/cG	Oshiro et al. (1989)
Goldfish	RSV/Neo	Yoon et al. (1990)
Goldfish	RSV/CAT	Hallerman et al. (1990)
Common carp	RSV/rtGH	Zhang et al. (1990)
Pike	RSV/bGH	Guise et al. (1991)
Chinese carp	mMT/hGH	Powers et al. (1991)
Catfish	RSV/rtGH	Powers et al. (1991)
Common carp	mMT/hGH	Hernandez et al. (1991)
Loach	mMT/hGH	Xie et al. (1989)
Common carp	mMT/hGH	Xie et al. (1993)

cattle (HGH, rGH, rtGH, and bGH, respectively); the chicken crystallin gene (cCR); the flounder antifreeze gene (fAFP); the hygromycin gene (Hygro); the β-galactosidase gene (β-Gal); the chloramphenicol acetyltransferase gene (CAT); the neomycin-resistance gene (neo); and the luciferase gene (Luc).

The method of microinjection has been proven to be successful and effective in producing transgenic fish. However, there is a need for the development of efficient mass transfer methods in the commercial aquaculture industry (Guise et al., 1991). Currently, three main systems for mass transfer are being investigated. These techniques include electroporation, sperm binding, and lipofection. Other methods of gene transfer presently in the preliminary stages include the use of embryonic stem cells, retroviral vectors, high-speed particle gun penetration, and muscle precursor cells. These strategies are discussed below.

B. Materials and Methods for Gene Insertion

The current methods for the production of transgenic fish involve the following strategies: (1) microinjection of fish embryos, (2) mass transfer systems for the production of transgenic fish, and (3) other technologies such as embryonic stem cells, retroviral vectors, high-speed particle gun gene transfer, and muscle precursor cells.

1. Microinjection

Microinjection has been a very successful technique used in the production of transgenic fish in a variety of species (Table 5). The most effective method of introducing foreign genes into the germ line of transgenic mice, rabbits, pigs, cows, and sheep has been the direct microinjection of DNA into the male pronuclei of fertilized eggs. However, the pronuclei of fertilized fish eggs cannot be visualized. Earlier studies using the toad *Xenopus laevis* showed that DNA injected directly into the cytoplasm of the *X. laevis* eggs was successfully integrated and expressed (Etkin *et al.*, 1984). Initial studies in the production of transgenic fish were conducted primarily by cytoplasmic injection of foreign DNA sequences. The following equipment and protocols are those that are used at the Edison Animal Biotechnology Center, Ohio University (Athens, OH), for the production of transgenic fish (see Fig. 2).

a. Essential Equipment   The essential equipment needed for the production of transgenic fish include a fish tank, inverted microscope, micromanipulator, glass microneedles, and a needle drawing apparatus.

b. Solutions   The solutions needed for microinjection of fish eggs include 0.25% (w/v) trypsin (for dechorionation of the fish eggs), Hanks' saline (137 mM NaCl, 5.4 mM KCl, 1.3 mM MgSO$_4$, 0.44 mM KH$_2$PO$_4$, 0.25 mM Na$_2$HPO$_4$, 4.2 mM NaHCO$_3$) and ST DNA injection solution (88 mM NaCl, 10 mM Tris-HCl, pH 8.0).

c. Routine Fish Care and Production of Fertilized Eggs   At the Edison Animal Biotechnology Center at Ohio University, we have initially carried out studies using the zebrafish as our model system (Bremiller *et al.*, 1989). Physical parameters that are optimized for the production of fertilized eggs include temperature, nutritional state, photoperiod, physiological condition of the fish, and age. The zebrafish are fed a variety of foods up until breeding. Ground dry or moist trout pellets and dry flake food can be found in most pet stores. Adults are fed at least

Breeding fish (7-18 months)

↓

Males transferred to female tank
1-2 hour before dark cycle

↓

Siphon bottom of tank and
Collect eggs 40 minutes after light cycle begins

· ↓

Filter through nylon mesh and rinse

↓

Remove chorion with 0.25 % trypsin

↓

Microinjection of embryos with 2-3 μl DNA

↓

Transfer embryos to 10 % Hanks' saline solution

↓

Embryos will hatch 3 days after microinjection

↓

Feed baby fish after 4 days

Figure 2. Schematic of microinjection procedure in zebrafish.

twice daily. When optimizing breeding of adult fish, they are fed adult live brine shrimp. A large supply of zebrafish embryos (1000 embryos per tank) can be produced once or twice a week if the zebrafish are kept under optimum conditions. The breeding fish should be between 7 and 18 months of age, and the females and males are kept in separate tanks. The tanks should be kept clean and the fish should be fed twice daily.

The fish are kept on a 14-hr light/10-hr dark cycle. On the day before embryos are desired, 1–2 hr before the dark cycle, the males are transferred into the female tank at a ratio of approximately 2 females to 1 male. A single layer of marbles is added to the bottom of the fish tank to prevent the fish from eating the eggs. The following light cycle stimulates females to release eggs and males to excrete milt. After 40 min into the light cycle, the eggs found on the bottom of the fish tank are collected by siphoning the bottom of the tank.

Another method by which fertilized eggs from some fish are produced is by hormonal induction (Zhu *et al.*, 1986). Overnight hormonal induction includes the use of 800–1000 units of chorionic gonadotrophin or 5 mg of common carp pituitary homogenate per kilogram body weight. The mature eggs are collected and placed in a petri dish. A drop of milt is immediately mixed in, and the eggs are allowed to stand for 5 min, during which time fertilization occurs.

d. Collection of Fertilized Fish Eggs and Dechorionation Embryos are collected into a beaker and filtered through a medium mesh nylon net. The embryos are then rinsed from the net into a petri dish with 10% Hanks' saline. The fertilized zebrafish eggs are placed in a 0.25% trypsin solution for 5 min to remove the chorion. The embryos are then transferred to a petri dish containing Hanks' saline solution and rinsed twice. The most difficult part of injecting fish eggs is the ability to penetrate the outer surface of the fertilized fish egg. This membrane is known as the chorion. Most fish eggs have a chorion that hardens on fertilization and exposure to water. Also, it has been found that some fish which spawn in turbulent water are more likely to have tougher chorions than fish which spawn in still water (Hallerman *et al.*, 1988). Therefore, before microinjection of DNA into fish eggs can be achieved, the chorion must be removed or softened. There are currently five different approaches to overcome this barrier.

1. Enzymatic dechorionation: Removal of the chorion by enzymatic digestion is the most common approach and has been successful in most fish including goldfish, carp, zebrafish, loach, and northern pike (Zhu *et al.*, 1985; Hallerman *et al.*, 1988). Enzymes used in the dechorionation are trypsin and protease type XXV (Sigma, St. Louis, MD).

2. Manual dissection of the chorion: In zebrafish, the chorion is manually dissected by use of forceps. The chorion is thin and the perivitelline space is large in the zebrafish zygote. On removal of the chorion, the zebrafish zygote has sufficient structural stability that it can survive without the mechanical support of the chorion (Stuart *et al.*, 1988).

3. Injection through the micropyle: In Atlantic salmon and tilapia, foreign sequences of DNA are injected through the micropyle, the entry tunnel used by the sperm during fertilization (Fletcher *et al.*, 1988; Brem *et al.*, 1988). This method requires clear visualization of the micropyle and correct positioning of the zygote before microinjection.

4. Piercing of the chorion: The chorion of rainbow trout and salmon are resistant to all tested enzymes. Therefore, to penetrate the chorion, a hole is cut through the chorion by the use of a short, strong needle. A broken end of a Pasteur pipette is inserted through the hole of the chorion. A second microneedle is then inserted in through the glass pipette and is used to introduce the foreign DNA.

5. Glutathione treatment: Rainbow trout eggs have been found to be resistant to hardening of the chorion by immersing them during fertilization in a solution containing 1 mM glutathione, pH 8.0 (Oshiro *et al.*, 1989).

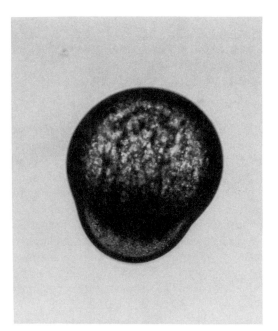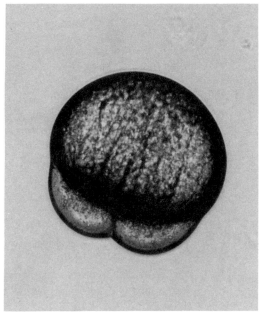

Figure 3. One- and two-cell fertilized zebrafish eggs after microinjection.

e. Microinjection of Fertilized Fish Eggs To prepare DNA to be used for microinjection, recombinant plasmid DNA is linearized and precipitated with ethanol. The DNA is then dissolved in ST buffer (88 mM NaCl and 10 mM Tris, pH 7.5) at a concentration of 100 μg/ml. This solution (2 nl) is microinjected into each fertilized egg.

Microinjection is carried out on fertilized eggs until the first cleavage occurs (Fig. 3). Under optimum conditions, up to 200 eggs can be injected per hour. Glass microneedles are pulled with a needle drawing apparatus to a 3–4 μm inner tip diameter. The needle is loaded with 2–3 μl of the DNA solution to be injected. Dechorionated eggs are lined up and positioned so that the germinal disk is facing upward. The needle is slowly lowered and the tip dipped down to one-third of the thickness of the germinal disk. This allows the DNA solution to be released around the area of the pronuclei. Routinely, 105–106 copies of DNA are injected per egg.

The concentration of DNA used to microinject fish embryos has been found to be of considerable importance if the concentration is greater than 100 μg/ml. In zebrafish, it was determined that by increasing the amount DNA injected into eggs at fertilization, the percentage of embryos surviving to 10 days of age decreased (Stuart *et al.*, 1988).

These initial studies on microinjection of exogenous DNA into fertilized eggs have shown that the fish embryo tolerates the injection of foreign DNA into the cytoplasm. The insertion of needles into the blastodisk does not appear to be significantly detrimental in the fish species studied. Also, the amount of DNA injected into the fish embryo is very important and needs to be optimized to ensure embryonic survival and adequate integration of the foreign DNA.

f. Embryo Recovery and Culturing   Fish embryos do not require complex manipulations of *in vitro* culturing nor subsequent transfer of embryos into foster mothers as is required for mammalian gene transfer experiments. The survival rate of injected fish embryos is also much higher than that of mammalian embryos, ranging from 35 to 80% (Chen and Powers, 1990). After microinjection, the fertilized eggs are carefully transferred into 10% Hanks' saline solution. There should be 25–50 embryos per 100 ml solution in a 250-ml beaker. The optimum temperature is 28.5°C. The embryos will usually hatch after 3 days of development and should be fed 4 days after fertilization. The baby fish can be fed live paramecia or other infusion materials which can be found in a pet shop. After a few weeks the baby fish can be fed baby brine shrimp and finally switched to adult type food.

2. Mass Transfer Systems for Production of Transgenic Fish

Although microinjection has been successful in the production of transgenic fish, relatively few eggs can conveniently be injected per day. Therefore, there is a need for mass transfer technology to be developed. Three main strategies for mass transfer currently being investigated are electroporation, sperm-mediated gene transfer, and lipofection.

a. Electroporation   Electroporation is a successful method of transferring genes into yeast, plant, and animal cells (Hashimoto *et al.*, 1985; Callis *et al.*, 1987; Chu *et al.*, 1987; Knutson and Yee, 1987). Brief electrical charges are applied to the cell membrane during the electroporation process, which allows the cellular membrane to become permeable and DNA ultimately to enter the cell. The use of electroporation to introduce foreign genes into fish eggs holds great potential. Recently, a fusion metallothionein–human growth hormone gene was successfully introduced into the fertilized eggs of loach and red crucian carp using electroporation (Xie *et al.*, 1992). The transfer efficiency was as high as 62.5%, and some individuals integrated over 100 copies of the foreign gene into the genome with electroporation parameters of 250 V and 22 μF. This methodology holds great potential for introduction of foreign genes into large batches of fertilized fish eggs.

b. Sperm-Mediated Gene Transfer   Fish provide an excellent system in which to study gene transfer. The fertilization process is unlike that of mammals in that the spermatozoa are immobile in seminal plasma, and even after activation they are only mobile for 30–60 sec. Fish eggs have similar properties to those of the spermatozoa. They remain inactive in ovarian fluid and can even be stored for several days with little loss of fertility. On combination of the egg and sperm, the addition of water activates the fertilization process, and the sperm is incorporated into the egg.

It has been reported that the spermatozoa may be capable of binding DNA and carrying it into a mouse egg (Lavitrano et al., 1989), although this procedure in mammalian systems is quite controversial and has yet to be reproduced. The fish fertilization process is relatively simplistic, and exogenous DNA is encapsulated in immunoliposomes and allowed to bind spermatozoa. In this way, the genes are transferred into the fish eggs (Cloud, 1990). Sperm-mediated gene transfer has been accomplished in loach and common carp using the metallothionein (MT)–hGH gene with as high as 50% efficiency (Y. Xie, personal communication, 1991).

c. Lipofection   The lipofection methodology requires the encapsulation of DNA within a phospholipid bilayer and subsequent delivery into a cell. Liposomes have been used successfully as gene delivery systems in mammalian cell culture (Felgner et al., 1987). By specifically targeting liposomes to egg or sperm cells, transgenic germ cells may be produced. To date, no successful reports of gene transfer into fish using lipofection have been documented.

Although all of the methods (electroporation, sperm-mediated gene transfer, and lipofection) show great promise, further studies are needed to develop these alternative methods for the introduction of genes into large numbers of eggs.

3. Other Technologies for Gene Transfer in Fish

a. Embryonic Stem Cells   Establishment of pluripotent cells from mouse and hamster embryos has been reported by a variety of investigators (Evans and Kaufman, 1981; Doetschman et al., 1988). These pluripotent cells can be cultured, manipulated, and transferred into blastocysts to produce transgenic animals. One of the major advantages to using embryonic stem cells is that the cells can be screened for exogenous DNA integration, location, and expression before insertion into the germ line. It has been shown that, by using homologous recombination, genes can be targeted to a specific locus in a chromosome (Smithies et al., 1985; Doetschman et al., 1987). There have been no reports to date which utilize this methodology in the production of transgenic fish. Initial studies have been conducted in which chimeric rainbow trout have been produced by injecting isolated blastomeres into recipient embryos (Nillson and Cloud, 1989). It has been suggested that the integrated blastomeres are pluripotent, but this assumption has yet to be substantiated.

b. Retroviral Vectors   Retroviruses are ideal gene delivery vehicles. This is due to their efficient integration and single copy insertion of genetic material into the host genome. In fish, there has been little research conducted on retroviruses.

Therefore, much research must be done to identify and characterize species-specific fish retroviruses before this technology can be used to deliver genes into fish.

c. Muscle Precursor Cells There is a current hypothesis that isolated fish myogenic stem cells can be used for gene transfer and subsequent transplantation into skeletal muscle of recipient fish (Cloud, 1991). Mononuclear precursor cells (satellite cells) have been used to insert allogeneic genes into mouse muscle (Watt *et al.*, 1984). Identification of satellite cells from a variety of fish species has been described, including rainbow trout (Powell *et al.*, 1989), shark (Kryvi, 1975), and hagfish (Sandset and Kornelliussen, 1978).

In the rainbow trout, skeletal muscle was enzymatically dispersed, and mononucleated cells were grown and maintained in culture. The satellite cells increased in number during culture and fused together in what appeared to be myotubes. The myotubes then went on to develop elongated, striated structures. These observations support the conclusion that the cells are satellite cells and can be maintained in culture. To use this methodology for gene insertion into fish, satellite cells from fish muscle will need to be isolated, transfected, and transplanted into the recipient fish skeletal muscle.

d. High-Speed Particle Gun Gene Insertion A relatively new methodology for introducing genes into cells involves the use of a high-speed particle gun. A particle gun is used to transfer foreign DNA, which has been adsorbed to 4-μm spherical tungsten particles. Successful gene transfer has been reported in plants by the use of this technique (Klein *et al.*, 1987). The plant cell wall and the fish chorion are similar in that they are tough barriers to overcome for gene insertion. The high-velocity microprojectiles have been able to penetrate the plant cell wall, and, therefore, this technique may prove to be very useful in the penetration of the fish chorion for gene insertion into the fish cell.

C. Economically Important Traits

Aquaculture and fishery-stocking programs are of increasing importance to the world economy owing to the increasing use and consumption of fish products. The successful production of faster growing, feed-efficient strains of fish would have a major economic impact as an important source of protein. Most groups investigating gene transfer in fish are utilizing the growth hormone gene controlled by mammalian or viral promoter/enhancer regions to achieve enhancement of growth in fish. It has been shown that the GH gene is successfully transferred into fish, integrated into the fish genome, genetically transferred to the F1 generation, and that in both the P1 and F1 generations the growth rate is increased in transgenic fish owing to expression of the foreign GH sequences (Chen *et al.*, 1989; Zhang *et al.*, 1990; Powers *et al.*, 1991).

Reporter genes such as the CAT gene, luciferase gene, and the *lacZ* gene have been used to study gene expression in specific tissues and during development. At the Edison Animal Biotechnology Center, the CAT gene under the control of the SV40 promoter and the MT promoter is being investigated. High levels of CAT expression were measured in the zebrafish embryo, with the SV–CAT fusion gene showing the highest levels of activity (Y. Xie and C. Wu, personal communication, 1991). Also, the luciferase gene under control of the RSV promoter was microinjected. High levels of luciferase activity were detected in the zebrafish 24 hr after microinjection. Further studies are underway to determine tissue-specific expression levels for these constructs.

Dominant selectable marker genes such as those encoding neomycin or hygromycin resistance have been used to produce transgenic fish. By the addition of selection marker genes into the fish genome, fish may be able to be put in a selection medium in which only the transgenic fish containing the selectable marker gene would survive.

The use of antifreeze genes to protect fish from freezing in cold waters is an important trait. The antifreeze proteins are found in some species that live in cold waters. These proteins prevent the blood from freezing and allow the fish to live in water which is $-2°C$. By binding to micro ice crystals, the antifreeze proteins have been found to lower the freezing point of fish blood (Lin and Gross, 1981; Davies *et al.*, 1982, 1984; Gourlie *et al.*, 1984; Scott *et al.*, 1986, 1988). The cloned winter flounder antifreeze protein (AFP) genes have been transferred and expressed in rainbow trout, bluegill, and salmon cell lines (Huang *et al.*, 1990). Current research is being performed to produce transgenic fish that are cold resistant. The winter flounder AFP genes have been transferred and integrated into the Atlantic salmon (Fletcher *et al.*, 1988); however, there was found to be no expression of the AFP genes. Continued research by Fletcher *et al.* is being done to achieve expression and protection of the transgenic fish against freezing.

To date, there are no commercially available vaccines or antiviral drugs to combat fish viral infections. In the aquaculture industry, the only solution to a virally infected fish population is destruction of the fish. This accounts for a major loss of revenue in the fisheries business. Two strategies involving the generation of transgenic immune fish have been suggested. One strategy uses antisense technology to block replication of virus. The other involves the transfer of the viral envelope gene into the fish to confer resistance against the virus. Both methods have been found to be successful in the avian system using retroviruses (Neiman *et al.*, 1988). Research is ongoing in this area of disease resistance.

D. Environmental Impact of Transgenic Fish

One of the crucial areas in the production of transgenic fish is the responsibility of the scientific community to prevent the release of transgenic fish into the native environment. Therefore, much time and effort has been spent to develop method-ologies in which sterile transgenic fish can be produced. There are several protocols that have been used to produce sterile transgenic fish. These techniques include polyploidization, interspecies hybridization, hormonal sterilization, and production of nonfunctional gonadal ducts.

1. Aneu-Polyploidization

In the rainbow trout it has been shown that by the induction of triploidy the triploid fish are sterile (Thorgaard and Gall, 1979; Lincoln and Scott, 1984). A $3N$ genome in fish is produced by temperature shock or pressure treatment of newly fertilized eggs (Thorgaard, 1983; Purdom, 1984). An alternative strategy to produce triploid fish is by crossing diploid fish stocks with tetraploid fish, and this procedure is currently being investigated (Chourrout *et al.*, 1986). Also, creating triploid hy-brids in which the diploid is lethal is another way in which to produce sterile fish (Chevassus *et al.*, 1983).

2. Interspecies Hybridization

Many fish species, when crossed to make interspecific hybrids, are found to be sterile. In this way, the specific brood stock can be protected by production of sterile fish hybrids.

3. Hormonal Sterilization

Sterility has been attained in adult fish that have been treated at early fry stages with androgens. It is an easy and effective method by which to sterilize fish and has been found to be successful in salmon (Donaldson and Hunter, 1983).

4. Nonfunctional Gonadal Ducts

The production of fish that have incomplete or nonfunction gonadal ducts would create a fish population which is infertile in the natural environment but could be made fertile in the hatchery. Two methods are currently being investigated. One study concerns the production of transgenic fish in which the gonadal duct would be completely nonfunctional, but the sperm would still have the ability to be used in *in vitro* fertilization. Another possible method is the development of a genetic strain of sterile fish which could, on exposure to a hormone or nutrient, be induced to reverse the infertile condition (Bye and Lincoln, 1986).

E. Future of Transgenic Fish

Gene transfer in fish has been accomplished. Methodology must now be developed for the improvement of mass gene transfer techniques, and research involving the targeting of foreign genes through homologous recombination as well as studies on the regulation and expression of genes at different stages of development need to be pursued. There is a need to search for genes that are important in aquaculture. Without such genes, the application of gene transfer technology in fish will remain a dream.

ACKNOWLEDGMENTS

We greatly appreciate the suggestions and assistance in the writing of this chapter provided by Bill Hayes, Xie Yuefang, Dr. Wu Chingjiang, and Dr. Tom Chen.

REFERENCES

Benyumov, A. O., Enikolopov, G. N., Barmintsev, V. A., Zelenia, I. A., Sleptsova, L. A., Doronin, Yu. K., Golichenkov, V. A., Grashchuk, M. A., and Georgiev, G. P. (1989). Integration and expression of human growth hormone gene in teleostei. *Genetika* **25**, 24–35.

Bosselman, R. A., Hsu, R. Y., Boggs, T., Hu, S., Bruszewski, J., Ou, S., Kosar, L., Martin, F., Green, C., Jacobsen, F., Nicolson, M., Schultz, J. A., Semon, K. M., Rishell, W., and Stewart, R. G. (1989a). Germline transmission of exogenous genes in the chicken. *Science* **243**, 533–535.

Bosselman, R. A., Hsu, R. Y., Boggs, T., Hu, S., Bruszewski, J., Ou, S., Kosar, L., Martin, F., Nicolson, M., Rishell, W., Schultz, J. A., Semon, K. M., and Stewart, R. G. (1989b). Replication-defective vectors of reticuloendotheliosis virus transduce exogenous genes into somatic stem cells of the unincubated chicken embryo. *J. Virol.* **63**, 2680–2689.

Brem, G., Brenig, B., Horstgen-Schwark, G., and Winnacker, E.-L. (1988). Gene transfer in tilapia (*Oreochromis niloticus*). *Aquaculture* **68**, 209–219.

Bremiller, R., Frost, D., Kimmel, C., Metcalfe, W., Myers, P., Nawrocki, L., Stuart, G., Sullivan, E., Walker, C., Warga, R., and Molven, A. (1989). *In* "The Zebrafish Book" (M. Westerfield, ed.), Univ. of Oregon Press, Eugene.

Bye, V. J., and Lincoln, R. F. (1986). Commercial methods for the control of sexual maturation in rainbow trout (*Salmo gairdneri* R.). *Aquaculture* **57**, 299–309.

Callis, J., Fromm, M., and Walbot, V. (1987). Expression of mRNA electroporated into plant and animal cells. *Nucleic Acids Res.* **15**, 5823.

Chang, C., and Lin, S.-J. (1991). An enkephalin analog stimulates growth of *Tilapia*. *In* "Fish, Physiology and Biochemistry," Vol. 9, No. 2, pp. 101–106. Kugler Publ., Amsterdam/Berkeley.

Chen, T. T., and Powers, D. A. (1990). Transgenic fish. *TibTech* **8**, 209–215.

Chen, T. T., Lin, C. M., Zhu, Z., Gonzalez-Villasenor, L. I., Dunham, R. A., and Powers, D. A.

(1989). Gene transfer, expression and inheritance of rainbow trout and growth hormone genes in carp and loach. *In* "UCLA Symposium on Transgenic Models in Medicine and Agriculture" (R. B. Church, ed.), Vol. 116, pp. 127–139. Wiley–Liss, New York.

Chen, H., Garber, E., Mills, E., Smith, J., Kopchick, J. J., DiLella, A. G., and Smith, R. G. (1990). Vectors, promoters, and expression of genes in chicken embryos. *J. Reprod. Fertil. Suppl.* **41,** 173–182.

Chevassus, B., Guyomard, R., Chourrout, D., and Quillet, E. (1983). Production of viable hybrids in salmonids by triploidization. *Genet. Sel. Evol.* **15,** 519–532.

Chourrout, D., Guyomard, R., and Houdebine, L.-M. (1986). High efficiency gene transfer in rainbow trout (*Salmo gairdneri* Rich.) by microinjection into egg cytoplasm. *Aquaculture* **51,** 143–150.

Chourrout, D., Guyomard, R., Leroux, C., Pourrain, F., and Houdebine, L. M. (1988). Integration and germ-line transmission of foreign genes in trout after injection into the egg cytoplasm. *J. Cell. Biochem.* **12B** (Suppl.), 188 (abstract).

Chu, G., Hayakawa, H., and Berg, P. (1987). Electroporation for the efficient transfection of mammalian cells with DNA. *Nucleic Acids Res.* **15,** 1311.

Cloud, J. G. (1990). Strategies for introducing foreign DNA into the germ line of fish. *J. Reprod. Fertil. Suppl.,* **41,** 107–116.

Crittenden, L. B., Salter, D. W., and Federspiel, M. J. (1989). Segregation of viral phenotype and proviral structure of 23 avain leukosis virus inserts in the germ line of chickens. *Theor. Appl. Genet.* **77,** 505–515.

Davies, P. L., Roach, A. H., and Hew, C. L. (1982). DNA sequence coding for an antifreeze protein precursor from winter flounder *Pseudopleuronectes americanus. Proc. Natl. Acad. Sci. U.S.A.* **79,** 335–339.

Davies, P. L., Hough, C., Scott, G. K., White, B. N., and Hew, C. L. (1984). Antifreeze protein genes of the winter flounder *Pseudopleuronectus americanus. J. Biol. Chem.* **259,** 9241–9247.

Doetschman, T., Gregg, R. G., Maeda, N., Hooper, M. L., Melton, D. W., Thompson, S., and Smithies, O. (1987). Targeted correction of a mutant HPRT gene in mouse embryonic stem cells. *Nature (London)* **330,** 576–578.

Doetschman, T., Williams, P., and Maeda, N. (1988). Establishment of hamster blastocyst-derived embryonic (ES) cells. *Dev. Biol.* **127,** 224–227.

Donaldson, E. M., and Hunter, G. A. (1983). Induced final maturation, ovulation, and spermiation in cultured fish. *In* "Fish Physiology" (W. S. Hoar, D. J. Randall, and E. M. Donaldson, eds.), Vol. 9B, pp. 405–435. Academic Press, New York.

Dunham, R. A., and Eash, J. (1987). Transfer of the metallothionein–human growth hormone fusion gene into channel catfish. *Trans. Am. Fish. Soc.* **116,** 87–91.

Emerman, M., and Temin, H. M. (1984). Genes with promoters in retrovirus vectors can be independently suppressed by an epigenic mechanism. *Cell (Cambridge, Mass.)* **39,** 459–467.

Etkin, L. D., Pearman, B., Roberts, M., and Bektesh, S. L. (1984). Replication, integration and expression of exogenous DNA injected into fertilized eggs of *Xenopus laevis. Differentiation (Berlin)* **26,** 194–202.

Evans, M. J., and Kaufman, M. (1981). Establishment in culture of pluripotent cells from mouse embryos. *Nature (London)* **292,** 154–156.

Eyal-Giladi, H., and Kochav, S. (1976). From cleavage to primitive streak formation: A complementary normal table and a new look at the first stages of the development of the chick. *Dev. Biol.* **49,** 321–337.

Felgner, P. L., Gadek, T. R., Holm, R., Chan, H. W., Wenz, M., Northrup, J. P., Ringold, G. M., and Danielson, M. (1987). Lipofection: A highly efficient, lipid-mediated DNA transfection procedure. *Proc. Natl. Acad. Sci. U.S.A.* **84,** 7413–7417.

Fletcher, G. L., Shears, M. A., King, M. J., Davies, P. L., and Hew, C. L. (1988). Evidence for antifreeze protein gene transfer in Atlantic salmon (*Salmo salar*). *Can. J. Aquat. Sci.* **45,** 352–357.

Ginsburg, M., and Eyal-Giladi, H. (1987). Primordial germ cells of the young chick blastoderm originate from the central zone of the area pellucida irrespective of the embryo-forming process. *Development (Cambridge, UK)* **101**, 209–219.

Ginsburg, M., and Eyal-Giladi, H. (1986). Temporal and spatial aspects of the gradual migration of primordial germ cells from the epiblast into the germinal crescent in the avian embryo. *J. Embryol. Exp. Morphol.* **95**, 53–71.

Gourlie, B., Lin, Y., Price, J., DeVries, A. L., Powers, D., and Hung, R. C. C. (1984). Winter flounder antifreeze proteins: A multigene family. *J. Biol. Chem.* **259**, 14960–14965.

Gruenbaum, Y., Revel, E., Yarus, S., and Fainsod, A. (1991). Sperm cells as vectors for the generation of transgenic chickens. *J. Cell. Biochem.* **15E** (Suppl.), 194 (abstract).

Guise, K. S., Kapuscinski, A., Hackett, P. B., Jr., and Faras, A. J. (1991). Gene transfer in fish. *In* "Transgenic Animals" (N. First and F. P. Haseltine, eds.), pp. 295–306. Butterworth–Heineman, Stoneham, Massachusetts.

Hallerman, E. M., Schneider, J. F., Gross, M. L., Faras, A. J., Hackett, P. B., Guise, K. S., and Kapuscinski, A. R. (1988). Enzymatic dechorionation of goldfish and northern pike eggs. *Trans. Am. Fish. Soc.* **117**, 456–460.

Hallerman, E. M., Schneider, J. F., Gross, M., Lui, Z., Yoon, S. J., He, L., Hackett, P. B., Faras, A. J., Kapuscinski, A. R., and Guise, K. S. (1990). Gene expression promoted by the RSV long terminal repeat elements in transgenic goldfish. *Anim. Biotechnol.* **1**, 79–93.

Hammer, R. E., Pursel, V. G., Rexroad, C. F., Wall, R. J., Bolt, D. J., Ebert, K. M., Palmiter, R. D., and Brinster, R. L. (1985). Production of transgenic rabbits, sheep, and pigs by microinjection. *Nature (London)* **315**, 680–683.

Hashimoto, H., Morikawa, H., Yamada, Y., and Kimura, A. (1985). A novel method for transformation of intact yeast cells by electroporation of plasmid DNA. *Appl. Microbiol. Biotechnol.* **21**, 336.

Hernandez, O., Castro, F. O., Aguilar, A., Uliyer, C., Perez, A., Herrera, L., and De La Fuente, J. (1991). Gene transfer in common carp cyprinus-carpio l. by microinjection. *Theriogenology* **35**, 625–632.

Hippenmeyer, P. J., Given, G. K., and Highkin, M. K. (1988). Transfer and expression of the bacterial NPT-11 gene in chick embryos using a Schmidt-Ruppin retrovirus vector. *Nucleic Acids Res.* **16**, 7619–7632.

Hu, S., Bruszewski, J., Nicolson, M., Tseng, J., Hsu, R. Y., and Bosselman, R. (1987). Generation of competent virus in the REV helper cell line C3. *Virology* **159**, 446–449.

Huang, R. C. C., Price, J. L., and Gourlie, B. (1990). Regulation of antifreeze gene expression in winter flounder and in transgenic fish cells. *In* "Transgenic Models in Medicine and Agriculture: UCLA Symposium on Molecular and Cellular Biology" (R. B. Church, ed.), Vol. 116, pp. 109–126. Wiley–Liss, New York.

Inoue, K., Yamashita, S., Hata, J., Kabeno, S., Asada, S., Nagahisa, E., and Fujita, T. (1990). Electroporation as a new technique for producing transgenic fish. *Cell Differ. Dev.* **29**, 123–128.

Klein, T. M., Wolf, E. D., Wu, R., and Sanford, J. C. (1987). High-velocity microprojectiles for delivering nucleic acids into living cells. *Nature (London)* **327**, 70–73.

Knutson, J. C., and Yee, D. (1987). Electroporation: Parameters affecting transfer of DNA into mammalian cells. *Anal. Biochem.* **164**, 44.

Kochav, S., Ginsburg, M., and Eyal-Giladi, H. (1980). From cleavage to primitive streak formation: A complementary normal table and a new look at the first stages of the development of the chick. *Dev. Biol.* **79**, 296–308.

Kopchick, J. J., Mills, E., Rosenblum, C., Taylor, J., Macken, F., Leung, F., Smith, J., and Chen, H. (1990). Methods for the introduction of recombinant DNA into chicken embryos. *In* "Transgenic Animals" (N. First and F. P. Haseltine, eds.), pp. 275–293. Butterworth-Heineman, Stoneham, Massachusetts.

Kopchick, J. J., Molavarca, R. H., Livelli, T. J., and Leung, F. C. (1985). Use of avian retroviral bovine growth hormone DNA recombinants to direct expression of biologically active growth hormone by cultured fibroblasts. *DNA* **4**, 23–31.

Kryvi, H. (1975). The structure of myosatellite cells in axial muscles of the shark *Galeus melastonus*. *Anat. Embryol.* **147**, 35–44.

Lannett, E. H., and Schmidt, N. J. (1980). "Diagnostic Procedures for Viral and Rickettsial Infections," 4th Ed., pp. 2–65. Aneum Public Health Association, New York.

Lavitrano, M., Camaioni, A., Fazio, V. M., Polci, S., Farace, M. G., and Spadafora, C. (1989). Sperm cells as vectors for introducing foreign DNA into eggs: Genetic transformation of mice. *Cell (Cambridge, Mass.)* **57**, 717–723.

Lin, Y., and Gross, J. K. (1981). Molecular cloning and characterization of winter flounder *Pseudopleuronectes* antifreeze complementary DNA. *Proc. Natl. Acad. Sci. U.S.A.* **78**, 2825–2829.

Lincoln, R. F., and Scott, A. P. (1984). Sexual maturation in triploid rainbow trout salmogaidneri. *J. Fish. Biol.* **25**, 385–392.

McEvoy, T., Stack, M., Keane, B., Barry, T., Sreenan, J. M., and Gannon, F. (1988). The expression of a foreign gene in salmon embryos. *Aquaculture* **68**, 27–38.

Maclean, N., Penman, D., and Zhu, Z. (1987). Introduction of novel genes into fish. *Bio/Technology* **5**, 257–261.

Markert, J. R., Higgs, D. A., Dye, H. M., and MacQuarrie, D. W. (1975). Influence of bovine growth hormone on growth rate, appetite and food conversion of yearling coho salmon (*Onchorhynchus kisutch*). *Gen. Comp. Endocrinol.* **27**, 240–253.

Mindur, C., Krawczyk, E., and Wezyk, S. (1985). Development of the uterine chick embryos after storage at 5°C for 15 hours. *Br. Poult. Sci.* **26**, 527–529.

Neiman, P. E., Booth, S. C., and To, R. Y. (1988). *In* "Antisense RNA and DNA" (D. A. Melton, ed.), pp. 153–158. Cold Spring Harbor Laboratory, Cold Spring Harbor, New York.

Nilsson, E., and Cloud, J. G. (1989). Production of chimeric embryos of trout (*Salmo gairdneri*) by introducing isolated blastomeres into recipient blastulae. *Biol. Reprod.* **40** (Suppl. 1), 109 (abstract).

Oshiro, T., Yoshizaki, G., and Takashina, F. (1989). *In* "Proceedings First International Marine Biotechnology Conference," Abstract No. 4–10.

Ozata, K., Kondoh, H., Inohara, H., Iwamatsu, T., Wakamatsu, Y., and Okada, T. S. (1986). Production of transgenic fish: Introduction and expression of chicken crystallin gene in medaka embryos. *Cell Differ.* **19**, 237–244.

Palmiter, R. D., and Brinster, R. L. (1986). Germline transformation of mice. *Annu. Rev. Gen.* **20**, 465–499.

Perry, M. M. (1987). A complete culture system for the chicken embryo. *Nature (London)* **331**, 70–72.

Petitte, J. N., Clark, M. E., Liu, G., Gibbons, A. M. V., and Etches, R. J. (1990). Production of somatic and germline chimeras in the chicken by transfer of early blastodermal cells. *Development (Cambridge, UK)* **108**, 185–190.

Powell, R. L., Dodson, M. V., and Cloud, J. G. (1989). Cultivation and differentiation of satellite cells from skeletal muscle of the rainbow trout *Salmo gairdneri*. *J. Exp. Zool.* **250**, 333–338.

Powers, D. A., Gonzalez-Villasenor, L. I., Zhang, P., Chen, T. T., and Dunham, R. A. (1991). Studies in transgenic fish: Gene transfer, expression, and inheritance. *In* "Transgenic Animals" (N. First and F. P. Haseltine, eds.), pp. 307–324. Butterworth–Heineman, Stoneham, Massachusetts.

Purdom, C. E. (1984). Atypical modes of reproduction in fish. *In* "Oxford Reviews of Reproductive Biology" (J. R. Clarke, ed.), Vol. 6, pp. 303–340. Oxford Univ. Press, Oxford.

Rokkones, E., Alestrom, P., Skjervold, H., and Gantvik, K. M. (1989). Microinjection and expression of a mouse metallothionein fusion gene in fertilized salmonid eggs. *J. Comp. Physiol.* **158**, 751–758.

Salter, D. W., and Crittenden, L. B. (1989). Artificial insertion of a dominant gene for resistance to avian leukemia virus into the germline of the chicken. *Theor. Appl. Genet.* **77**, 457–461.

Salter, D. W., Smith, E. J., Hughes, S. H., Wright, S. E., and Crittendon, L. B. (1987). Transgenic chickens: Insertion of retroviral genes into the chicken germline. *Virology* **157**, 236–240.

Sandset, P. M., and Kornelliussen, H. (1978). Myosatellite cells associated with different muscle fibre types in the Atlantic hagfish (*Myxine glutinosa* L.). *Cell Tissue Res.* **195**, 17–27.

Savatier, P., Bagnis, C., Thoroval, P., Poncet, D., Belakebi, M., Mallet, F., Legras, C., Cossett, F. L., Thomal, J. L., Chebloune, Y., Faure, C., Verdick, G., Samarut, J., and Nigon, V. (1989). Generation of a helper cell line for packaging avain leukosis virus-based vectors. *J. Virol.* **63**, 513–522.

Scott, G. K., Fletcher, G. L., and Davies, P. L. (1986). Fish antifreeze proteins: Recent gene evolution. *Can. J. Fish. Aquat. Sci.* **43**, 1028–1034.

Scott, G. K., Davies, P. L., Kao, M. H., and Fletcher, G. L. (1988). Differential amplification of antifreeze protein genes in the Pleuronectinae. *J. Mol. Evol.* **27**, 29–35.

Simkiss, K., Vick, L., Luke, G., Page, N., and Savva, D. (1990). Infection of primordial germ cells with defective retrovirus and their transfer to the developing embryo. *4th World Congress Genetics Applied Livestock Production*, abstract.

Smithies, O., Gregg, R. G., Boggs, S. S., Koralewski, M. A., and Kucherlapati, R. S. (1985). Insertion of DNA sequences into the human chromosomal-globin locus by homologous recombination. *Nature (London)* **317**, 230–234.

Souza, L., Boone, T. C., Murdock, D., Langley, K., Wypych, J., Fenton, D., Johnson, S., Lai, P. H., Everett, R., and Itsy, R. Y. (1989). Application of recombinant DNA technologies to studies on chicken growth hormone. *J. Exp. Zool.* **232**, 465–473.

Spratt, N. T., and Haas, H. (1961). *J. Exp. Zool.* **147**, 57–93.

Stoker, A. W., and Bissell, J. (1988). Development of avain sarcoma and leukosis virus-based vector-packaging cell lines. *J. Virol.* **62**, 1008–1015.

Stuart, G. W., McMurray, J. V., and Westerfield, M. (1988). Replication, integration, and stable germ-line transmission of foreign sequences injected into early zebrafish embryos. *Development (Cambridge, UK)* **103**, 403–412.

Stuart, G. W., Vielkind, J. R., McMurray, J. V., and Westerfield, M. (1990). Stable lines of transgenic zebrafish exhibit reproducible patterns of transgene expression. *Development (Cambridge, UK)* **109**, 577–584.

Tamiya, E., Sugiyama, T., Masaki, K., Hirose, A., Okoshi, T., and Karube, I. (1990). Spacial imaging of luciferase gene expression in transgenic fish. *Nucleic Acids Res.* **18**, 1072.

Thorgaard, G. H., and Gall, G. A. E. (1979). Adult triploids in a rainbow trout family. *Genet. Princeton* **93**, 961–973.

Thorgaard, G. H. (1983). Chromosome set manipulation and sex control in fish. *In* "Fish Physiology" (W. S. Hoar, D. J. Randall, and E. M. Donaldson, eds.), Vol. 9B, pp. 405–435. Academic Press, New York.

Thorgaard, G. H., Scheerer, P. D., and Parsons, J. E. (1985). Residual paternal inheritance in gynogenetic rainbow trout: Implication for gene transfer. *Theor. Appl. Genet.* **71**, 119–121.

Watanabe, S., and Temin, H. M. (1982). Sequences for spleen necrosis virus, an avian retrovirus, are between the 5′ long terminal repeat and the start of the *gag* gene. *Proc. Natl. Acad. Sci. U.S.A.* **79**, 5986–5990.

Watanabe, S., and Temin, H. M. (1983). Construction of a helper cell line for avian reticuloendotheliosis virus cloning vectors. *Mol. Cell. Biol.* **3**, 2241–2249.

Watt, D. J., Morgan, J. E., and Partridge, T. A. (1984). Use of mononuclear precursor cells to insert allogenic genes into growing mouse muscle. *Muscle Nerve* **7**, 741–750.

Xie, Y., Liu, D., Zou, J., and Zhu, Z. (1989). Novel gene transfer in the fertilized eggs of loach via electroporation. *Acta Hydrobiol. Sin. (Shuisheng Shengwu Xuebao)* **13**(40), 387–389.

Xie, Y., Lui, D., Zou, J., Li, G., and Zhu, Z. (1993). Gene transfer via electroporation in fish. *Aquaculture* (abstract).

Yoon, S. J., Lui, Z., Kapuscinski, A. R., Hackett, P. B., Faras, A., and Guise, K. S. (1988). Successful gene transfer in fish. *J. Cell. Biochem.* **12B** (Suppl.), 190.

Yoon, S. J., Hallerman, E. M., Gross, M. L., Lui, Z., Schneider, J. F., Faras, A. J., Hackett, P. B., Kapuscinki, A. R., and Guise, K. S. (1990). Transfer of the gene for neomycin resistance into goldfish (*Carassius auratus*). *Aquaculture* **85**, 21–34.

Zhang, P., Hayat, M., Joyce, C., Gonzelez-Villasenor, L. I., Lin, C. M., Dunham, R. A., Chen, T. T., and Powers, D. A. (1990). Gene transfer, expression and inheritance of pRSV–rainbow trout GH complementary DNA in the common carp *Cyprinus carpio* Linnaeus. *Mol. Reprod. Dev.* **25,** 13–25.

Zhu, Z., Li, G., He, L., and Chen, S. (1985). Novel gene transfer into the fertilized eggs of goldfish (*Carassius auratus* L. 1758). *Z. Agnew. Ichthyol.* **1,** 31–34.

Zhu, Z., Xu, K., Li, G., Xie, Y., and He, L. (1986). Biological effects of human growth hormone gene microinjected into fertilized eggs of loach *Misgurnus anguillicaudatus* (Cantor). *Kexue Tongbao Acad. Sin. (Engl. Transl.)* **31,** 988–990.

Production of Transgenic Swine

Michael J. Martin
DNX Corporation
Swine Research Group
Princeton, New Jersey 08540

Carl A. Pinkert
Department of Comparative Medicine
Schools of Medicine and Dentistry
The University of Alabama at Birmingham
Birmingham, Alabama 35294

I. INTRODUCTION AND DISCUSSION

A. Background

The production of transgenic mice that grew twice as large as their nontransgenic littermates (Palmiter *et al.*, 1982) was the first evidence that genetic engineering could be used to greatly modify the phenotype of an animal. With U.S. animal agriculture representing a multibillion dollar a year industry (Pursel *et al.*, 1989), livestock species (i.e., cattle, swine, sheep, goats, and poultry) are obvious candidates for the application of genetic engineering. Genetic engineering is currently being used to enhance livestock performance in two major areas: growth/development (Pursel *et al.*, 1989) and disease resistance (Pinkert *et al.*, 1989a,b; Lo *et al.*, 1991; Weidle *et al.*, 1991). A third application of genetic engineering concerns the conversion of livestock into bioreactors which produce human proteins that can eventually be harvested from blood or milk (Clark *et al.*, 1987; Van Brunt, 1988; Swanson *et al.*, 1992).

Genes that have been introduced into swine are listed in Table 1. The majority of genes utilized in swine experiments have been growth hormone (GH) or growth hormone releasing factor (GRF) fusion constructs. In contrast to mice that express foreign GH genes, neither sheep (Murray *et al.*, 1989; Rexroad *et al.*, 1989, 1990) nor swine (Ebert *et al.*, 1988; Miller *et al.*, 1989; Pursel *et al.*, 1990a,b; Wieghart *et al.*, 1990) grew at an accelerated rate when compared to nontransgenic control

TABLE 1
Genes That Have Been Introduced into Swine[a]

Gene[a]	Eggs injected	Transgenic offspring	Functional transgenics	Reference
mMT/hGH	268	1	b	Brem *et al.* (1985)
	2035	20	11/18	Hammer *et al.* (1985)
hMT/bGH	423	6	1/6	Vize *et al.* (1988)
mMT/bGH	2198	11	8/11	Pursel *et al.* (1987)
bPRL/bGH	289	4	2/4	Polge *et al.* (1989)
rPEPCK/bGH	1057	7	5/7	Wieghart *et al.* (1990)
MLV/rGH	59	1	1/1	Ebert *et al.* (1988)
mMT/hGRF	2627	8	2/8	Pinkert *et al.* (1987), Pursel *et al.* (1990b)
	1041	6	b	Brem *et al.* (1988)
hALB/hGRF	968	5	3/3	Pursel *et al.* (1990a)
mMT/hIGF-1	387	4	1/4	Pursel *et al.* (1990b)
mMT/Mx	1083	6	—	Brem *et al.* (1988)
mWAP	850	5	3/3	Shamay *et al.* (1991)
hLCR/$\alpha\alpha\beta$	709	3	3/3	Swanson *et al.* (1992)
MSV/c-ski	1091	29	10/29	Pursel *et al.* (1992)
mμIg	119	0		Pinkert (1990)
m$\kappa\gamma$Ig	b	3	1/3	Weidle *et al.* (1991)
mαIg	542	2	2/2	Lo *et al.* (1991)
mMT/hβIFN	848	2	2[c]/2	Pinkert (1990)

[a] For chimeric constructs, a slash separates a promoter or enhancer sequence from the structural gene. The species derivation is indicated by a lowercase letter before the abbreviation of the gene: b, bovine; c, chicken; h, human; m, mouse; o, ovine; p, porcine; r, rat. ALB, albumin; $\alpha\alpha\beta$ (α, α, and β) globin chains; GH, growth hormone; GRF, growth hormone-releasing factor; Ig, immuno-globulin (κ, α, γ and/or μ chain); βIFN, β-interferon; LCR, human β-globin gene control locus; MLV, Moloney murine leukemia virus; MSV, mouse sarcoma virus; MT, metallothionein; Mx, myxovirus resistant; PEPCK, phosphoenolpyruvate carboxy kinase; PRL, prolactin; WAP, whey acidic protein. A functional transgene indicates the number of founder transgenic animals (or their offspring) expressing a transgene encoded mRNA or protein product divided by the total number of founder animals (or offspring within a line) evaluated.

[b] Data incomplete.

[c] Pigs died at birth.

animals. Vize *et al.* (1988), however, reported the production of several transgenic pigs containing a porcine growth hormone (pGH) construct, one of which grew substantially faster than littermate controls. Serum concentrations of GH in this animal were twice those observed in nontransgenic littermates. These researchers have speculated that the enhanced growth exhibited by the pig was due to the incorporation of a gene construct which expressed a homologous (pGH) as opposed to heterologous [bovine (bGH) or human (hGH)] growth hormone gene. In contrast, Ebert *et al.* (1990) found no difference in phenotype between transgenic swine which expressed a homologous versus heterologous growth hormone construct.

A comparison of endocrine profiles between transgenic swine and sheep producing bGH or hGH and respective nontransgenic littermate controls revealed the presence of elevated plasma concentrations of insulin and insulin-like growth factor I (IGF-I) in the transgenic individuals (Miller *et al.*, 1989; Murray *et al.*, 1989; Rexroad *et al.*, 1989; Pinkert, 1991). The physiological consequences of circulating hGH and bGH in transgenic swine and sheep are similar to those observed following the exogenous administration of growth hormone, for example, body fat is reduced (Hammer *et al.*, 1986; Pursel *et al.*, 1987; Murray *et al.*, 1989) while feed efficiency is enhanced (Campbell *et al.*, 1988; Evock *et al.*, 1988; Pursel *et al.*, 1989).

Unfortunately, transgenic livestock that express a heterologous GH construct demonstrate a host of health problems. Transgenic swine expressing GH genes exhibited lameness, peptic ulcers, lethargy, and impaired reproductive performance (Pursel *et al.*, 1989, 1990a,b; Wieghart *et al.*, 1990). The occurrence of multiple health problems in hGH, bGH, and rat GH (rGH) transgenic swine may result from the continuous exposure of the animal to elevated concentrations of growth hormone (Pursel *et al.*, 1990a). This hypothesis applies to transgenic swine that express a pGH construct as well (Ebert *et al.*, 1990). The identification of promoters that cause expression of the GH transgene exclusively during the rapid growth phase or in an episodic fashion may reduce the health concerns associated with GH fusion constructs (Pursel *et al.*, 1990a).

The hypothesis that farm animals may be utilized as bioreactors received major support with the production of transgenic mice which expressed the sheep β-lactoglobulin gene in the mammary gland (Simons *et al.*, 1987). β-Lactoglobulin concentrations in the milk of the mice were 5-fold higher than those estimated in sheep milk. Gordon *et al.* (1990) have further shown that two human proteins, tissue plasminogen activator and protein C, can be produced in the mammary gland of lactating transgenic mice.

The first livestock species to be utilized as a bioreactor was sheep (Simons *et al.*, 1988). Using the β-lactoglobulin gene as a promoter, Simons *et al.* (1988) produced transgenic sheep that secreted human clotting factor IX or α_1-antitrypsin in the milk. Researchers have now been able to create transgenic swine (Wall *et al.*, 1991), sheep (Wright *et al.*, 1991), and goats (Denman *et al.*, 1991; Ebert *et al.*, 1991) that secrete heterologous milk proteins as well. Although the mammary gland has been the major target for the so-called biofarming of heterologous proteins, the production of transgenic pigs which synthesize and secrete human hemoglobin (Swanson *et al.*, 1992) indicates that erythroid tissues can be utilized in a similar manner.

B. Production of Transgenic Swine: Current Efficiencies

In contrast to the mouse, the efficiency associated with the production of transgenic livestock, including swine, is low (Hammer *et al.*, 1985; Pursel *et al.*, 1989, 1990a,b; Wieghart *et al.*, 1990). Since the first transgenic swine were reported in 1985, DNA microinjection has been the only successful method identified to produce transgenic pigs. Although other technologies are in development, this chapter describes the use and refinement of DNA microinjection technology.

An initial problem encountered during the creation of transgenic swine and other farm animal species concerned the visualization of the pronuclei or nuclei within the ova. Wall *et al.* (1985) found that centrifugation of pig ova at 15,000 *g* for 3 to 5 min results in stratification of the cytoplasm that renders the pronuclei visible using differential interference contrast microscopy (Fig. 1). Unfortunately, centrifugation failed to reveal the pronuclei in 15% (Wall *et al.*, 1985) to 33% (Brem *et al.*, 1989) of the fertilized ova. After experiencing a similar difficulty in observing pronuclei following centrifugation, Hammer *et al.* (1986) attempted to produce transgenic swine by injecting DNA into the cytoplasm of 1- and 2-cell ova. None of the fetuses derived from the ova were transgenic. Introduction of the same gene construct into the pronuclei or nuclei of centrifuged 1- and 2-cell ova, however, yielded an integration frequency in offspring of 10%. These findings indicate that genomic incorporation of the transgene takes place only when the gene is introduced directly into the pronucleus or nucleus of the ovum.

The proportion of transferred microinjected ova that develop into viable offspring in swine varies from 6 to 11.7% (Pursel *et al.*, 1989, 1990b). The survival to term of microinjected porcine ova is related to several factors, including the developmental stage of ova injected (Pursel *et al.*, 1987), the duration of *in vitro* culture (Davis, 1985; Brem *et al.*, 1989), synchrony of donors and recipients (Polge, 1982; Brem *et al.*, 1989), the number of ova transferred (Brem *et al.*, 1989), and the age of the donor (Pinkert *et al.*, 1989a; Brem *et al.*, 1989). Other factors that have been shown to influence the development of microinjected mouse and sheep ova, for example, proficiency of the microinjectionist, DNA concentration and form (Brinster *et al.*, 1985), and injection pipette diameter (Walton *et al.*, 1987), may affect the viability of microinjected swine ova as well.

The proportion of transgenic swine that develop from microinjected ova is also lower than that observed in rodents (0.31–1.73% for swine versus 3% for mice) (Pursel *et al.*, 1990b). The low frequency at which integration of the transgene occurs in the pig genome may be associated with the age of the host ovum or the

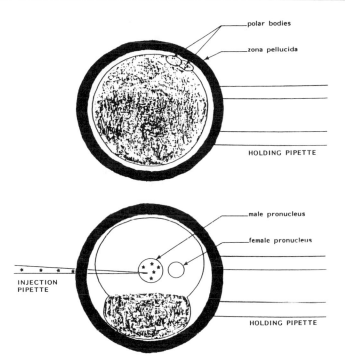

polar bodies

zona pellucida

HOLDING PIPETTE

male pronucleus

female pronucleus

INJECTION
PIPETTE

HOLDING PIPETTE

Figure 1. Microinjection of a pig zygote. Although the cytoplasm in murine eggs does not inhibit visualization of the pronuclei, lipid-rich pig eggs are relatively opaque (top). To facilitate visualization of nuclear structures, 1- and 2-cell eggs are centrifuged to stratify lipids in the cytoplasm (bottom). After centrifugation, the eggs are placed in a microdrop of medium overlaid with silicone oil and held in place, sequentially, using the large-bore holding pipette. A small-bore injection pipette containing DNA in a buffer solution is inserted through the zona pellucida and plasma membrane into a pronucleus (or nuclei of 2-cell eggs), which readily expands (~50% greater in volume) as the DNA solution is delivered. The diameter of the egg is approximately 125 μm. If 2-cell eggs are used, generally both blastomeres are filled with DNA. [Reproduced with permission from Pinkert (1987) and Pinkert et al. (1990).]

composition of its chromosomes (Pursel et al., 1989). If integration does occur, the number and orientation (head to head versus head to tail) of transgene copies inserted into the genome vary greatly as well (Hammer et al., 1985; Vize et al., 1988; Miller et al., 1989; Polge et al., 1989; Wieghart et al., 1990).

In swine, approximately 70% of the transgenic pigs containing an exogenous growth hormone construct expressed the integrated protein product of the transgene (Pursel et al., 1990b). Failure of the remaining 30% of transgenic swine to express the foreign gene has been attributed to integration of the transgene into an inactive chromosomal locus or alteration of the transgene sequence during its integration (Pursel et al., 1990a). Once the gene has been integrated, the level of expression appears to vary greatly among individuals regardless of the source of the growth hormone construct (bGH, pGH, or hGH) (Hammer *et al.*, 1985; Miller *et al.*, 1989; Pursel *et al.*, 1989). According to Pursel *et al.* (1990b), transgene transcription rates are probably regulated by the level of activity present at the locus, or site of integration, and the properties of the enhancer sequences located in genes flanking the fusion gene. The existence of an interaction between gene copy number and level of expression is most likely dependent on the particular construct or transgene. Hammer *et al.* (1985) saw no relationship between the number of gene (hGH) copies present per cell and plasma concentrations of hGH. Miller *et al.* (1989), however, found a positive correlation between plasma concentrations of bGH and the number of bGH copies per cell in transgenic swine.

II. METHODS

A. Donor and Recipient Management

1. Synchronization of Estrus

The successful transfer of ova in swine depends greatly on synchronization of estrus between the donor and recipient. Webel *et al.* (1970) found no difference in the pregnancy rate between recipients whose estrous periods were 24 hr ahead or behind those of the donors. Polge (1982), however, noted that the pregnancy rate was highest (70–86%) among recipients which had expressed estrus 24 to 40 hr later than the donors and lowest (10–56%) among recipients which had expressed estrus earlier than the donors. Blum-Reckow and Holtz (1991) observed a similar increase in both the pregnancy and embryo survival rates when ova were transferred to recipients whose estrous cycles were 24 hr behind that of the donors. The current recommendation for embryo transfer in swine is to utilize recipient females whose onset of estrus is at least synchronous or 12 to 24 hr later than the occurrence of estrus in the donors.

The majority of methods used for synchronizing estrus in sexually mature gifts or sows involves manipulation of the luteal phase of the estrous cycle. These procedures include pseudopregnancy induction followed by exogenously induced luteal regression, induction of accessory or secondary corpora lutea (CL) followed by exogenously induced CL regression, and maintenance of an artificial luteal phase through the administration of an orally active progestin (e.g., allyl trenbolone).

2. Pseudopregnancy

Pseudopregnancy can be induced by the administration of estradiol benzoate (EB) or estradiol valerate (EV) during the period of pregnancy recognition. Injections of 5 or 10 mg of EB (Guthrie, 1975) or EV (Zavy *et al.*, 1984) on days 11 to 15 of the estrous cycle have proven to be effective in significantly prolonging luteal function. The administration of EB for 20 days also prolonged luteal function and suppressed estrus (Kraeling and Rampacek, 1977). The return of pseudopregnant gilts to estrus is achieved through prostaglandin $F_{2\alpha}$ ($PGF_{2\alpha}$)-induced luteal regression. The time of $PGF_{2\alpha}$ treatment is apparently unimportant, as its administration on days 1, 5, 10, or 20 following cessation of EB treatment appears to be equally effective in causing gilts to exhibit synchronized estrus 4 to 6 days later (Guthrie, 1975).

3. Accessory Corpora Lutea Induction

Ovulation induction and subsequent formation of accessory CL during the luteal phase of the estrous cycle have been accomplished in the pig through the use of pregnant mare's serum gonadotropin (PMSG) and human chorionic gonadotropin (HCG) (Neill and Day, 1964; Caldwell *et al.*, 1969). Because $PGF_{2\alpha}$ is not luteolytic in the pig until at least day 11 or 12 of the estrous cycle (Diehl and Day, 1974; Guthrie and Polge, 1976), estrus induction in animals possessing accessory CL is most effective when $PGF_{2\alpha}$ is given on days 12 and 13 (Guthrie and Polge, 1967) or days 13 and 14 (Guthrie, 1979) after HCG administration (day 0). The incidence of estrus expression can be increased by administering a second injection of PMSG alone on the second day of $PGF_{2\alpha}$ treatment (Guthrie and Polge, 1967) or by administering PMSG followed 96 hr later by HCG (Guthrie, 1979). Accessory CL induction does not appear to be detrimental to fertilization, as 80% conception rates have been achieved following a single insemination (Guthrie and Polge, 1967).

4. Orally Active Progestagen

Perhaps the most effective estrus synchronization method involves the feeding of an orally active synthetic progestin [allyl trenbolone (AT), 15 mg/head/day] for 14 to 21 days (Webel, 1976; Knight *et al.*, 1976; O'Reilly *et al.*, 1979; Davis *et al.*, 1979; Redmer and Day, 1981; Pursel *et al.*, 1981) to sexually mature females. Estrus synchronization rates (i.e., the proportion of females which exhibit estrus between 4 and 7 days following AT withdrawal) of 89% or better have been reported (Knight *et al.*, 1976; Redmer and Day, 1981; Pursel *et al.*, 1981). Allyl trenbolone also appears to have no detrimental effect on ovulation, fertilization, or gestation. In fact, AT consumption may enhance the ovulation rate (Davis *et al.*, 1979).

B. Superovulation

Superovulation of donors in a transgenic swine program provides two major benefits: it increases the number of ova that can be obtained from each donor and it establishes the onset of ovulation. The latter benefit enables one to determine the optimum time at which to recover pronuclear stage ova.

The likelihood of producing mosaic individuals increases when the nuclei of 2-cell ova are microinjected with DNA, and thus it is advantageous to recover a high percentage of pronuclear stage ova. One-cell ova can be obtained from swine that have ovulated naturally or from donors that have been superovulated using PMSG and HCG. In the former case, the start of ovulation and the optimum time at which to collect pronuclear stage ova are based solely on the onset of estrus.

Natural ovulation in swine begins between 30 and 34 hr after the onset of estrus (Signoret et al., 1972; Pope et al., 1989) and takes at least 3 to 6.5 hr to complete (Burger, 1952; Betteridge and Raeside, 1962). Unfortunately, the relationship between the onset of estrus and the start of ovulation in swine is quite variable. Pope et al. (1989) found that the majority of gilts begin to ovulate sometime between 30 and 38 hr after the onset of estrus. Furthermore, when ovaries from gilts at 34 hr after the onset of estrus were examined, ovulation appeared to have been completed in some cases, whereas in others ovulation had not yet begun. Additional data indicate that neither ovulation nor early embryonic development are synchronous processes in swine (Pope et al., 1989; Didion et al., 1990; Martin et al., 1990; Xie et al., 1990a,b). It is clear that a great amount of variation exists throughout the chronology of events associated with estrus, ovulation, and early embryonic development in swine.

In contrast to natural ovulation, a superovulatory regimen that includes PMSG and HCG enables one to program the onset and reduce the asynchronicity of ovulation (Dziuk and Baker, 1972; Pope et al., 1989). These benefits should increase the likelihood of selecting the proper time at which to recover ova at the pronuclear stage of development.

Superovulation of sexually mature gilts can be achieved through either the administration of PMSG alone (Hunter, 1964, 1966) or in combination with HCG (Day et al., 1965, 1967). Gilts given PMSG alone following luteal regression (day 15 to 16 of the estrous cycle) show estrus 4 to 7 days later following treatment and ovulate without receiving HCG (Day et al., 1965). Ovulation can be more precisely controlled when HCG is administered between 72 and 80 hr following PMSG (Wall et al., 1985; Hammer et al., 1985; Pope et al., 1989). Gilts should exhibit estrus within 24 hr following HCG administration.

The dose of PMSG to administer depends on both the target ovulation rate and hormone source one chooses. As the ovulatory response to PMSG/HCG approaches or exceeds 40 ova, the probability of obtaining immature, degenerate, or otherwise abnormal ova may increase as well (Holtz and Schlieper, 1991). The PMSG should be administered at a dosage that results in an average ovulation rate of between 25 and 35. This dosage must be determined for each new source of PMSG (Martin et al., 1989).

Exogenous gonadotropin treatment can stimulate ovulation in prepuberal gilts as well (Casida, 1935; Dziuk and Gehlbach, 1966; Baker and Coggins, 1968; Amet *et al.,* 1991). However, Pinkert *et al.* (1989a) found that fertilized ova recovered from superovulated prepuberal gilts exhibited reduced development *in vitro* as compared to zygotes recovered from sexually mature gilts. French *et al.* (1991) further demonstrated that both the pregnancy rate and litter size were lower following the transfer of microinjected ova obtained from prepuberal as opposed to puberal gilts. In contrast, no reduction in the *in vitro* development of 1-cell ova following microinjection was noted in German studies (Brem *et al.,* 1989) which utilized prepuberal gilts exclusively as ova donors. Differences in the developmental capacity of 1-cell ova recovered from prepuberal gilts, illustrated in the above studies, may be due to variation in the genetic composition of the donor animals, namely, European versus U.S. versus Australian breeds of swine.

C. Recipient Management

Prospective recipients must be structurally and reproductively sound. Females that fall into this category include sows and puberal gilts. Prepuberal swine should not be utilized as recipients of microinjected ova. Both Rampacek *et al.* (1979) and Segal and Baker (1973) found that the pregnancy rate was low among prepuberal gilts that had been induced to ovulate unless HCG or EB and progesterone were administered after breeding. Luteinizing hormone receptor studies of CL from prepuberal versus puberal swine (Estienne *et al.,* 1988) further suggest that CL formation and function following induced ovulation may be abnormal in the prepuberal pig.

An alternate approach to the two-pool system (one pool of donor animals and one pool of recipients) for creating transgenic swine is the use of a single animal pool where ovum donors also serve as recipients of microinjected ova. This dual-purpose approach reduces the number of animals required for the production of transgenic swine and appears to have no detrimental effect on pregnancy rate or litter size (Pursel and Wall, 1991).

D. Breeding Management

A factor critical to the production of transgenic swine is the recovery of fertilized ova from donor animals. Because fertilization of ova is dependent on the presence of sperm in the reproductive tract of the female shortly before the onset of ovulation, the timing and frequency of insemination performed during the estrous period is important. Inseminations performed too early or too late during estrus have been shown to result in low fertility (Boender, 1966). If females are exposed to a mature boar once daily, gilts/sows should be bred each day they exhibit standing estrus. If estrus detection is conducted twice daily, females should be bred at 12 and 24 hr after the onset of estrus (Diehl *et al.,* 1990).

TABLE 2
Boar Semen Extenders[a]

Ingredient	BL-1 (1 quart)	Egg yolk (1 quart)
Egg yolk	—	317.0 ml
Distilled water	b	739.0 ml
Glucose	27.4 g	31.7 g
Potassium chloride	0.3 g	—
Sodium bicarbonate	1.9 g	1.6 g
Sodium citrate	9.5 g	—
Penicillin	1.0×10^6 IU	1.1×10^6 IU
Streptomycin sulfate	1.0 g	1.6 g

[a]Reprinted with permission from Diehl et al. (1979).
[b]Put salts in a clean quart container and fill to the line with distilled water.

When artificial insemination (AI) is utilized, females should be inseminated with at least 3 billion live sperm in a total volume of 100 ml (semen plus extender). As a mature boar will produce an average of 100 to 300 million sperm/ml (Foote, 1980), a typical ejaculate should contain a sufficient number of sperm to breed six to eight females when properly extended (Diehl et al., 1979).

Semen collected from a boar of proven fertility can usually be extended at a ratio of 1 part semen to 4 to 5 parts extender and still yield acceptable conception rates. If a 1:10 ratio of semen to extender is required, one should determine the actual sperm concentration before the semen is extended in order to make sure each female receives at least 3 billion live sperm. Several commercial extenders, some of which maintain sperm viability for up to 1 week, are available. The formulas for two commonly used semen extenders are presented in Table 2 (Diehl et al., 1979).

E. Ova Recovery

To obtain ova at the pronuclear or 2-cell stage for microinjection, surgical recovery is performed between 60 and 66 hr after the administration of HCG (Hammer et al., 1985; Wall et al., 1985; Pinkert et al., 1989a). General anesthesia may be induced in swine by administering one of the following drug combinations through a peripheral ear vein:

- 1 g sodium thiopental/100 kg body weight (bw)
- 1.3 mg xylazine/kg bw plus 1.3 mg ketamine/kg bw
- 0.4 mg acepromazine/kg bw plus 7.5 mg ketamine/kg bw plus 0.1 ml 0.9% saline/kg bw

The use of sodium thiopental or xylazine/ketamine requires that anesthesia be maintained by a closed circuit system of O_2 (600 to 1000 ml/min) and halothane (4–5%, v/v; Webel et al., 1970).

Once the animal has been anesthetized, the reproductive tract is exteriorized via a midventral laparotomy. A drawn glass cannula (o.d. 5 mm, length 8 cm) is then inserted into the ostium of the oviduct a distance of 3 cm and anchored in place using a single 2-O silk tie through the mesosalpinx (Vincent et al., 1964; Day, 1979). The oviduct is flushed toward the infundibulum by inserting a 20-gauge needle into the lumen of the oviduct at a point just superior to the uterotubal junction and infusing 10 ml of sterile, warm (30°C) Dulbecco's phosphate buffered saline (PBS) supplemented with 2% (w/v) bovine serum albumin (BSA). The medium is collected in 15-ml sterile plastic tubes. Flushings are transferred to 15×100 mm petri dishes and searched at low power ($50\times$) using a stereomicroscope. After washing the ova twice in G medium (Pinkert et al., 1989a), the ova are transferred to microdrops of G medium that have been overlaid with silicone oil. The ova are maintained in this manner at 38°C under 5% CO_2, 5% O_2 (by volume) until just prior to microinjection.

F. Ova Culture

To create transgenic swine, 1- and/or 2-cell ova must be recovered, injected with DNA, and subsequently transferred back to suitable recipients. The developmental capacity of porcine 1- and 2-cell ova in various media has been reviewed by Wright and Bondioli (1981). Until recently, however, the ability of 1- and 2-cell ova to undergo more than one or two cleavages in vitro has been difficult to achieve. Culture of 1- and 2-cell ova to the 2- to 4-cell stage (short-term culture, 12–24 hr) can be accomplished using modified Krebs–Ringer bicarbonate medium (mKRB) or modified BMOC-3 supplemented with BSA and EDTA (G medium; Pinkert et al., 1989a; Wieghart et al., 1990), oviductal fluid (Archibong et al., 1989), or hypotaurine and taurine (NCSU 23 medium; Reed et al., 1992). Long-term culture ($\geq$72 hr) of 1- and 2-cell ova to the blastocyst stage has also been demonstrated using Whitten's medium supplemented with 1.5% BSA (Beckman et al., 1990; Beckman and Day, 1991) and NCSU 23 medium. The birth of live pigs from 1- or 2-cell ova cultured in G, Whitten's, and NCSU 23 media indicates that any of these media may be used to culture porcine ova prior to and immediately following microinjection.

The use of a bicarbonate buffer to maintain the pH of a culture medium such as mKRB, Whitten's, and NCSU 23 between 7.2 and 7.4 requires that the medium be stored under an atmosphere which includes at least a 5% (v/v) CO_2 gas phase. The atmospheres commonly used are 5% CO_2 in air or a mixture of 5% CO_2, 5% O_2, and 90% N_2. Wright (1977) found that a reduced O_2 atmosphere was superior to a 5% CO_2 in air atmosphere in supporting embryonic development, whereas Niemann et al. (1983) observed just the opposite. A review of several studies in which transgenic swine were successfully produced indicates that the most viable atmosphere for maintaining the pH of bicarbonate-buffered media is 5% CO_2, 5% O_2.

Coculture (Allen and Wright, 1984; Krisher *et al.*, 1989) or xenogeneic (Papaioannou and Ebert, 1988) culture systems have also been shown to support development of 1-cell porcine ova. The use of such elaborate systems, however, is not necessary for the production of transgenic swine, since the recovery, microinjection, and transfer processes should require only a short ova culture period.

The duration of the culture period is especially important as it greatly affects ova viability. James *et al.* (1980, 1983) found that the proportion of transferred embryos which develop to term decreases rapidly as the duration of *in vitro* storage increases even though the embryos appear morphologically normal at the time of transfer. In a recent study, Blum-Reckow and Holtz (1991) examined the effects of long- and short-term culture in PBS on embryo viability. Pregnancy and embryo survival rates of 9 and 1.8%, respectively, were obtained following the transfer of morulae/blastocysts cultured from the 4-cell stage (a period of 72 hr). When freshly collected blastocysts were transferred, the pregnancy and embryo survival rates increased to 60 and 23.9%, respectively. These observations support our recommendation that ova be microinjected and returned to recipients as soon as possible after recovery.

G. Microinjection

1. Equipment

The equipment used to create transgenic swine is similar to that used to create transgenic mice (see Chapter 2 for a detailed breakdown). A brief list of instruments and supplies currently used in our laboratories to create transgenic swine is identified in Table 3.

TABLE 3
Specialty Equipment and Supplies Used for Gene Transfer in Swine

Interference contrast microscope (e.g., Leitz Laborlux; Leica Inc., Deerfield, IL)
Micromanipulators (2@; Leitz mechanical manipulators)
Micromanipulator baseplate and antivibration table (Micro-G, Woburn, MA, or ServaBench)
Eppendorf centrifuge and microcentrifuge tube carriers (Brinkman)
Microforge (de Fonbrune; TPI, Inc.)
Pipette puller (Sutter, San Raphael, CA; P-87)
Stereomicroscope (e.g., Wild M3 or M8)
P-3 microsyringe assemblies (2@; 100-μl syringe attached to a micrometer head)
Threaded plunger syringe (100 μl)
Glass capillary tubing (e.g., i.d. 0.027, o.d. 0.037; Garner Glass, La Jolla, CA, or World Precision
  Instruments, Sarasota, FL)
Silicone oil
Incubator with 5% CO_2, 5% O_2 balance N_2 environment
Surgical suite (with surgical supplies for multiple laparotomies)

2. DNA Preparation

Because of variation associated with the injection ability of the technician and the dimensions of the injection pipette tip, the volume of DNA solution introduced into the pronucleus or nucleus is difficult to regulate initially. Most investigators estimate that 1 to 2 pl is injected per nucleus for mouse ova (Hogan *et al.*, 1986); we estimate that 4 pl is injected per pig ovum according to micrometer-based estimates. Two factors which appear to affect embryo viability are DNA concentration and injection buffer composition. In mouse ova, optimal integration was obtained when DNA was introduced in a linear form at a concentration of 2 ng/μl in Tris-HCl buffer which contained between 0.1 and 0.3 mM EDTA (Brinster *et al.*, 1985). DNA integration and embryo survival were significantly reduced when the DNA and EDTA concentrations were increased beyond 10 ng/ml and 5 mM, respectively.

In swine, the relationship of DNA concentration, buffer composition, and DNA form (linear versus circular) to gene integration and embryo survival are unknown. However, Williams *et al.* (1992) found that the detrimental effect of pronuclear microinjection on subsequent development of porcine 1-cell ova was caused by the injection of DNA and/or contaminants, not by the mechanical process associated with microinjection per se. The most recently reported protocols for the production of transgenic swine suggest that one should microinject DNA at a concentration between 1 and 2 ng/μl (Shamay *et al.*, 1991; Swanson *et al.*, 1992). This concentration can be achieved by resuspending the DNA fragment in injection buffer (TE, which is 5–10 mM Tris-HCl, pH 7.5, 1 mM EDTA) and diluting this solution with TE, Dulbecco's PBS, or HEPES-buffered medium. One to two picoliters of this solution should contain between 10 and 2000 copies of the DNA fragment.

3. Ova Preparation

One- and/or two-cell ova are initially recovered using Dulbecco's PBS plus 0.4% BSA. Ova are then transferred to 10 × 15 mm petri dishes that contain microdrops (25 μl) of G medium (Pinkert *et al.*, 1989a) overlaid with silicone oil. The ova are stored under a humidified atmosphere of 5% CO_2, 5% O_2. Although paraffin or silicone oil can be equilibrated with culture medium prior to its use as an overlay, we have observed no detrimental effect of unequilibrated oil on ova viability during short-term culture under a humidified atmosphere.

Prior to microinjection, 10 to 50 ova are transferred to a 2-ml Eppendorf tube that contains 1 ml of G medium + 20.85 mM HEPES (G-HEPES) substituted for sodium bicarbonate (Fig. 2). The ova are centrifuged at 15,000 g for 6 min (Wall *et al.*, 1985). As previously mentioned, nuclear structures will remain hidden in

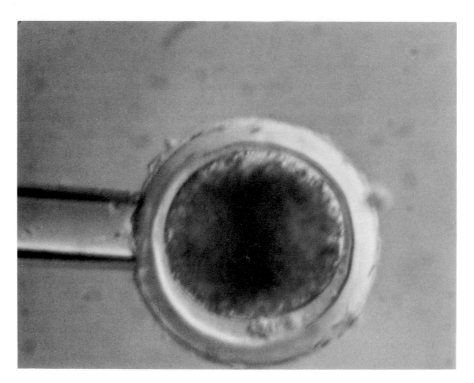

Figure 2.   One-cell pig egg before centrifugation. Magnification, 250 ×. Note opaque cytoplasm.

one-third of the ova following initial centrifugation. This figure can be reduced (Martin *et al.*, 1992) by culturing the ova an additional 1–4 hr and subjecting them to recentrifugation.

4. Microinjection Procedure

Initial steps include attachment of the holding and injection pipettes to their respective lines followed by the uptake of silicone or paraffin oil. Care must be taken to ensure that no air bubbles are left in any portion of the microinjection apparatus. The holding pipette tip should have outer and inner diameters of 25 to 50 μm and 10 to 15 μm, respectively. The injection pipette tip should be 1 to 2 μm in diameter (o.d.) and initially closed. The tip is opened by breaking it on the edge of the holding pipette. Bevelled tip injection pipettes can also be made by grinding the ends with a diamond dust-coated wheel. However, pipettes prepared in this manner do not appear to be as sharp as those with broken tips (M. J. Martin and C. A. Pinkert, personal observation, 1991).

Ova to be microinjected are held in a microdrop of G-HEPES medium (10 to 15 ova per drop) that has been placed on the middle of a depression slide. Another drop (2–5 μl) containing the DNA–buffer solution is placed above or below the ova-containing drop. Both drops are covered with silicone or paraffin oil. The injection pipette is loaded by drawing up HEPES medium first followed by the DNA–buffer solution.

Once the DNA solution has stopped flowing into the injection pipette, the syringe plunger is screwed down until the DNA begins to flow out of the tip slowly. The injection pipette is then transferred to the microdrop containing the ova. The pipette tip is inserted through the zona pellucida and vitelline membrane and into the pronucleus of a 1-cell ovum or nucleus of a 2-cell ovum using one continuous motion. When the pipette tip has penetrated the pronuclear or nuclear membrane, the pronucleus or nucleus should immediately begin to expand (Figs. 3 and 4). The instant expansion ceases, the injection pipette is withdrawn, again using one continuous motion. DNA should be injected into both nuclei of 2-cell ova in order to enhance the rate of transgenic animal production (Rexroad et al., 1988).

In addition to standard hollow capillary tubes, filament fiber capillary tubes (World Precision Instruments) can also be used for DNA microinjection (see Chapter 2). Although filament fiber capillary tubes negate the need for a DNA drop on the microinjection slide, a greater volume of DNA for microinjection is required over the course of an experiment.

H. Ova Transfer

Ova collected at the 1- or 2-cell stage for microinjection are subsequently transferred to the oviducts of sexually mature recipients (Dziuk et al., 1964; Pope and Day, 1977). Following anesthetization, the reproductive tract is exposed via a midventral laparotomy. Oviductal transfer entails aspiration of the ova along with 1 to 2 ml of G medium into the tubing obtained from a 21-gauge 3/4 in. butterfly infusion set. The tube is then fed through the ostium of the oviduct until it reaches the lower third or isthmus of the oviduct. The ova are expelled as the tubing is slowly withdrawn.

The pregnancy rate and the number of ova transferred appear to be directly related. Pope et al. (1972), using noninjected ova, observed that the pregnancy rate at 26 to 29 days of gestation improved when the number of ova transferred was increased from 12 to 24 (pregnancy rates were 71.4 and 100%, respectively). A similar finding was noted by Brem et al. (1989) following the transfer of various numbers of microinjected ova to recipients (Table 4). To maximize the pregnancy rate, we suggest the transfer of at least 35 to 40 microinjected ova per recipient.

The transplantation of noninjected or control ova has proved useful in ensuring pregnancy maintenance (C. A. Pinkert and M. J. Martin, unpublished data). In addition, the transfer of control ova from a strain of swine that exhibits a unique

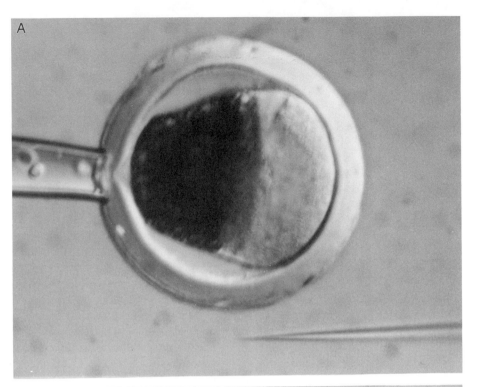

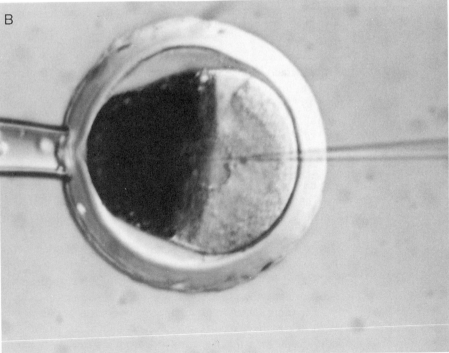

Figure 3. One-cell pig egg after centrifugation. Magnification, 250 ×. (A) After 6 min, the cytoplasmic lipids are stratified and the pronuclei are visible. A nucleolus and pronuclear membrane are evident at the center of the lipid–cytoplasmic interface. The injection pipette is focused in the same plane as the visible pronucleus, then brought in line to pierce the egg. (B) After impaling the pronucleus, expansion is seen as the pronucleus fills with the DNA solution.

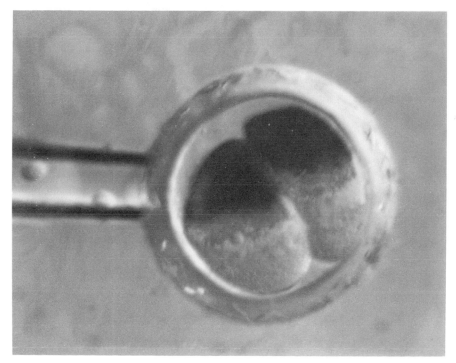

Figure 4. Two-cell pig egg after centrifugation. Magnification, 250 ×. Even after maneuvering the egg, the nuclei are not always in focus in the same plane. However, both nuclei are injected with DNA and show similar enlargement (~50%) as the DNA is delivered.

TABLE 4
Influence of Number of Embryos Transferred per Recipient on Success Rate[a]

Number of embryos transferred per recipient	Number of transfers	Pregnancy rate (%)	Piglets per	
			Litter	Transfer
<30	48	25	5.0	1.3
31–40	57	32	4.6	1.4
>41	77	47	4.4	2.0

[a]Reprinted with permission from Brem *et al.* (1989).

coat color pattern allows one to differentiate easily between offspring developing from control versus microinjected ova. However, this procedure does increase the labor and overall cost associated with the production system.

III. SUMMARY

The procedures outlined in this chapter represent the current state-of-the-art technology for the production of transgenic swine (Fig. 5). It is envisioned that these procedures will be modified and enhanced as data from future transgenic pig research become available. Central to short-term enhancement will be (1) develop-

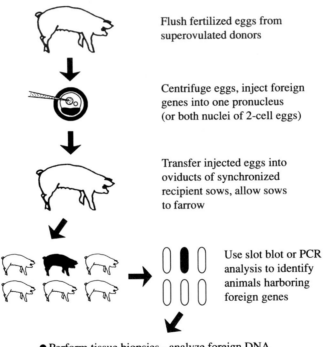

Flush fertilized eggs from superovulated donors

Centrifuge eggs, inject foreign genes into one pronucleus (or both nuclei of 2-cell eggs)

Transfer injected eggs into oviducts of synchronized recipient sows, allow sows to farrow

Use slot blot or PCR analysis to identify animals harboring foreign genes

● Perform tissue biopsies - analyze foreign DNA integration, mRNA transcription, and protein production

● Establish transgenic lines to study gene regulation in progeny

Figure 5. Scheme for production of transgenic swine. The methodology employed in the production and evaluation of transgenic pigs is illustrated. [Adapted with permission from Pinkert (1987) and Pinkert et al. (1990).]

ment of alternate DNA delivery systems, (2) refinement of existing porcine embryo culture systems, and (3) the utilization of PCR amplification of blastomere DNA in order to reduce the number of ova required for transfer per recipient female and per experiment.

REFERENCES

Allen, R. L., and Wright, R. W., Jr. (1984). *In vitro* development of porcine embryos in coculture with endometrial cell monolayers or culture supernatants. *J. Anim. Sci.* **59,** 1657–1661.

Amet, T. M., Li, J., and Markert, C. L. (1991). Estrus, ovulation, and fertilization in immature gilts treated with PG 600. *J. Anim. Sci.* **69** (Suppl. 1), 444 (abstract).

Archibong, A. E., Petters, R. M., and Johnson, B. H. (1989). Development of porcine embryos from one- and two-cell stages to blastocysts in culture medium supplemented with porcine oviductal fluid. *Biol. Reprod.* **41,** 1076–1083.

Baker, R. D., and Coggins, E. G. (1968). Control of ovulation rate and fertilization in prepuberal gilts. *J. Anim. Sci.* **27,** 1607–1610.

Beckmann, L. S., and Day, B. N. (1991). Culture of the one-and two-cell porcine embryo: Effects of varied osmolarity in Whitten's and Krebs Ringer bicarbonate media. *Theriogenology* **35,** 184 (abstract).

Beckmann, L. S., Cantley, T. C., Rieke, A. R., and Day, B. N. (1990). Development and viability of one- and two-cell porcine embryos cultured through the "four-cell block." *Theriogenology* **33,** 193 (abstract).

Betteridge, K. J., and Raeside, J. I. (1962). Observation of the ovary by peritoneal cannulation in pigs. *Res. Vet. Sci.* **3,** 390–398.

Blum-Reckow, B., and Holtz, W. (1991). Transfer of porcine embryos after 3 days of *in vitro* culture. *J. Anim. Sci.* **69,** 3335–3342.

Boender, J. (1966). The development of AI in pigs in the Netherlands and the storage of boar semen. *World Rev. Anim. Prod.* **2,** 29–44.

Brem, G., Brenig, B., Goodman, H. M., Selden, R. C., Graf, F., Kruff, B., Springmann, K., Hondele, J., Meyer, J., Winnacker, E.-L. and Kräusslich, H. (1985). Production of transgenic mice, rabbits and pigs by microinjection into pronuclei. *ZF Zuchthygiene* **20,** 251–252.

Brem, G., Brenig, B., Müller, M., Kräusslich, H., and Winnacker, E.-L. (1988). Production of transgenic pigs and possible application to pig breeding. *Occ. Publ. Br. Soc. Anim. Prod.* **12,** 15–31.

Brem, G., Springmann, K., Meier, E., Krausslich, H., Brenig, B., Muller, M., and Winnacker, E.-L. (1989). Factors in the success of transgenic pig programs. *In* "Transgenic Models in Medicine and Agriculture" (R. B. Church, ed.), Vol. 116, pp. 61–72. Wiley–Liss, New York.

Brinster, R. L., Chen, H. Y., Trumbauer, M. E., Yagle, M. K., and Palmiter, R. D. (1985). Factors affecting the efficiency of introducing foreign DNA into mice by microinjecting DNA. *Proc. Natl. Acad. Sci. U.S.A.* **82,** 4438–4442.

Burger, J. F. (1952). Sex physiology of pigs. *Onderspoort J. Vet. Res.* **25** (Suppl. 1), 22–131.

Caldwell, B. V., Moor, R. M., Wilmut, I., Polge, C., and Rowson, L. E. A. (1969). The relationship between day of formation and functional lifespan of induced corpora lutea in the pig. *J. Reprod. Fertil.* **18,** 107–113.

Campbell, R. G., Steele, N. C., Caperna, T. J., McMurtry, J. P., Solomon, M. B., and Mitchell, A. D. (1988). Interrelationships between energy intake and endogenous porcine growth hormone administration on the performance, body composition and protein and energy metabolism of growing pigs weighing 25 to 55 kilograms live weight. *J. Anim. Sci.* **66,** 1643–1655.

Casida, L. E. (1935). Prepuberal development of the pig ovary and its relation to stimulation with gonadotropic hormones. *Anat. Rec.* **61,** 389–396.

Clark, A. J., Simons, P., Wilmut, I., and Lathe, R. (1987). Pharmaceuticals from transgenic livestock. *Trends Biotechnol.* **5**, 20–24.

Davis, D. L. (1985). Culture and storage of pig embryos. *J. Reprod. Fertil. Suppl.* **33**, 115–124.

Davis, D. L., Knight, J. W., Killian, D. B., and Day, B. N. (1979). Control of estrus in gilts with a progestogen. *J. Anim. Sci.* **49**, 1506–1509.

Day, B. N. (1979). Embryo transfer in swine. *Theriogenology* **11**, 27–31.

Day, B. N., Oxenreider, S. L., Waite, A. B., and Lasley, J. F. (1965). Use of gonadotropins to synchronize estrus in swine. *J. Anim. Sci.* **24**, 1075–1079.

Day, B. N., Longnecker, D. E., Jaffey, S. C., Gibson, E. W., and Lasley, J. F. (1967). Fertility in swine following superovulation. *J. Anim. Sci.* **21**, 697–699.

Denman, J., Hayes, M., O'Day, C., Edmunds, T., Bartlett, C., Hirani, S., Ebert, K. M., Gordon, K., and McPherson, J. (1991). Transgenic expression of a variant of human tissue-type plasminogen activator in goat milk: Purification and characterization of the recombinant enzyme. *Bio/Technology* **6**, 839–843.

Didion, B. A., Martin, M. J., and Markert, C. L. (1990). Characterization of fertilization and early embryonic development of naturally ovulated pig ova. *Theriogenology* **33**, 284 (abstract).

Diehl, J. R., and Day, B. N. (1974). Effect of prostaglandin $F_{2\alpha}$ on luteal function in swine. *J. Anim. Sci.* **39**, 392–396.

Diehl, J. R., Day, B. N., Stevermer, E. J., Pursel, V., and Holden, K. (1979). Artificial insemination in swine. *Iowa State Univ. Pork Ind. Handb.* No. 64 (reproduction).

Diehl, J. R., Danion, J. R., and Thompson, L. H. (1990). Managing sows and gilts for efficient reproduction. *Iowa State Univ. Pork Ind. Handbook,* Reproduction Fact Sheet No. 8.

Dziuk, P. J., and Baker, R. D. (1972). Induction and control of ovulation in swine. *J. Anim. Sci.* **21**, 697–699.

Dziuk, P. J., and Gehlbach, G. D. (1966). Induction of ovulation and fertilization in the immature gilt. *J. Anim. Sci.* **25**, 410–413.

Dziuk, P. J., Polge, C., and Rowson, L. E. A. (1964). Intra-uterine migration and mixing of embryos in swine following egg transfer. *J. Anim. Sci.* **23**, 37–42.

Ebert, K. M., Low, M. J., Overstrom, E. W., Buonomo, F. C., Baile, C. A., Roberts, T. M., Mandel, G., and Goodman, R. H. (1988). A Moloney MLV–rat somatotropin fusion gene produces biologically active somatotropin in a transgenic pig. *Mol. Endocrinol.* **2**, 227–283.

Ebert, K. M., Smith, T. E., Buonomo, F. C., Overstrom, E. W., and Low, M. J. (1990). Porcine growth hormone gene expression from viral promoters in transgenic swine. *Anim. Biotechnol.* **1**, 145–159.

Ebert, K. M., Selgrath, J. P., DiTullio, P., Denman, J., Smith, T. E., Memon, M. A., Schindler, J. E., Monastersky, G. M., Vitale, J. A., and Gordon, K. (1991). Transgenic production of a variant of human tissue-type plasminogen activator in goat milk: Generation of transgenic goats and analysis of expression. *Bio/Technology* **6**, 835–838.

Estienne, C. E., Rampacek, G. B., Kraeling, R. R., Estienne, M. J., and Barb, C. R. (1988). Luteinizing hormone receptor number and affinity in corpora lutea from prepuberal gilts induced to ovulate and spontaneous corpora lutea of mature gilts. *J. Anim. Sci.* **66**, 917–922.

Evock, C. M., Etherton, T. D., Chang, C. S., and Ivy, R. E. (1988). Pituitary porcine growth hormone (pGH) and a recombinant pGH analog stimulate pig growth performance in a similar manner. *J. Anim. Sci.* **66**, 1928–1941.

Foote, R. H. (1980). Artificial insemination. *In* "Reproduction in Farm Animals" (E. S. E. Hafez, ed.), p. 525. Lea & Febiger, Philadelphia, Pennsylvania.

French, A. J., Zviedrans, P., Ashman, R. J., Heap, P. A., and Seamark, R. F. (1991). Comparison of prepubertal and postpubertal young sows as a source of one-cell embryos for microinjection. *Theriogenology* **35**, 202 (abstract).

Gordon, K., Vitale, Roberts, B., Monastersky, G., DiTullio, P., and Moore, G. (1990). Expression of foreign genes in the lactating mammary gland of transgenic animals. *In* "Transgenic Models in Medicine and Agriculture" (R. B. Church, ed.), pp. 55–59. Wiley–Liss, New York.

Guthrie, H. D. (1975). Estrus synchronization and fertility in gilts treated with estradiol benzoate and prostaglandin F$_{2\alpha}$. *Theriogenology* **4**, 69–78.

Guthrie, H. D. (1979). Fertility and estrous cycle control using gonadotropin and prostaglandin F$_{2\alpha}$ treatment of sows. *J. Anim. Sci.* **49**, 158–162.

Guthrie, H. D., and Polge, C. (1967). Control of oestrus and fertility in gilts treated with accessory corpora lutea by prostaglandin analogues, ICI 29,939 and ICI 80,996. *J. Reprod. Fertil.* **48**, 427–430.

Guthrie, H. D., and Polge, C. (1976). Luteal function and oestrus in gilts treated with a synthetic analogue of prostaglandin F$_{2\alpha}$ (ICI-79,939) at various times during the oestrus cycle. *J. Reprod. Fertil.* **48**, 423–425.

Hammer, R. E., Pursel, V. G., Rexroad, C. E., Jr., Wall, R. J., Bolt, D. J., Ebert, K. M., Palmiter, R. D., and Brinster, R. L. (1985). Production of transgenic rabbits, sheep and pigs by microinjection. *Nature (London)* **315**, 680–683.

Hammer, R. E., Pursel, V. G., Rexroad, C. E., Jr., Wall, R. J., Bolt, D. J., Palmiter, R. D., and Brinster, R. L. (1986). Genetic engineering of mammalian embryos. *J. Anim. Sci.* **63**, 269–278.

Hogan, B., Costantini, F., and Lacy, E. (1986). "Manipulating the Mouse Embryo," p. 158. Cold Spring Harbor Laboratory, Cold Spring Harbor, New York.

Holtz, W., and Schlieper, B. (1991). Unsatisfactory results with the transfer of embryos from gilts superovulated with PMSG and hCG. *Theriogenology* **35**, 1237–1249.

Hunter, R. H. F. (1964). Superovulation and fertility in the pig. *Anim. Prod.* **6**, 189–194.

Hunter, R. H. F. (1966). The effect of superovulation on fertilization and embryonic survival in the pig. *Anim. Prod.* **8**, 457–465.

James, J. E., Reeser, P. D., Davis, D. L., Straiton, E. C., Talbot, A. C., and Polge, C. (1980). Culture and long-distance shipment of swine embryos. *Theriogenology* **14**, 463–469.

James, J. E., James, D. M., Martin, P. A., Reed, D. E., and Davis, D. L. (1983). Embryo transfer for conserving valuable genetic material from swine herds infested with pseudorabies. *J. Am. Vet. Med. Assoc.* **183**, 525–528.

Knight, J. W., Davis, D. L., and Day, B. N. (1976). Estrus synchronization in gilts with a progestogen. *J. Anim. Sci.* **42**, 1358 (abstract).

Kraeling, R. R., and Rampacek, G. B. (1977). Synchronization of estrus and ovulation in gilts with estradiol and prostaglandin F$_{2\alpha}$. *Theriogenology* **8**, 103.

Krisher, R. L., Petters, R. M., Johnson, B. H., Bavister, B. D., and Archibong, T. E. (1989). Development of porcine embryos from the one-cell stage to blastocyst in mouse oviducts maintained in organ culture. *J. Exp. Zool.* **249**, 235–239.

Lo, D., Pursel, V., Linton, P. J., Sandgren, E., Behringer, R., Rexroad, E., Palmiter, R. D., and Brinster, R. L. (1991). Expression of mouse IgA by transgenic mice, pigs and sheep. *Eur. J. Immunol.* **21**, 1001–1006.

Martin, M. J., Didion, B. A., and Markert, C. L. (1989). Effect of gonadotropin administration on estrus synchronization and ovulation rate following induced abortion in swine. *Theriogenology* **32**, 929–937.

Martin, M. J., Didion, B. A., and Markert, C. L. (1990). Characterization of fertilization and early embryonic development of naturally ovulated pig ova. *Theriogenology* **33**, 284.

Martin, M. J., Merriman, C., Freeman, B., Merriman, J., and Kearns, J. (1992). Effect of extended centrifugation on visualization of nuclear structures in 1- and 2-cell porcine ova. *J. Anim. Sci.* **70** (Suppl. 1), 266 (abstract).

Miller, K. F., Pursel, V. G., Hammer, R. E., Pinkert, C. A., Palmiter, R. D., and Brinster, R. L. (1989). Expression of human or bovine growth hormone gene with a mouse metallothionein-1 promoter in transgenic swine alters the secretion of porcine growth hormone and insulin-like growth factor-I. *J. Endocrinol.* **120**, 481–488.

Murray, J. D., Nancarrow, C. D., Marshall, J. T., Hazelton, I. G., and Ward, K. A. (1989). Production of transgenic merino sheep by microinjection of ovine metallothionein–ovine growth hormone fusion genes. *Reprod. Fertil. Dev.* **1**, 147–155.

Neill, J. D., and Day, B. N. (1964). Relationship of developmental stage to regression of the corpus luteum in swine. *Endocrinology (Baltimore)* **74**, 355–360.

Niemann, H., Illera, M. J., and Dziuk, P. J. (1983). Developmental capacity size and number of nuclei in pig embryos cultured *in vitro*. *Anim. Reprod. Sci.* **5**, 311–321.

O'Reilly, P. J., McCormack, R., O'Mahoney, K., and Murphy, C. (1979). Estrus synchronization and fertility in gilts using a synthetic progestagen (allyl trenbolone) and inseminated with fresh stored or frozen semen. *Theriogenology* **12**, 131–137.

Palmiter, R. D., Brinster, R. L., Hammer, R. E., Trumbauer, M. E., Rosenfeld, M. G., Birnberg, N. C., and Evans, R. M. (1982). Dramatic growth of mice that develop from eggs microinjected with metallothionein–growth hormone fusion genes. *Nature (London)* **300**, 611–615.

Papaioannou, V. E., and Ebert, K. M. (1988). The preimplantation pig embryo: Cell number and allocation to trophectoderm and inner cell mass of the blastocyst *in vivo* and *in vitro*. *Development (Cambridge, UK)* **102**, 793–803.

Pinkert, C. A. (1990). Transgenic livestock development in the year 2000. *Proc. 7th FAVA Congr.*, pp. 20–30.

Pinkert, C. A. (1991). Transgenic animals as models for metabolic and growth research. *J. Anim. Sci.* **69** (Suppl. 3), 49–55.

Pinkert, C. A., Pursel, V. G., Miller, K. F., Palmiter, R. D. and Brinster, R. L. (1987). Production of transgenic pigs harboring growth hormone (MTbGH) or growth hormone releasing factor (MThGRF) genes. *J. Anim. Sci.* **65** (Suppl. 1), 260.

Pinkert, C. A., Kooyman, D. L., Baumgartner, A., and Keisler, D. H. (1989a). *In vitro* development of zygotes from superovulated prepuberal and mature gilts. *J. Reprod. Fertil.* **87**, 63–66.

Pinkert, C. A., Manz, J., Linton, P. J., Klinman, N. R., and Storb, U. (1989b). Elevated PC responsive B cells and anti-PC antibody production in transgenic mice harboring anti-PC immunoglobulin genes. *Vet. Immunol. Immunopathol.* **23**, 321–332.

Polge, C. (1982). Embryo transplantation and preservation. *In* "Control of Pig Reproduction" (D. J. A. Cole and G. R. Foxcroft, eds.), pp. 283–285. Butterworth, London.

Polge, E. J. C., Barton, S. C., Surani, M. H. A., Miller, J. R., Wagner, T., Elsome, K., Davis, A. J., Goode, J. A., Foxcroft, G. R., and Heap, R. B. (1989). Induced expression of a bovine growth hormone construct in transgenic pigs. *In* "Biotechnology of Growth Regulation" (R. B. Heap, C. G. Prosser, and G. E. Lamming, eds.), pp. 189–199. Butterworth, London.

Pope, C. E., and Day, B. N. (1977). Transfer of preimplantation pig embryos following *in vitro* culture for 24 or 48 h. *J. Anim. Sci.* **44**, 1036–1040.

Pope, C. E., Christenson, R. K., Zimmerman-Pope, V. A., and Day, B. N. (1972). Effect of number of embryos on embryonic survival in recipient gilts. *J. Anim. Sci.* **35**, 805–808.

Pope, W. F., Wilde, M. H., and Xie, S. (1989). Effect of electrocautery of nonovulated day 1 follicles on subsequent morphological variation among day 11 porcine embryos. *Biol. Reprod.* **39**, 882–887.

Pursel, V. G., and Wall, R. J. (1991). Use of donor gilts as recipients of microinjected ova. *J. Anim. Sci.* **67** (Suppl. 1), 375 (abstract).

Pursel, V. G., Elliot, D. O., Newman, C. W., and Staigmiller, R. B. (1981). Synchronization of estrus in gilts with allyl trenbolone: Fecundity after natural service and insemination with frozen semen. *J. Anim. Sci.* **52**, 130–133.

Pursel, V. G., Miller, K. F., Pinkert, C. A., Palmiter, R. D., and Brinster, R. L. (1987). Development of 1-cell and 2-cell pig ova after microinjection of genes. *J. Anim. Sci.* **65** (Suppl. 1), 402 (abstract).

Pursel, V. G., Pinkert, C. A., Miller, K. F., Bolt, D. J., Campbell, R. G., Palmiter, R. D., Brinster, R. L., and Hammer, R. E. (1989). Genetic engineering of livestock. *Science* **244**, 1281–1288.

Pursel, V. G., Hammer, R. G., Bolt, D. J., Palmiter, R. D., and Brinster, R. L. (1990a). Integration, expression and germ-line transmission of growth-related genes in pigs. *J. Reprod. Fertil. Suppl.* **41**, 77–87.

Pursel, V. G., Bolt, D. J., Miller, K. F., Pinkert, C. A., Hammer, R. E., Palmiter, R. D., and Brinster, R. L. (1990b). Expression and performance in transgenic pigs. *J. Reprod. Fertil. Suppl.* **40**, 235–245.

Pursel, V. G., Sutrave, P., Wall, R. J., Kelly, A. M., and Hughes, S. H. (1992). Transfer of c-ski gene into swine to enhance muscle development. *Theriogenology* **37**, 278 (abstract).

Rampacek, G. B., Kraeling, R. R., Kiser, T. E., Barb, C. R., and Benyshek, L. L. (1979). Prostaglandin F concentrations in uterovarian vein plasma of prepuberal and mature gilts. *Prostaglandins* **18**, 247–255.

Redmer, D. A., and Day, B. N. (1981). Ovarian activity and hormonal patterns in gilts fed allyl trenbolone. *J. Anim. Sci.* **53**, 1089–1094.

Reed, M. L., Illera, M. J., and Petters, R. M. (1992). *In vitro* culture of pig embryos. *Theriogenology* **37**, 95–109.

Rexroad, C. E., Jr., Pursel, V. G., Hammer, R. E., Bolt, D. J., Miller, K. F., Mayo, K. E., Palmiter, R. D., and Brinster, R. L. (1988). Gene insertion: Role and limitations of technique in farm animals as a key to growth. *In* "Biomechanisms Regulating Growth and Development" (G. L. Steffens and T. S. Rumsey, eds.), Vol. 12, pp. 87–97. Kluwer, Dordrecht, The Netherlands.

Rexroad, C. E., Jr., Hammer, R. E., Bolt, D. J., Mayo, K. E., Frohman, L. A., Palmiter, R. D., and Brinster, R. L. (1989). Production of transgenic sheep with growth regulating genes. *Mol. Reprod. Dev.* **1**, 164–169.

Rexroad, C. E., Jr., Hammer, R. E., Behringer, R. R., Palmiter, R. D., and Brinster, R. L. (1990). Insertion, expression and physiology of growth-regulating genes in ruminants. *J. Reprod. Fertil. Suppl.* **41**, 119–124.

Segal, D. H., and Baker, R. D. (1973). Maintenance of corpora lutea in prepuberal gilts. *J. Anim. Sci.* **37**, 762–767.

Shamay, A., Sabina, S., Pursel, V. G., McKnight, R. A., Alexander, L., Beattie, C., Hennighausen, L., and Wall, R. J. (1991). Production of the mouse whey acidic protein in transgenic pigs during lactation. *J. Anim. Sci.* **69**, 4552–4562.

Signoret, J. P., du Mesnil du Buisson, F., and Mauleon, P. (1972). Effect of mating on the onset and duration of ovulation in the sow. *J. Reprod. Fertil.* **31**, 327–330.

Simons, J. P., McClenaghan, M., and Clark, J. A. (1987). Alteration of the quality of milk by expression of sheep β lactoglobulin in transgenic mice. *Nature (London)* **328**, 530–532.

Simons, J. P., Wilmut, I., Clark, A. J., Archibald, A. L., Bishop, J. O., and Lathe, R. (1988). Gene transfer into sheep. *Bio/Technology* **6**, 179–183.

Swanson, M. E., Martin, M. J., O'Donnell, J. K., Hoover, K., Lago, W., Huntress, V., Parsons, C. T., Pinkert, C. A., Pilder, S., and Logan, J. S. (1992). Production of functional human hemoglobin in transgenic swine. *Bio/Technology* **10**, 557–559.

Van Brunt, J. (1988). Molecular farming: Transgenic animals as bioreactors. *Bio/Technology* **6**, 1149–1154.

Vincent, C. K., Robison, O. W., and Ulberg, L. C. (1964). A technique for reciprocal embryo transfer in swine. *J. Anim. Sci.* **23**, 1084–1088.

Vize, P. D., Michalska, A. E., Ashman, R., Lloyd, B., Stone, B. A., Quinn, P., Wells, J. R. E., and Seamark, R. F. (1988). Introduction of a porcine growth hormone fusion gene into transgenic pigs promotes growth. *J. Cell Sci.* **90**, 295–300.

Wall, R. J., Pursel, V. G., Hammer, R. E., and Brinster, R. L. (1985). Development of porcine ova that were centrifuged to permit visualization of pronuclei and nuclei. *Biol. Reprod.* **32**, 645–651.

Wall, R. J., Pursel, V. G., Shamay, A., McKnight, R. A., Pettius, C. W., and Hennighausen, L. (1991). High-level synthesis of a heterologous protein in the mammary glands of transgenic swine. *Proc. Natl. Acad. Sci. U.S.A.* **88**, 1696–1700.

Walton, J. R., Murray, J. D., Marshall, T. T., and Nancarrow, C. D. (1987). Zygote viability in gene transfer experiments. *Biol. Reprod.* **37**, 957–967.

Webel, S. K. (1976). Estrus control in swine with a progestogen. *J. Anim. Sci.* **42**, 1358.

Webel, S. K., Peters, J. B., and Anderson, L. L. (1970). Synchronous and asynchronous transfer of embryos in the pig. *J. Anim. Sci.* **30**, 565–568.

Weidle, U. H., Lenz, H., and Brem, G. (1991). Genes encoding a mouse monoclonal antibody expressed in transgenic mice, rabbits and pigs. *Gene* **98**, 185–191.

Wieghart, M., Hoover, J. L., McGrane, M. M., Hanson, R. W., Rottman, F. M., Holtzman, S. H.,

Wagner, T. E., and Pinkert, C. A. (1990). Production of transgenic swine harboring a rat phos-phoenolpyruvate carboxykinase–bovine growth hormone fusion gene. *J. Reprod. Fertil. Suppl.* **41,** 89–96.

Williams, B. L., Sparks, A. E. T., Canseco, R. S., Knight, J. W., Johnson, J. L., Velander, W. H., Page, R. L., Drohan, W. N., Young, J. M., Pearson, R. W., Wilkins, T. D., and Gwazdauskas, F. C. (1992). *In vitro* development of zygotes from prepubertal gilts after microinjection of DNA. *J. Anim. Sci.* **70,** 2207–2211.

Wright, C., Carver, A., Cottom, D., Reeves, D., Scott, A., Simons, P., Wilmut, I., Garner, I., and Colman, A. (1991). High level expression of active human α_1-antitrypsin in the milk of transgenic sheep. *Bio/Technology* **9,** 830–834.

Wright, R. W., Jr. (1977). Successful culture *in vitro* of swine embryos to the blastocyst stage. *J. Anim. Sci.* **44,** 854–858.

Wright, R. W., and Bondioli, K. R. (1981). Aspects of *in vitro* fertilization and embryo culture in domestic animals. *J. Anim. Sci.* **53,** 702–729.

Xie, S., Broermann, D. M., Nephew, K. P., Bishop, M. D., and Pope, W. F. (1990a). Relationships between oocyte maturation and fertilization on zygote diversity in swine. *J. Anim. Sci.* **68,** 2027–2033.

Xie, S., Broermann, D. M., Nephew, K. P., Ottobre, J. S., and Pope, W. F. (1990b). Changes in follicular endocrinology during maturation of porcine oocytes. *Dom. Anim. Endocrinol.,* **7,** 75–82.

Zavy, M. T., Buchanan, D., and Geisert, R. D. (1984). The combination of estrogen-induced prolonged luteal function (PLF) and $PGF_{2\alpha}$ administration as a means of estrus synchronization in swine. *J. Anim. Sci.* **59** (Suppl. 1), 327 (abstract).

Production of Transgenic Ruminants

Caird E. Rexroad, Jr., and Harold W. Hawk

U.S. Department of Agriculture

Agricultural Research Service

Beltsville Agricultural Research Center

Livestock and Poultry Sciences Institute

Gene Evaluation and Mapping Laboratory

Beltsville, Maryland 20705

I. INTRODUCTION

Transgenic sheep, cows, and goats have been produced by the microinjection of DNA into the pronucleus of zygotes (Hammer *et al.,* 1985; Roschlau *et al.,* 1989; Ebert *et al.,* 1991). The number of lines of transgenic ruminants produced to date has been small compared to the number of lines of mice and even of pigs. A number of factors limits the efficiency of production of transgenic ruminants. The greatest limiting factor is the low rate of incorporation of transgenes into the genome of microinjected embryos that survive to term. In several studies, transgenic offspring numbered less than 1% of the zygotes microinjected (for review, see Rexroad,

1992). Efficiency is further reduced by a rate of expression of the transgenes of less than 50%. The low efficiencies, small litter size (1–3), and relatively long generation interval make it expensive to conduct experiments to produce transgenic ruminants. Nonetheless, the potential value of genetically modified ruminants for production of pharmaceuticals and for use in agriculture has ensured continued research efforts to produce transgenic ruminants.

This chapter covers the methodology involved in producing transgenic ruminants by pronuclear microinjection, with particular attention to those methods peculiar to ruminants, especially the cow. Methods include embryo production by *in vitro* maturation (IVM), fertilization (IVF), and embryo culture or by culture in temporary surrogate hosts until the final embryo transfer.

II. TRANSGENIC CATTLE

A. Embryo Donor Management

1. General

The production of transgenic calves depends on the availability of a large number of one-cell embryos to be microinjected. Collection of one-cell embryos from live cattle is expensive and difficult, requiring either surgical collection or collection at necropsy. Transgenic cattle have been produced by the microinjection of embryos obtained from superovulated donors (Bondioli *et al.*, 1991; Roschlau *et al.*, 1989), but the most common technique for the production of large numbers of synchronously staged embryos is by IVM–IVF of oocytes from slaughterhouse ovaries (Krimpenfort *et al.*, 1991).

2. Superovulation of Donors

a. Pregnant Mare's Serum Gonadotropin   Roschlau *et al.* (1989) induce superovulation in beef heifers by a single injection of 2500–3000 IU of pregnant mare's serum gonadotropin (PMSG) given at 0800 hr on day −4 of treatment, followed 55 hr later (at 1500 hr on day −2 of treatment) by the injection of a prostaglandin $F_{2\alpha}$ (PGF) analog (cloprostenol, 0.5 mg) to induce luteolysis (Fig. 1). Onset of estrus (day 0) is expected about 40 hr after the administration of PGF. Superovulation protocols for cattle are usually initiated between days 9 and 14 postestrus. Cattle are subjected to artificial insemination (AI) at 0800 and 1500 hr on the expected day of estrus and at 0800 hr on the following day (day 1). Anti-PMSG serum (600 IU) is injected at 1500 hr on day 0. Embryos are collected by flushing

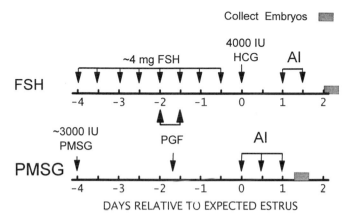

Figure 1. Protocols for treatments to superovulate cows to produce single-cell embryos.

the oviducts from reproductive tracts collected at ovohysterectomy at 1400 hr on day 1. Collection time actually ranges from 78–82 hr after PGF (Fig. 1). With this protocol, 75.5% of the cows ovulated more than two follicles, for an average ovulation rate of 12.6 and embryo recovery rate of 8.2.

b. Follicle-Stimulating Hormone Bondioli *et al.* (1991) produced transgenic calves from embryos that had been obtained after superovulation of cattle with a crude follicle-stimulating hormone (FSH) preparation (Fig. 1). They administer either 28 mg FSH (crossbred cows) or 37 mg FSH (dairy cows; Westhusin *et al.*, 1992) split into eight equal doses and given over 4 days. (Approximately the same amount of FSH is often given in decreasing doses over 4 days.) Also, PGF (25 mg) is given concurrently with the fifth and sixth FSH injections on day −2 relative to the expected day of estrus (day 0). Human chorionic gonadotropin (HCG, 4000 IU) is given on the morning of the day of expected estrus, 48 hr after the first PGF injection. HCG is intended to improve the synchrony of ovulation and provide proper timing for the collection of single-cell embryos. Cows are artificially inseminated in the morning and evening of the sixth day of the treatment regime, and embryos are collected from excised oviducts on the morning of the seventh day, or day 2 postestrus.

B. In Vitro *Embryo Production*

Oocytes can be collected from live cattle via laparoscopy, but, for research purposes, large numbers of oocytes can be collected from ovaries obtained from a slaughterhouse. Several protocols exist for the production of IVM–IVF embryos. The following description is largely from Kay *et al.* (1991).

1. Oocyte Collection

Collected ovaries are usually kept between 20 and 30°C and have been processed as late as 8 hr or more after slaughter. Ovaries should be washed prior to oocyte collection to reduce the chances of contamination in subsequent culture. The ovaries can be washed in tap water using a dilute solution of bacteriostatic soap. Oocytes may be collected from follicles either by aspiration or by cutting the surface of follicles while the ovary lies in a solution. Follicles 2 to 6 mm in diameter can be aspirated using a needle attached to a syringe or connected by tubing to a vacuum pump with a tube in the line to trap the follicular fluid and oocytes. Collected oocytes can be aspirated from the bottom of tubes after they settle or concentrated by filtration in an ovum concentrator (Immuno Systems, Spring Valley, WI).

2. Oocyte Maturation

Oocytes to be matured are generally selected on the basis of having at least 50% of their zona pellucida covered with compact cumulus cells and having a uniformly granular cytoplasm. Oocytes are generally passed through several rinses of a medium such as TALP–HEPES (Bavister *et al.*, 1983). The oocytes must be matured in a maturation medium, usually TCM-199 or Ham's F10, either in oil-covered drops of about 50 μl or in open wells of culture plates in about 0.5 ml of medium. The medium contains luteinizing hormone, FSH, or both, and often with added estradiol. Incubation is usually at 39°C in a humidified atmosphere of 5% (v/v) CO_2 in air for about 24 hr.

3. Sperm Preparation

Sperm for fertilization treatment have usually been prepared by the swim-up method of Parrish *et al.* (1986). A more rapid procedure that results in high fertilization rates is that of X. Yang (Hawk *et al.*, 1992). Frozen semen is thawed at 34°C, and the semen is added to Dulbecco's phosphate-buffered saline (PBS) containing D-glucose (1 g/liter) and sodium pyruvate (1 mg/liter). The sperm are washed by centrifugation at 500 g, the supernatant aspirated, and the sperm resuspended in fertilization medium to provide a final concentration of 1.6×10^6 motile sperm/ml.

4. Fertilization Medium

The critical component of fertilization medium is heparin to prepare the sperm for fertilization. At the Beltsville Agricultural Research Center, the fertilization medium is a modification of that described by Fukui (1990). The fertilization medium is freshly prepared from stock and supplemented with 2.5 mM fatty acid-free bovine serum albumin (BSA), 0.25 mM sodium pyruvate, and 340 units/liter (2 μg/ml) of heparin. The maturation medium is carefully aspirated from wells after 24 hr of maturation followed by the addition of 0.5 ml of fertilization medium containing 0.8 $\times$ 10^6 motile spermatozoa. Pipetting of the oocytes prior to either maturation or fertilization to remove most of the adhering cumulus cells improves the fertilization rate (Hawk *et al.*, 1992). The fertility of bulls used for IVF is variable; if genetics of the bull is not an overriding consideration, several bulls should be tested to obtain one that will give high fertilization rates.

About 15 hr after exposure of oocytes to sperm, the ova are placed in PBS in a 15-ml tube and vortexed to remove adhering cumulus and corona cells. The number of oocytes containing two pronuclei is near a maximum by 13 hr after addition of the fertilization medium to oocytes (Kay *et al.*, 1991).

5. Embryo Microinjection

In cattle, the pronuclei to be microinjected with DNA are not visible under differential interference contrast microscopy (DIC), but the pronuclei can be visualized following centrifugation to stratify the granular cytoplasm (Wall *et al.*, 1985). The ova are centrifuged in 2-ml microcentrifuge tubes for up to 8 min at 15,600 g (Wall and Hawk, 1988; Kay *et al.*, 1991).

The procedures for microinjection of cow embryos are similar to those of other species (see Chapters 2 and 11). Using IVM–IVF procedures and *in vitro* culture, Krimpenfort *et al.* (1991) produced two transgenic calves from an initial collection of 2470 oocytes. Only 1154 of the oocytes had visible pronuclei and were injected, and only 687 embryos cleaved (28% of the initial oocytes). Of 129 embryos transferred to 99 recipient cows, 19 calves were born and only two were transgenic (1.55% of the transferred embryos). Hill *et al.* (1992) produced four live transgenic calves after starting with 19,336 oocytes, of which 11,206 were microinjected and 193 calves were born. Thus, the procedure is highly inefficient, and large numbers of recipients are needed for the production of transgenic calves even by total *in vitro* technology.

C. *Embryo Culture*

The low incidence of successful gene transfer after microinjection makes it costly to transfer each microinjected embryo immediately into a recipient cow. Two methods have been utilized to culture bovine embryos to the blastocyst stage, at which time viable embryos can be transferred to recipients. One method makes use of an intermediate host, including rabbit does, ewes, and heifers. The second is *in vitro* coculture.

1. Intermediate Hosts

Hawk *et al.* (1989) transfer microinjected bovine embryos into the oviducts of rabbit does. Oviducts are ligated at the tubo-uterine junction prior to transfer. Embryos are recovered by flushing the oviducts at necropsy from 7 to 9 days after embryo transfer; 56% of the transferred zygotes were recovered and 56% of the recovered embryos had developed to the blastocyst stage. Transfer of morulae or blastocysts to recipient heifers resulted in over half of the embryos forming elongated blastocysts when recovered 7 days after transfer. Thus, the total development rate to the elongated blastocyst stage of embryos recovered from does was about 30%. Bondioli *et al.* (1991) obtained a 17% survival rate of microinjected cow embryos that were cultured temporarily in sheep oviducts, transferred to bovine recipients, and recovered as fetuses at 60 to 65 days of gestation. Roschlau *et al.* (1989) used heifers as intermediate hosts for embryos, but the lack of a strong species–specific requirement for temporary host makes it feasible to use smaller animals as temporary hosts.

2. *In Vitro* Culture

The development of *in vitro* coculture systems for bovine embryos using granulosa cells (Goto *et al.*, 1992), oviductal cells (Bondioli *et al.*, 1991), oviduct cell-conditioned medium (Eyestone and First, 1989), or buffalo rat liver (BRL) cells (Voelkel and Hu, 1992) has reduced the number of recipient animals needed. Coculture can be conducted on an oviductal cell monolayer or with motile oviduct cell vesicles (Xu *et al.*, 1992). If the reproductive tract is not removed from the cow aseptically, the tract should be cleaned with soapy water and thoroughly rinsed with 70% (v/v) alcohol. The oviduct is removed from the supporting ligaments by dissection. Cells for culture can be removed from the oviduct by extrusion by squeezing the oviduct beneath a glass side and pulling from the isthmus toward the fimbria. Cells can be suspended in medium (e.g., TCM-199 with 10% heat-treated fetal bovine serum) and aliquoted into culture wells or drops. Cells are added at sufficient density to result in a confluent monolayer in 2 to 3 days or in numerous epithelial cell vesicles the next day. Another approach is to add 2 to 3 μl of cells to two 20-μl drops of culture medium to which embryos are added (Bondioli *et al.*, 1991). Embryos are cultured in microdrops under silicone oil at 38 or 39°C. Embryos are cocultured for 5 to 7 days and are transferable at the blastocyst stage.

Eyestone and First (1989) culture bovine embryos in conditioned TCM-199. Oviductal cells extruded from the oviduct are diluted 1:50 in TCM-199 supplemented with 10% fetal bovine serum and cultured 1 to 7 days. Conditioned medium from the culture provided support for embryonic development to the blastocyst stage for more than 20% of IVM–IVF embryos. Conditioned medium was used in the studies of Krimpenfort *et al.* (1991) in the production of transgenic calves.

D. Recipient Management

1. Estrus Synchronization

Optimal fertility using bovine embryo transfer is achieved by transfer of embryos to hosts that are within 36 hr synchrony of estrus with donors (Coleman *et al.*, 1987). Whether this optimum is modified by transfer of *in vitro* cultured embryos has not been established.

Use of cows in natural estrus as recipients would require a large group from which to select. The 18- to 24-day estrous cycle of cattle would indicate that more than 40 cows would need to be maintained to obtain two recipients with close synchrony on most days. However, synchronization of estrus can be achieved by the administration of a single injection of PGF between days 6 and 14 of the estrous cycle to cause regression of the corpus luteum. Most cows will be in estrus 3 or 4 days after PGF treatment, and a majority of treated cows would be expected to be in estrus within a 36-hr period.

2. Embryo Transfer

Bovine embryos are transferred nonsurgically by use of standard embryo transfer techniques. Embryos are usually transferred fresh rather than freeze–thawed. Cryopreservation significantly reduces the viability of *in vitro*-derived embryos.

3. Pregnancy Detection

Pregnancy can be diagnosed by detecting the embryo or its membranes. The absence of pregnancy can be detected by observing some indication of the failure of a potential pregnancy. Means of detecting pregnancy or the absence of pregnancy include palpation of the ovaries and uterus per rectum, analysis of milk or blood for progesterone concentration, and ultrasonography. A definite aid to pregnancy diagnosis is a sound estrus detection program; an animal returning to estrus after insemination or after being the recipient of an embryo will not be pregnant.

Palpation of the ovaries per rectum can provide evidence of lack of pregnancy by revealing a regressing corpus luteum, and palpation of the uterus at about 35 days of pregnancy or later will reveal the presence of fluid-filled chorionic membranes and the embryo-containing amniotic vesicle. Palpation of corpora lutea or the amniotic vesicle requires experience on the part of the palpator but requires little time or expense.

High progesterone concentrations in blood or milk at 20–25 days after insemination or 13–18 days after receiving an embryo will indicate maintenance of the corpus luteum and likely presence of an embryo, but other means will be required subsequently to confirm the pregnancy. Low progesterone concentrations indicate regression of the corpus luteum and impending return to estrus. In one study, milk progesterone concentration was 80% accurate in indicating pregnancy at 24 days after insemination and 100% accurate in indicating lack of pregnancy (Heap *et al.*, 1976). Use of progesterone analyses requires the necessary analytical equipment or access to it.

A valuable addition to methods of diagnosing pregnancy has been application of ultrasound techniques that can detect the presence of embryonic membranes in cattle as early as the second week of pregnancy (Kastelic *et al.*, 1988). Ultrasonography can be used to follow the course of pregnancy and ascertain the loss or maintenance of pregnancy through the critical 14- to 35-day period of embryo development. Use of ultrasonography requires ultrasound equipment and expertise in its use.

III. TRANSGENIC SHEEP AND GOATS

A. General

Small ruminants in transgenic experiments can serve as models for the genetically related bovine and in some instances can be used as models for humans. In addition, they can produce sufficient quantities of milk to merit consideration as bioreactors in the production of pharmaceuticals, they are relatively inexpensive to purchase and maintain, and their generation interval is short enough to accomplish lactation experiments in 2 years. The first transgenic livestock reported in the literature was a lamb (Hammer *et al.*, 1985). Litters of sheep and goats are small (1–3 for most breeds), necessitating large groups for transgenic experimentation. Sheep requirements for a transgenic experiment, depending on the number of transgenic lambs to be produced, might include 150 ewes to be donors and recipients, four intact rams for semen, and four vasectomized rams for detection of estrus. The following discussion of transgenic small ruminants will focus on sheep. Most of the work reported thus far has been in that species.

B. Donor Management

The selection of a breed of sheep to be used in transgenic experiments may in part depend on the nature of the experiment. Sheep breeds vary significantly in meat

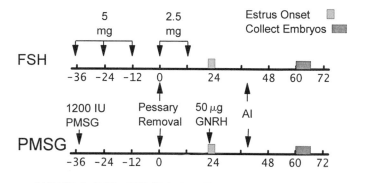

Figure 2. Protocols for treatments to superovulate ewes to produce single-cell embryos.

production, wool production, milk production, fecundity, and response to seasonal changes. Sheep are seasonally polyestrous. Reproductive cycles of most breeds begin in late August/early September and cease between February and April in the Northern Hemisphere. Seasonality is limiting for male fertility and for recipient management, although embryos can be collected throughout the year. Depending on the duration of an experiment, breeds might be selected that have a prolonged reproductive season. Lighter breeds are easier to work with and less expensive to maintain if other considerations are not important.

Sheep embryos are usually collected *in vivo* and microinjected at the 1-cell stage. The natural ovulation rate of sheep varies from one to three ovulations per estrous cycle, but sheep respond well to superovulatory techniques. Two basic treatments can be used to induce superovulation in sheep: PMSG or FSH (Fig. 2). Either hormone may be used in conjunction with gonadotropin-releasing hormone (GnRH) treatment near the time of expected estrus to regulate the time of ovulation precisely.

1. Estrus Synchronization

For efficient experiments, donors should ovulate in synchrony with one another and with recipients. Estrus can be synchronized by using either progestin pessaries or PGF. The advantage of progestin pessaries is that they synchronize nearly 100% of ewes that retain them after a single treatment (pessary insertion). However, some sheep lose the pessaries. In addition, pessaries are not available in all parts of the world, although they can be made in the laboratory using a number of synthetic progestins.

Pessaries can be inserted into the vagina of ewes at any stage of the estrous cycle. Pessaries are usually compacted into a hollow tube of about 2.0 cm inner diameter that has rounded edges. The tube is inserted into the vagina of restrained ewes, and the pessary expelled from the insertion device with a rod. The insertion device is cleansed prior to use in each ewe. Leaving the pessaries in place for at least 12 to 14 days ensures that most ewes will have undergone luteal regression. Leaving pessaries in for longer intervals increases the chance of reproductive tract infection. Some vaginal bleeding from a mild level of infection is common at the time of pessary removal.

PGF causes luteal regression in ewes after day 5 of the luteal phase and thus must be used twice, about 8 days apart, to regulate all ewes if a random group is allocated to experiment. If ewes are checked routinely for estrus, then a single treatment of PGF can be used by selecting ewes between days 5 and 14 of their estrous cycle. PGF can be administered as a single injection of 15 mg of prostaglandin $F_{2\alpha}$ or the equivalent of analogs or as two injections of 5 mg, 3 hr apart.

2. Superovulation

Superovulation is most often induced by treatment with multiple injections of crude FSH preparations (Hammer *et al.*, 1985). The ovulation rate can be adjusted by changing the dosage or number of injections. Five injections with a total dosage of 20 mg of crude FSH has been used at Beltsville to produce about 10 ovulations per ewe. FSH is administered as injections of 5, 5, 5, 2.5, and 2.5 mg, with the next to last injection at a 5 p.m. pessary removal (Rexroad and Powell, 1991a; Fig. 2). Most ewes (>90%) display estrus at 18 to 24 hr after pessary removal. A simpler protocol for FSH administration is the subcutaneous injection of 2.15 mg equine FSH at -30 and at 0 hr relative to pessary removal (Simons *et al.*, 1988). The long half-life of equine FSH permits it to be used in the same manner as PMSG. Superovulation can also be induced by FSH when PGF is used to cause luteal regression. FSH treatments are begun during the midluteal phase of the estrous cycle. Prostaglandin can be administered as a single injection at the time of the next to last FSH injection (Rexroad and Powell, 1991a).

3. Estrus Detection

Estrus is detected in ewes by the observation of mating. If fertility is not desired, for example, in recipient ewes, estrus should be detected using vasectomized or aproned rams (Evans and Maxwell, 1987). The duration of estrus in ewes is frequently less than 25 hr, so ewes should be checked for estrus at least twice daily if direct observation is to be used. An alternative is to use rams wearing harnesses with chalk attached that marks the rumps of ewes that were mated. Estrus detection is probably not necessary for superovulated donors after the technique has been established for a flock; estrus usually begins 18 to 24 hr after pessary removal or PGF injection in superovulated ewes. Injection of 50 μg of GnRH intramuscularly at the time of estrus is a common technique to improve the synchronization of ovulation among ewes (Murray *et al.*, 1989). The efficacy of GnRH is not clearly demonstrated; it may be more efficacious after PMSG treatment than after FSH treatment.

Most studies indicate that optimum fertility after embryo transfer is ensured by close synchrony of estrus between donors and recipients; however, such a study has not been performed for microinjected embryos. The time of estrus after pessary removal or prostaglandin treatment in recipients is variable and probably should be established in each flock. Without superovulation, estrus usually begins between 36 and 54 hr after either PGF injection or progestin pessary removal.

4. Insemination

Insemination of ewes can be performed satisfactorily either by natural service or artificial insemination. Natural service in conjunction with superovulation and drug-regulated control of the estrous cycle results in variable fertility (Rexroad and Powell, 1991a). Artificial insemination, with sperm being deposited in the uterus (Rexroad and Powell, 1991a), overcomes low fertility associated with reduced sperm transport through the cervix after regulation of estrus (Hawk and Conley, 1975). Intrauterine AI has been performed by laparoscope (Murray *et al.*, 1989; Robinson *et al.*, 1989) and by midventral laparotomy.

Laparoscopy has the advantage of requiring analgesia rather than anesthesia and may be less traumatic to the uterus, depending on the operator. Semen is drawn into a glass pipette of 4 to 5 mm diameter drawn to a fine point. The pipette is passed through a laparoscopic cannula, and the semen (about 50 μl equivalent of neat semen) is deposited into each uterine horn (Murray *et al.*, 1989; Robinson *et al.*, 1989). Transgenic lambs have been produced from embryos collected from sheep that had been inseminated at 44 hr after pessary removal via laparoscope or midventral laparotomy. Robinson *et al.* (1989) obtained high fertility and good embryo recovery after laparoscopic insemination at 60 hr after pessary removal in superovulated ewes.

For insemination by midventral laparotomy, uterine horns are exteriorized while ewes are under sodium pentobarbital anesthesia, and 0.4 ml of semen, diluted 1:1 with phosphate-buffered saline containing 1% glucose, is injected into the lumen of each uterine horn. AI at laparotomy requires correct timing because barbiturate anesthetics can interfere with the ovulatory process and block ovulation. Laparotomy is usually performed after the onset of estrus (Rexroad and Powell, 1991a).

5. Embryo Collection

The time from beginning of estrus to ovulation in ewes is reported to vary from 24 to 30 hr. From the time of pessary removal, ewes are in estrus at 18 to 24 hr and ovulate at 42 to 54 hr. Most oocytes should be fertilized within a few hours after ovulation, so embryo collection can follow shortly after the expected time of fertilization. We routinely begin embryo collection at 64 hr after pessary removal, which results in embryos that are 10 to 24 hr postovulation, with the time after fertilization somewhat less. These times are in part validated by the collection of predominantly pronuclear-stage 1-cell embryos following this protocol, although the proportion of 2-cell embryos can vary from 10 to 20%.

Embryos are collected at laparotomy by retrograde flushing of the oviducts via a polyethylene tube. Ewes are anesthetized with pentobarbital and the uterus exteriorized via midventral laparotomy. The polyethylene tube, with an internal diameter of about 1.2 mm, has had the cut edges of one end rounded by placing the end near a flame. The rounded end of the tube is inserted through the fimbria and about 2 cm into the oviduct. A blunt-ended 22-gauge needle is passed through the uterine wall near the uterotubal junction. If possible, the needle is threaded through the junction. With digital pressure (or by a grasping device such as a child's play clothespin) around the polyethylene tubing, 3 to 6 ml of flushing medium is passed through the 22-gauge needle toward the tubing. The free end of the tubing is usually in a sterile 35-mm petri dish. Flushings are immediately inspected under a dissecting microscope to determine recovery and stage of the embryos. Inspection also permits the determination of the presence of a reproductive tract infection, in which case the embryos should be washed thoroughly or handled separately from those of other ewes. A single flush is satisfactory as embryos are rarely found on subsequent flushes. Embryos are then transferred to sterile culture vials for transport to the laboratory.

Removal of embryos from transport vials can be assured by inverting the tubes a few times and quickly pouring the contents into a culture dish. Any embryos left behind can be removed by sharply rapping the edge of the inverted tube on the dish. In the laboratory, embryos can be stored, handled, and microinjected in the same medium as that used for collection.

6. Microinjection of Sheep Embryos

Sheep embryos in most studies have been held at room temperature during microinjection. They have been handled in either a phosphate-buffered saline or a HEPES-containing buffer to prevent pH changes. Sheep pronuclei are visible, although not sharply outlined, by differential interference contrast microscopy (Hammer *et al.*, 1985). Centrifugation is not needed. The percentage of collected embryos that are fertilized and have visible pronuclei is reported to be 80 to 90% (Hammer *et al.*, 1985; Murray *et al.*, 1989). Microinjection procedures for sheep embryos are not unique, and different systems have been used. About 10% (C. E. Rexroad and others, unpublished) to about 30% (Walton *et al.*, 1987) of microinjected sheep eggs lyse within 1 hr after microinjection. Variation in the extent of lysis is probably caused by variation in methods of injecting the solution.

7. Embryo Transfer

Microinjected sheep embryos are usually transferred into recipients immediately, which necessitates a large number of recipients. The number of embryos transferred into each recipient depends on the likelihood of viability of each embryo. Two to five embryos are commonly transferred to each recipient, and 20 to 50% of transferred embryos have produced offspring in different studies (Rexroad, 1992). We commonly transfer three embryos per recipient. Estrus in recipient ewes is synchronized with that of donor ewes with progestin pessaries, which are removed 48 hr before desired time of estrus. Vasectomized rams are utilized for detection of estrus at 0800 and 1600 hr daily. In addition, detection of estrus in an untreated flock can be conducted to provide recipients. Embryo transfer is performed while ewes are under sodium pentobarbital anesthesia. Embryos to be transferred are aspirated into a positive displacement glass micropipette in a volume of approximately 10 μl. The pipette is inserted 1 to 3 cm into the fimbriated end of the oviduct of recipient ewes, and embryos are expelled.

C. Embryo Culture

The ability to maintain embryos *in vitro* to assess development and possible incorporation of a transgene would enhance the efficiency of transgenic experiments greatly. Culture systems now exist for sheep embryos. High rates of development are achieved by a 5-day culture of sheep embryos in synthetic oviduct fluid (SOF) supplemented with human serum (Walker *et al.*, 1992). Culture in SOF results in excellent viability and good development after transfer. The most serious restraints are the use of fresh human sera and an increased incidence of abnormally large lambs. In addition, sheep embryos can be cultured for 3 to 5 days on monolayers of sheep oviductal cells or STO cells.

1. Oviduct Cell Coculture

In our laboratory, oviducts are collected aseptically from naturally cycling ewes at necropsy on day 2 after estrus, trimmed free of ligaments, and slit lengthwise on the side of the mesosalpinx. Cells from two oviducts are collected by lightly scraping the lumenal surface with a sterile scalpel, and the cells are suspended in 5 ml of M199. One-half milliliter of the cell suspension is pipetted into 5 ml of M199 medium containing fetal calf serum in 35-mm wells of a six-well culture plate. The oviductal cells are cultured at 37.5°C in a humidified atmosphere of 95% air, 5% CO_2 (by volume). Medium is completely replaced once at 24 hr after the start of culture, which removes nonadhering cells. Embryos are placed on the monolayer after 3 days of cell culture, when the monolayer is about 50% confluent. About 50% of cocultured nonmicroinjected embryos develop to late morulae or blastocysts after 5 days of coculture.

If oviduct cells are not convenient to use for coculture, STO cells can be purchased from the American Type Culture Collection (Rockville, MD). STO cells are thawed according to the directions provided and placed in M199 supplemented with fetal calf serum. Cells are seeded densely in 35-mm culture dishes. Culture is the same as with oviductal monolayers. The percentage of superovulated embryos that become blastocysts exceeds 30% (Rexroad and Powell, 1991b, 1993).

2. Transfer of Cultured Embryos

Embryos are transferred via midventral laparotomy or via laparoscopy. At laparotomy, uteri are exteriorized, and a small puncture wound is made bluntly along the outer curvature of a uterine horn. Embryos are picked up in a positive displacement glass micropipette, the pipette is inserted into the lumen toward the oviduct, and the embryos are expelled. After 5 days of culture, only embryos that have reached the blastocyst stage are likely to produce offspring. Use of long-term cultures for production of transgenic lambs has not been reported, so the likelihood of success is unknown, although the technique is similar to those used for cattle. The success of transferring cocultured embryos can be improved by using recipients that were in estrus 1 day after embryo donors were in estrus; this timing compensates for the slower development of embryos *in vitro* than *in vivo* (Rexroad and Powell, 1991b).

D. Recipient Management

1. Pregnancy Diagnosis

Recipient ewes are bled by puncture of the jugular vein at 15 and 18 days after estrus. Plasma is assayed for progesterone concentration by radioimmunoassay (RIA). Ewes are considered pregnant if plasma progesterone concentrations exceed 1 mg/ml on both days. Ewes are also checked daily for estrus.

2. Transgene Diagnosis

The techniques for diagnosis of gene insertion in sheep are the same as in other species. Tail docking is a routine husbandry practice in sheep for hygienic reasons. We typically dock the tails using electrically heated cutters at 2 to 3 days after birth.

Cross sections of tail tissue are frozen on dry ice to be used for DNA isolation. All lambs born in a transgenic experiment are tattooed to ensure identification. There are no provisions in law or regulatory codes that would allow a transgenic animal to be disposed into the food chain. This currently increases significantly the costs of transgenic experiments in ruminants.

IV. SUMMARY

Figure 3 summarizes the steps involved in generating transgenic cattle. Each step has been subjected to intensive research. *In vitro* technologies have enhanced the

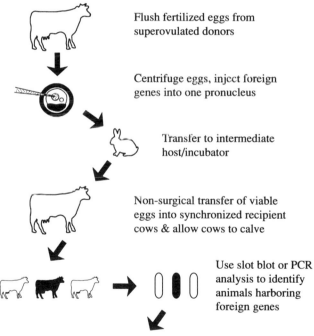

Flush fertilized eggs from superovulated donors

Centrifuge eggs, inject foreign genes into one pronucleus

Transfer to intermediate host/incubator

Non-surgical transfer of viable eggs into synchronized recipient cows & allow cows to calve

Use slot blot or PCR analysis to identify animals harboring foreign genes

● Perform tissue biopsies - analyze foreign DNA integration, mRNA transcription, and protein production

● Establish transgenic lines to study gene regulation in progeny

Figure 3. Scheme for production of transgenic cattle. (Courtesy of C. A. Pinkert).

possibilities for the production of transgenic cattle. Until techniques are devised for the insertion of genes into the nucleus with high efficiency, the limiting step in production of transgenic ruminants will be the cost of maintaining recipients. The development of technologies to select embryos containing incorporated genes would greatly enhance the value of transgenic ruminants for the study of gene expression as it relates to the physiology of animals and to their improvement as livestock. Until then, the production of transgenic ruminants will likely be limited to efforts to produce valuable products such as pharmaceuticals (Clark *et al.*, 1989).

REFERENCES

Bavister, B. D., Leibfried, M. L., and Lieberman, G. (1983). Development of preimplantation embryos of the Golden Hamster in a defined culture medium. *Biol. Reprod.* **28,** 235–247.

Bondioli, K. R., Biery, K. A., Hill, K. G., Jones, K. B., and De Mayo, F. J. (1991). Production of transgenic cattle by pronuclear injection. *In* "Transgenic Animals" (N. First and F. P. Haseltine, eds.), pp. 265–273. Butterworth–Heinemann, London.

Clark, A. J., Bessos, H., Bishop, J. O., Brown, P., Harris, S., Lathe, R., McClenaghan, M., Prowse, C., Simons, J. P., Whitelaw, C. B. A., and Wilmut, I. (1989). Expression of human anti-hemophilic factor IX in the milk of sheep. *Bio/Technology* **7,** 487–492.

Coleman, D. A., Dailey, R. A., Leffel, R. E., and Baker, R. D. (1987). Estrous synchronization and establishment of pregnancy in bovine embryo transfer recipients. *J. Dairy Sci.* **70,** 858–866.

Ebert, K. M., Selgrath, J. P., DiTullio, P., Denman, J., Smith, T. E., Memon, M. A., Schindler, J. E., Monastersky, G. M., Vitale, J. A., and Gordon, K. (1991). Transgenic production of a variant of human tissue-type plasminogen activator in goat milk. *Bio/Technology* **9,** 835–838.

Evans, G., and Maxwell, W. M. C. (1987). Preparation of teaser males. *In* "Salamon's Artificial Insemination of Sheep and Goats," pp. 75–78. Butterworth, London.

Eyestone, W. H., and First, N. L. (1989). Co-culture of early cattle embryos to the blastocyst stage with oviductal tissue or in conditioned medium. *J. Reprod. Fertil.* **85,** 715–720.

Fukui, Y. (1990). Effect of follicle cells on the acrosome reaction, fertilization, and developmental competence of bovine oocytes matured *in vitro*. *Mol. Reprod. Dev.* **26,** 40–46.

Goto, K., Iwai, N., Takuma, Y., and Nakanishi, Y. (1992). Co-culture of *in vitro* fertilized bovine embryos with different cell monolayers. *J. Anim. Sci.* **70,** 1449–1453.

Hammer, R. E., Pursel, V. G., Rexroad, C. E., Jr., Wall, R. J., Bolt, D. J., Ebert, K. M., Palmiter, R. D., and Brinster, R. L. (1985). Production of transgenic rabbits, sheep and pigs by microinjection. *Nature (London)* **315,** 680–683.

Hawk, H. W., and Conley, H. H. (1975). Involvement of the cervix in sperm transport failures in the reproductive tract of the ewe. *Biol. Reprod.* **13,** 322–328.

Hawk, H. W., Nel, N. D., Waterman, R. A., and Wall, R. J. (1992). Investigation of means to improve rates of fertilization in *in vitro* matured/*in vitro* fertilized bovine oocytes. *Theriogenology* **38,** 989–998.

Heap, R. B., Holdsworth, R. J., Gadsby, J. E., Lang, J. A., and Walters, D. E. (1976). Pregnancy diagnosis in the cow from milk progesterone concentration. *Br. Vet. J.* **132,** 445–464.

Hill, K. G., Curry, J., DeMayo, F. J., Jones-Diller, K., Slapak, J. R., and Bondioli, K. R. (1992). Production of transgenic cattle by pronuclear injection. *Theriogenology* **37,** 222.

Kastelic, J. P., Curran, S., Pierson, R. A., and Ginther, O. J. (1988). Ultrasonic evaluation of the bovine conceptus. *Theriogenology* **29,** 39–54.

Kay, G. W., Hawk, H. W., Waterman, R. A., and Wall, R. J. (1991). Identification of pronuclei in *in vitro* fertilized cow embryos. *Anim. Biotechnol.* **2**, 45–59.

Krimpenfort, P., Rademakers, A., Eyestone, W., Van der Schans, A., Van den Broek, S., Kooiman, P., Kootwijk, E., Plantenburg, G., Pieper, F., Strijker, R., and de Boer, H. (1991). Generation of transgenic dairy cattle using "*in vitro*" embryo production. *Bio/Technology* **9**, 844–847.

Murray, J. D., Nancarrow, C. D., Marshall, J. T., Hazelton, I. G., and Ward, K. A. (1989). The production of transgenic Merino sheep by microinjection of ovine metallothionein–ovine growth hormone fusion genes. *Reprod. Fertil. Dev.* **1**, 147–155.

Parrish, J. J., Susko-Parrish, J. L., Leibfried-Rutledge, M. L., Critser, E. S., Eyestone, W. H., and First, N. L. (1986). Bovine *in vitro* fertilization with frozen thawed semen. *Theriogenology* **25**, 591–600.

Pinkert, C. A., Dyer, T. J., Kooyman, D. L., and Kiehm, D. J. (1990). Characterization of transgenic livestock production. *Domest. Anim. Endocrinol.* **7**, 1–18.

Rexroad, C. E., Jr. (1992). Transgenic technology in animal agriculture. *Anim. Biotechnol.* **3**, 1–13.

Rexroad, C. E., Jr., and Powell, A. M. (1991a). FSH injections and intrauterine insemination protocols for superovulation of ewes. *J. Anim. Sci.* **69**, 246–251.

Rexroad, C. E., Jr., and Powell, A. M. (1991b). Effect of serum-free co-culture and synchrony of recipients on development of cultured sheep embryos to fetuses. *J. Anim. Sci.* **69**, 2066–2072.

Rexroad, C. E., Jr., and Powell, A. M. (1993). Development of ovine embryos co-cultured on oviductal cells or STO cell monolayers. *Biol. Reprod.*, 609. In press.

Rexroad, C. E., Jr., Mayo, K., Bolt, D. J., Elsasser, T. H., Miller, K. F., Behringer, R. R., Palmiter, R. D., and Brinster, R. L. (1991). Transferrin- and albumin-directed expression of growth-related peptides in transgenic sheep. *J. Anim. Sci.* **69**, 2995–3004.

Robinson, J. J., Wallace, J. M., and Aitken, R. P. (1989). Fertilization and ovum recovery rates in superovulated ewes following cervical insemination or laparoscopic intrauterine insemination at different times after progestogen withdrawal and in one or both uterine horns. *J. Reprod. Fertil.* **87**, 771–782.

Roschlau, K., Rommel, P., Andreewa, L., Zackel, M., Roschlau, D., Zackel, B., Schwerin, M., Huhn, R., and Gazarjan, K. G. (1989). Gene transfer experiments in cattle. *J. Reprod. Fertil. Suppl.* **38**, 153–160.

Simons, J. P., Wilmut, I., Clark, A. J., Archibald, A. L., Bishop, J. O., and Lathe, R. (1988). Gene transfer into sheep. *Bio/Technology* **6**, 179–183.

Voelkel, S. H., and Hu, Y-X. (1992). Effect of gas atmosphere on the development of one-cell bovine embryos in two culture systems. *Theriogenology* **37**, 1117–1131.

Walker, S. K., Heard, T. M., and Seamark, R. F. (1982). *In vitro* culture of sheep embryos without co-culture: Successes and perspectives. *Theriogenology* **37**, 111–126.

Wall, R. J., and Hawk, H. W. (1988). Development of centrifuged cow zygotes cultured in rabbit oviducts. *J. Reprod. Fertil.* **82**, 673–680.

Wall, R. J., Pursel, V. G., Hammer, R. E., and Brinster, R. L. (1985). Development of porcine ova that were centrifuged to permit visualization of pronuclei and nuclei. *Biol. Reprod.* **32**, 645–651.

Walton, J. R., Murray, J. D., Marshall, J. T., and Nancarrow, C. D. (1987). Zygote viability in gene transfer experiments. *Biol. Reprod.* **37**, 957–967.

Xu, K. P., Yadav, B. R., Rorie, R. W., Plante, L., Betteridge, K. J., and King, W. A. (1992). Development and viability of bovine embryos derived from oocytes matured and fertilized *in vitro* and co-cultured with bovine oviducal epithelial cells. *J. Reprod. Fertil.* **94**, 33–43.

Index